THE CHEMISTRY BOOK

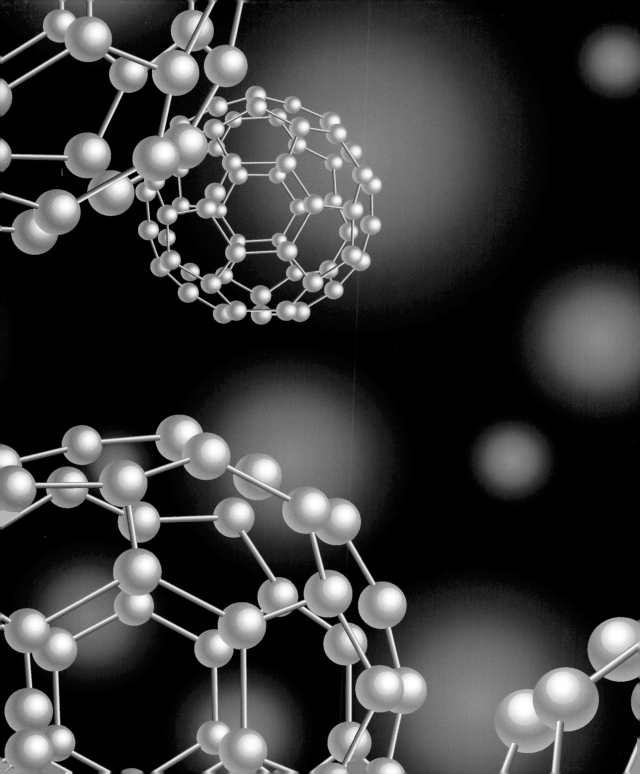

THE CHEMISTRY BOOK

FROM GUNPOWDER TO GRAPHENE,
250 MILESTONES IN THE HISTORY OF CHEMISTRY

Derek B. Lowe

STERLING
New York

STERLING
New York

An Imprint of Sterling Publishing
1166 Avenue of the Americas
New York, NY 10036

STERLING and the distinctive Sterling logo are registered trademarks of Sterling Publishing Co., Inc.

Text © 2016 by Derek B. Lowe

All rights reserved. No part of this publication may be reproduced, stored in a retrieval system, or transmitted, in any form or by any means (including electronic, mechanical, photocopying, recording, or otherwise) without prior written permission from the publisher.

ISBN 978-1-4549-1180-7

Distributed in Canada by Sterling Publishing
c/o Canadian Manda Group, 664 Annette Street
Toronto, Ontario, Canada M6S 2C8
Distributed in the United Kingdom by GMC Distribution Services
Castle Place, 166 High Street, Lewes, East Sussex, England BN7 1XU
Distributed in Australia by Capricorn Link (Australia) Pty. Ltd.
P.O. Box 704, Windsor, NSW 2756, Australia

For information about custom editions, special sales, and premium and corporate purchases, please contact Sterling Special Sales at 800-805-5489 or specialsales@sterlingpublishing.com.

Manufactured in China

2 4 6 8 10 9 7 5 3 1

www.sterlingpublishing.com

Opposite: The element bismuth melts easily (like its chemical neighbors, mercury and lead) and forms these characteristic "staircase" crystals as it cools. The rainbow colors come from a microscopically thin layer of bismuth oxide on the surface.

To my wife, Tanaz

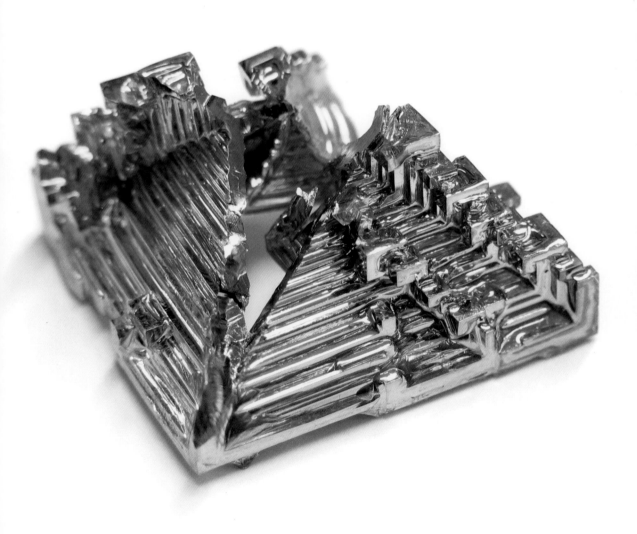

Contents

Introduction 10

c. 500,000 BCE Crystals 14
c. 3300 BCE Bronze 16
c. 2800 BCE Soap 18
c. 1300 BCE Iron Smelting 20
c. 1200 BCE Purification 22
c. 550 BCE Gold Refining 24
c. 450 BCE The Four Elements 26
c. 400 BCE Atomism 28
210 BCE Mercury 30
c. 60 CE Natural Products 32
c. 126 Roman Concrete 34
c. 200 Porcelain 36
c. 672 Greek Fire 38
c. 800 The Philosopher's Stone 40
c. 800 Viking Steel 42
c. 850 Gunpowder 44
c. 900 Alchemy 46
c. 1280 Aqua Regia 48
c. 1280 Fractional Distillation 50
1538 Toxicology 52
1540 Diethyl Ether 54
1556 *De Re Metallica* 56
1605 The Advancement of Learning 58
1607 Yorkshire Alum 60
1631 Quinine 62
1661 *The Sceptical Chymist* 64
1667 Phlogiston 66
1669 Phosphorus 68
1700 Hydrogen Sulfide 70

c. 1706 Prussian Blue 72
1746 Sulfuric Acid 74
1752 Hydrogen Cyanide 76
1754 Carbon Dioxide 78
1758 Cadet's Fuming Liquid 80
1766 Hydrogen 82
1774 Oxygen 84
1789 Conservation of Mass 86
1791 Titanium 88
1792 Ytterby 90
1804 Morphine 92
1805 Electroplating 94
1806 Amino Acids 96
1807 Electrochemical Reduction 98
1808 Dalton's Atomic Theory 100
1811 Avogadro's Hypothesis 102
1813 Chemical Notation 104
1814 Paris Green 106
1815 Cholesterol 108
1819 Caffeine 110
1822 Supercritical Fluids 112
1828 Beryllium 114
1828 Wöhler's Urea Synthesis 116
1832 Functional Groups 118
1834 Ideal Gas Law 120
1834 Photochemistry 122
1839 Polymers and Polymerization 124
1839 Daguerreotype 126
1839 Rubber 128
1840 Ozone 130
1842 Phosphate Fertilizer 132

1847 Nitroglycerine 134	1893 Borosilicate Glass 202
1848 Chirality 136	1893 Coordination Compounds 204
1852 Fluorescence 138	1894 The Mole 206
1854 Separatory Funnel 140	1894 Asymmetric Induction 208
1856 Perkin's Mauve 142	1894 Diazomethane 210
1856 Mirror Silvering 144	1895 Liquid Air 212
1859 Flame Spectroscopy 146	1896 Greenhouse Effect 214
1860 Cannizzaro at Karlsruhe 148	1897 Aspirin 216
1860 Oxidation States 150	1897 Zymase Fermentation 218
1861 Erlenmeyer Flask 152	1897 Hydrogenation 220
1861 Structural Formula 154	1898 Neon 222
1864 Solvay Process 156	1900 Grignard Reaction 224
1865 Benzene and Aromaticity 158	1900 Free Radicals 226
1868 Helium 160	1900 Silicones 228
1869 The Periodic Table 162	1901 Chromatography 230
1874 Tetrahedral Carbon Atoms 164	1902 Polonium and Radium 232
1876 Gibbs Free Energy 166	1905 Infrared Spectroscopy 234
1877 Maxwell-Boltzmann Distribution 168	1907 Bakelite® 236
1877 Friedel-Crafts Reaction 170	1907 Spider Silk 238
1878 Indigo Synthesis 172	1909 pH and Indicators 240
1879 Soxhlet Extractor 174	1909 Haber-Bosch Process 242
1881 Fougère Royale 176	1909 Salvarsan 244
1883 Claus Process 178	1912 X-Ray Crystallography 246
1883 Liquid Nitrogen 180	1912 Maillard Reaction 248
1884 Fischer and Sugars 182	1912 Stainless Steel 250
1885 Le Châtelier's Principle 184	1912 Boranes and the Vacuum-Line Technique 252
1886 Isolation of Fluorine 186	1912 Dipole Moments 254
1886 Aluminum 188	1913 Mass Spectrometry 256
1887 Cyanide Gold Extraction 190	1913 Isotopes 258
1888 Liquid Crystals 192	1915 Chemical Warfare 260
1891 Thermal Cracking 194	1917 Surface Chemistry 262
1892 Chlor-Alkali Process 196	1918 Radithor 264
1892 Acetylene 198	1920 Dean-Stark Trap 266
1893 Thermite 200	

Year	Topic	Page
1920	Hydrogen Bonding	268
1921	Tetraethyl Lead	270
1923	Acids and Bases	272
1923	Radioactive Tracers	274
1925	Fischer-Tropsch Process	276
1928	Diels-Alder Reaction	278
1928	Reppe Chemistry	280
1930	Chlorofluorocarbons	282
1931	Sigma and Pi Bonding	284
1931	Deuterium	286
1932	Carbonic Anhydrase	288
1932	Vitamin C	290
1932	Sulfanilamide	292
1933	Polyethylene	294
1934	Superoxide	296
1934	The Fume Hood	298
1935	Transition State Theory	300
1935	Nylon	302
1936	Nerve Gas	304
1936	Technetium	306
1937	Cellular Respiration	308
1937	Elixir Sulfanilamide	310
1937	Reaction Mechanisms	312
1938	Catalytic Cracking	314
1938	Teflon®	316
1939	The Last Element in Nature	318
1939	*The Nature of the Chemical Bond*	320
1939	DDT	322
1940	Gaseous Diffusion	324
1942	Steroid Chemistry	326
1942	Cyanoacrylates	328
1943	LSD	330
1943	Streptomycin	332
1943	Bari Raid	334
1944	Birch Reduction	336
1944	Magnetic Stirring	338
1945	Penicillin	340
1945	Glove Boxes	342
1947	Antifolates	344
1947	Kinetic Isotope Effects	346
1947	Photosynthesis	348
1948	Donora Death Fog	350
1949	Catalytic Reforming	352
1949	Molecular Disease	354
1949	Nonclassical Ion Controversy	356
1950	Conformational Analysis	358
1950	Cortisone	360
1950	Rotary Evaporator	362
1951	Sanger Sequencing	364
1951	The Pill	366
1951	Alpha-Helix and Beta-Sheet	368
1951	Ferrocene	370
1951	Transuranic Elements	372
1952	Gas Chromatography	374
1952	Miller-Urey Experiment	376
1952	Zone Refining	378
1952	Thallium Poisoning	380
1953	DNA's Structure	382
1953	Synthetic Diamond	384
1955	Electrophoresis	386
1956	The Hottest Flame	388
1957	Luciferin	390
1958	DNA Replication	392
1960	Thalidomide	394
1960	Resolution and Chiral Chromatography	396
1961	NMR	398
1962	Green Fluorescent Protein	400
1962	Noble Gas Compounds	402

1962	Isoamyl Acetate and Esters	*404*	
1963	Ziegler-Natta Catalysis	*406*	
1963	Merrifield Synthesis	*408*	
1963	Dipolar Cycloadditions	*410*	
1964	Kevlar®	*412*	
1965	Protein Crystallography	*414*	
1965	Cisplatin	*416*	
1965	Lead Contamination	*418*	
1965	Methane Hydrate	*420*	
1965	Woodward-Hoffman Rules	*422*	
1966	Polywater	*424*	
1967	HPLC	*426*	
1968	BZ Reaction	*428*	
1969	Murchison Meteorite	*430*	
1969	Gore-Tex®	*432*	
1970	Carbon Dioxide Scrubbing	*434*	
1970	Computational Chemistry	*436*	
1970	Glyphosate	*438*	
1971	Reverse-Phase Chromatography	*440*	
1972	Rapamycin	*442*	
1973	B_{12} Synthesis	*444*	
1974	CFCs and the Ozone Layer	*446*	
1975	Enzyme Stereochemistry	*448*	
1976	PET Imaging	*450*	
1977	Nozaki Coupling	*452*	
1979	Tholin	*454*	
1980	Iridium Impact Hypothesis	*456*	
1982	Unnatural Products	*458*	
1982	MPTP	*460*	
1983	Polymerase Chain Reaction	*462*	
1984	Electrospray LC/MS	*464*	
1984	AZT and Antiretrovirals	*466*	
1984	Quasicrystals	*468*	
1984	Bhopal Disaster	*470*	
1985	Fullerenes	*472*	
1985	MALDI	*474*	
1988	Modern Drug Discovery	*476*	
1988	PEPCON® Explosion	*478*	
1989	Taxol®	*480*	
1991	Carbon Nanotubes	*482*	
1994	Palytoxin	*484*	
1997	Coordination Frameworks	*486*	
1998	Recrystallization and Polymorphs	*488*	
2001	Click Triazoles	*490*	
2004	Graphene	*492*	
2005	Shikimic Acid Shortage	*494*	
2005	Olefin Metathesis	*496*	
2006	Flow Chemistry	*498*	
2006	Isotopic Distribution	*500*	
2009	Acetonitrile	*502*	
2010	Engineered Enzymes	*504*	
2010	Metal-Catalyzed Couplings	*506*	
2013	Single-Molecule Images	*508*	
2025	Hydrogen Storage	*510*	
2030	Artificial Photosynthesis	*512*	

Notes and Further Reading *514*
Index *525*
Image Credits *528*

Introduction

Electrons, protons, and neutrons form atoms—that's physics. But atoms bond together to form molecules, and that's chemistry. Beginning textbooks for chemistry students tend to say something about the "Central Science" to emphasize chemistry's role in scientific progress—and in the process make their readers feel better about taking the course—but that characterization is truly accurate. Chemistry really does occupy the middle ground between physics and biology, claiming territory from each of those sciences as well as its own. A look through this book shows what that means in practice. There are entries that straddle the borderline between physical chemistry and chemical physics, and others that land somewhere between biological chemistry and chemical biology. (And yes, those are all real names for fields of study, although even their own practitioners might disagree about what falls where!)

The study of chemistry is older than human writing. Only archaeologists could tell us when and where the early chemical experiments might have taken place, and the very first stirrings surely left no traces at all. When some distant ancestor of ours wondered about fire and its effects, thought about the colors of rocks and pigments, or ground up plants for medicine, they were making the sorts of chemical explorations that continue today. A modern chemist has connections throughout human history, to Bronze Age metalsmiths and Egyptian priests, to Chinese scholars and Persian alchemists. We can look back on many of these people and remark on all the things that they got wrong, but what's important is what they got right, because that built the science we have now.

It's worth remembering, too, that science itself is a very recent thing. Note the concentration of this book's entries on the historical timeline: There is a long, slow buildup, with practical discoveries in things like metals, building materials, and weaponry. A less practical pursuit (to our eyes) was alchemy, where searches for how to transmute metals and brew elixirs of life went on for centuries with no success. Along the way, though, the alchemists learned how to distill, purify, and classify the substances they worked with, and they laid the foundations for modern chemistry without realizing it. Sometime in the 1600s, in the twilight of alchemy, the sun began to come up on what we would recognize as modern science. Discovery built on discovery as the new breed of natural scientists learned how to do systematic, reproducible experiments. The 1700s eclipse everything before them, but the 1800s easily outdo them in turn.

The entries in this book don't have to be read in order, but here's a brief tour of what you'll encounter if you do. Experiments with gases of all sorts were cutting-edge science in the 1700s and early 1800s, and studying them proved an ideal way to learn how elements combined into compounds. Electricity then provided a way to make new chemical reactions happen that had never been seen before, and the field struggled to make sense of all the new elements and transformations that were being found so quickly. Organic chemists were busy isolating new substances from plants and other natural sources, and attempts to understand their structures gradually led to the realization that chemical compounds formed complex three-dimensional shapes.

The nineteenth century was also the era when some of the simplest questions finally began to be answered: Why are some chemicals so brightly colored, while others are clear? Why are some of them silvery metals that can only be melted in the hottest furnaces, while others are gases, some lighter than air? What makes some of them give off light, or even burst into flames, if opened to the air? Before the 1800s, these questions must have seemed nearly impossible to reconcile into theories that made sense, but a huge amount of work and several key advances began to make that possible.

By the early twentieth century, it was becoming clear that many substances were *polymers*—startlingly long chains of simpler molecules joined end to end. Living cells themselves use several of these, and polymer chemists found themselves creating everything from rubber and cornstarch to polyethylene in a quest to understand how these compounds worked. Meanwhile, organic chemists and inorganic chemists found themselves unexpectedly joining forces to produce a huge array of organometallic compounds—never before seen—and analytical chemistry moved into territories that no one had even realized were possible, with techniques like mass spectrometry revealing the weights of individual types of molecules.

World War II had an extraordinary effect on all technological fields. The war began with biplanes but finished with jet engines and guided missiles, and chemistry underwent similar changes. Petroleum chemistry, radioisotopes, and antibiotics were just three fields that advanced almost beyond recognition, but all parts of the science sped up dramatically. Then, by the end of the 1950s, DNA and protein sequences were recognized as being the keys to living systems and were deciphered for the first time during the 1960s. Analytical chemists were changing the science with new kinds of chromatography and NMR (nuclear magnetic resonance) machines, and medicinal chemists were taking the compounds of nature (antibiotics and steroids) and modifying their very structures.

The 1970s and 1980s saw the beginnings of molecular biology, a field that has moved biologists ever closer to chemistry's point of view. Chromatography and mass spectrometry began to merge into the most powerful analytical techniques ever developed, and the revolution in computer processing power turned X-ray crystallography calculations into just another afternoon in the lab.

The last twenty years have seen the rise of nanotechnology, with chemists starting to design and build molecular tools and scaffolds that would have been impossible to figure out in an earlier era. There has been a similar explosion of effort in what's now called *chemical biology*, using the techniques of chemistry to alter, probe, and understand proteins and other molecules of life. New organic chemistry reactions, new analytical equipment, and new computational power have all come together to make the field what it is today. If we're going to take carbon dioxide out of the air and turn it back into useful compounds and fuels in order to keep new pollutants from even being used while cleaning up the old ones, if we're going to synthesize new medicines or make new exotic materials stronger and lighter than anything before, we'll need these latest breakthroughs in chemistry.

It's easy, in this age, to take all our chemical knowledge for granted, but remember, what might seem mundane to us, our ancestors would have regarded as miracles (or as clear evidence of witchcraft). The story of chemistry is the story of mankind learning to write the missing instruction manuals for the physical world. It has taken perseverance, bravery, all the intelligence we can bring to bear, and no small amount of borderline craziness to get us to where we are today. And I've been very glad, through writing this book, to salute all the people who have made it happen.

The story continues. I myself am a professional chemist and wrote this book on nights and weekends. During the day, I'm in the lab, as are thousands of chemists around the world, helping to write the next chapters. Readers are welcome to visit my website, In the Pipeline (pipeline.corante.com), where I talk about the kind of work I do and give updates on the field's cutting-edge advances.

About This Book

Note that the dates given are often the dates of discovery, but in other instances they indicate the year in which a discovery or concept gained wide acceptance in the scientific community. Benzene, to pick one example, had been known for decades before 1865, but that's when its real structure was first worked out—a discovery that set off many others in turn. Many discoveries do not have precise origins, being spread out over a wide time frame (or among many different people). Spider silk was first analyzed chemically in 1907, but even after over a century of work by untold numbers of chemists, we still don't quite understand how it does what it does. In some cases I picked another landmark as the year, such as the date a Nobel Prize was awarded or other large development was made. For instance, chemical reactions that can take carbon dioxide out of the air (on a small scale) have been known since the 1800s, but a dramatic practical example of this knowledge came when this technique saved the lives of the *Apollo 13* astronauts in 1970. On a larger scale, the amount of carbon dioxide in the atmosphere has been a huge topic over the last twenty-five years, but the "greenhouse effect" was first understood back in 1896. You may be surprised at how early (or how late) some discoveries show up.

Crystals

A great number of chemical compounds will form crystals under the right conditions. Temperature is crucial—many things that we think of as liquids (or even gases) under "ordinary" conditions will crystallize if they're cooled down enough. Generally, a compound has to be fairly pure and fairly concentrated to crystallize well, and it has to have a regular enough structure to arrange itself into a repeated pattern. Compounds made of long disorderly chains—such as paraffin or fatty acids—turn into waxy solids, instead.

Crystallization also depends on how quickly a solution is cooled down and how it is stirred. Two spectacular examples of crystallization can be found in Mexican caves discovered by a mining operation. In 1910, Cueva de las Espadas (Cave of Swords), with impressive one-meter-long gypsum (calcium sulfate) crystals, was found around four hundred feet below sea level. It wasn't until the year 2000 that miners uncovered the majestic Cueva de los Cristales (Cave of Crystals) at a depth of one thousand feet. The best theory for what led to these colossal formations (the largest being around forty feet tall and weighing fifty-five tons) involves processes that can only happen on geological timescales. The cave formed on the Naica fault line in the Chihuahuan Desert of north-central Mexico and filled with ground water, which a magma chamber kept heated for hundreds of thousands of years. The water became totally saturated with calcium sulfate, which dissolves well in hot water, and then the whole mixture spent another extended period of at least five hundred thousand years slowly cooling down. These are perfect conditions for growing huge crystals, and it's not easy to find any larger than these.

Gypsum itself is a common mineral that crystallizes in several different forms, depending on the conditions. It's one of the main ingredients in plaster (with the common name *plaster of paris* referring to the ancient gypsum mines found in the Montmartre district of Paris), but no other deposits around the world are as spectacular as those found at the Cave of Crystals.

SEE ALSO X-Ray Crystallography (1912), Quasicrystals (1984), Coordination Frameworks (1997), Recrystallization and Polymorphs (1998)

In what looks like a scene from a science fiction movie, a cave explorer stands among some of the largest crystals on Earth in the Cave of Crystals.

Bronze

Bronze is the first metal that gets its own age, which began around 3300 BCE in Mesopotamia. Other metals were certainly in use before it—especially copper—but the addition of a small amount of tin to existing copper technology changed everything. Bronze was a step up in hardness, durability, and resistance to corrosion. Unfortunately, tin and copper ores generally aren't found together, which meant that an area rich in one ingredient had to trade for the other. Beginning around 2000 BCE, tin from Cornwall (southwest Britain) was in such demand that it turned up in many eastern Mediterranean archaeological sites, thousands of miles away.

We don't know much about these early chemists and metallurgists, but it's clear that they experimented with whatever they had on hand. Bronze alloys have turned up with all sorts of other metals in them—lead, arsenic, nickel, antimony, and even precious metals like silver. Those must have taken especially large amounts of nerve to add to the mix, since it was almost certain at the time that you would never see them again (the techniques to repurify such metals would not arrive for many centuries).

And thus, the long human adventure with metallurgy began—one that is nowhere near over. Bronze itself has been improved over the years—the Greeks added more lead to make the resulting alloy easier to work with, and the addition of zinc takes you into the various forms of brass. Modern bronzes often have aluminum or silicon in them, which were completely unknown to the ancients. If you want to see real, old-fashioned bronze of a kind that would have been recognized thousands of years ago, take a close look at a drum kit. Bronze has been the preferred metal for bells and cymbals for hundreds of years. The more tin in the mix, the lower the timbre, but there is no record of what adding arsenic or silver might do to the sound.

SEE ALSO Iron Smelting (c. 1300 BCE), *De Re Metallica* (1556)

This ancient, Chinese bronze bell may have been part of a larger set, tuned and shaped to produce different notes. Casting bronze instruments to such specific tolerances is a serious technical challenge.

Soap

Soapmaking may sound like a humble craft, but it's the first chemical preparation that we have any record of. Sumerian tablets dating from 2800 BCE make mention of soap-like materials, and soap's use in washing wool was described three hundred years later. Another Sumerian clay tablet from 2200 BCE offered a formula that would still work today: water, alkali from ashes, and oil.

Recipes of many kinds are found throughout the Egyptian, Roman, and Chinese records, but they all have the same chemistry behind them. No matter the plant or animal source, oils and fats are all triglycerides—a molecule of glycerol, also called *glycerin*, with three long-chain fatty acids attached to it as esters. The ester groups can be broken (hydrolyzed) in the presence of a strong alkali in water. In preindustrial times, the most reliable source of alkaline compounds was extract of wood ash (which we now know contained potassium carbonate). Treating this compound with the mineral slaked lime (calcium hydroxide) led to an even more alkaline potassium hydroxide, known as "lye" or "caustic soda," which is excellent soapmaking material.

After the hydrolysis reaction (combining with water), what's left are free glycerol molecules and the potassium salts of the fatty acids that were attached to it. Those molecules straddle the fence when it comes to their behavior in water: The acid/salt ends are completely water-soluble, but the long carbon chains behind them are not. They attract other greasy substances, however, which are pulled along into the water by the polar salt ends—effective for degreasing wool in a Sumerian stream.

That dissolves-both-ways trick turns out to be extremely useful in a world where almost everything can be divided into the categories "dissolves in water" and "dissolves in oil." In the twentieth century it was discovered that similar molecules (along with **cholesterol**) make up the membranes of every living cell. These molecules form a bilayer, with polar water-soluble ends interacting with water-based fluids inside and outside the cell, while the long nonpolar chains in the middle of the membrane form a protective barrier. This ensures that cell contents don't leak out and unwanted material from outside the cell doesn't soak in.

SEE ALSO Cholesterol (1815), pH and Indicators (1909), Isoamyl Acetate and Esters (1962)

Bright blues and yellows mean that a soap bubble's thickness is down to only two hundred to three hundred nanometers (billionths of a meter), which is shorter than the wavelengths of visible light.

Iron Smelting

The Iron Age definitively replaced the **Bronze** Age, so you would assume that the newly available iron must have been clearly superior. Not so—good bronze was harder and much more corrosion-resistant. However, major disturbances and population movements in the Mediterranean and Near East around 1300 BCE may have disrupted the metal trade that bronze-working depended on. Iron ore was much easier to come by, but higher-temperature furnaces were needed to smelt it, and these often depended on forced air. Iron production was thus, sometimes, a seasonal event, with furnaces built to take advantage of monsoons and other dependable winds. Objects made of iron from before 1300 BCE are known but uncommon, and many of these are not even from our own planet—produced from solid nickel-iron meteorites, they must have been very valuable objects indeed.

Given a chance, iron will react with oxygen to produce rust (iron oxide), and smelting iron ore is basically the reverse process. The early iron-smelting device, a clay or stone furnace with air inlet tubes, was called a *bloomery*. Charcoal and iron ore were heated, producing a lump of crude smelted iron (the *bloom*) in the bottom of the furnace. This was a laborious process, since the bloom needed further heating, and lumps of impurities had to be beaten out before it could be useful. Still, iron technology spread rapidly, and it seems to have been discovered independently in several locations, including India and sub-Saharan Africa. Ancient wind-driven iron furnaces evolved into the modern blast furnace—in which ore is fed in continuously from the top and has its oxygen stripped away by contact with carbon monoxide gas of ferocious temperatures—as early as the first or second century BCE in China.

Iron's properties change dramatically depending on what is mixed into it. Careful addition of some of the charcoal's carbon produces steel—a superior metal in every way—but this was a job for experienced craftsmen: too little carbon produced soft wrought iron, while too much carbon yielded a very hard metal that is too brittle for most uses. Now, the varieties of iron alloys and steels in modern metallurgy are almost too many to count.

SEE ALSO Bronze (c. 3300 BCE), Viking Steel (c. 800), *De Re Metallica* (1556), Aluminum (1886), Stainless Steel (1912).

A modern blast furnace can produce molten iron on a scale that ancient craftsmen could only dream of. But by any route, ironworking has always been a very energy-intensive process.

Purification

Who was the first chemist whose name we know? That record is currently held by Tapputi, a palace overseer and perfume maker mentioned on a Babylonian tablet from about 1200 BCE. The cuneiform text describes her using various scented raw materials (myrrh, balsam, and the like), filtering off impurities, and heating the results in water to collect the vapors. That also makes this tablet the oldest reference to distillation and filtration—processes that are still familiar to every working chemist today.

The science of scent has been the engine of more chemical discovery than you might think. Human civilization has cared about attractive smells for a very long time, and we've learned a lot about **natural-product** chemistry in the process. Long before anyone knew how to use chemistry to produce useful medicines, ancient chemists produced high-value perfumes (which were often considered medicinal as well). Many techniques were developed to extract and concentrate essences from flowers, barks, seeds, and other sources. Some of these extracts can stand up to heat (as with distillation) and be concentrated that way, but delicate extracts have to be purified at lower temperatures. This led to experiments with different kinds of solvents and separations, such as soaking aromatic plants in oils or alcoholic solutions.

It seems likely that Tapputi would quickly grasp the concepts behind a modern **rotary evaporator** and could surely put one to good use. We know almost nothing about her besides what is found on this single tablet. But it's enough.

An entire Bronze Age perfume factory has been excavated in Cyprus, showing how valuable this sort of work was in the ancient world. Modern perfume factories largely use synthetic molecules (starting with the groundbreaking **Fougère Royale** blend) with costly natural extracts reserved for the most expensive items in the catalog.

SEE ALSO Natural Products (c. 60 CE), Fractional Distillation (c. 1280), Separatory Funnel (1854), Fougère Royale (1881), Chromatography (1901), Zone Refining (1952), Rotary Evaporator (1950), Isoamyl Acetate and Esters (1962).

Earlier than Tapputi's time, this Fourth Dynasty (c. 2500 BCE) Egyptian tomb decoration shows the making of lily-scented perfume using methods the Babylonians would have known well.

Gold Refining

Croesus (595–c. 547 BCE)

The desire for metals has been a strong motivator in applied chemistry for thousands of years. While **bronze** and steel were needed for weapons and tools, it's generally been the case that gold could buy you a lot of tin, copper, and iron to make them with. Gold has been recognized since prehistoric times for its vivid color, resistance to all corrosion, and malleability that allows metalsmiths to fashion almost any shape in any thickness. All the gold that has ever been mined—beginning in the Caucasus region (present-day Georgia) before 3000 BCE—could be stored in a small, unimpressive warehouse. Many ancient civilizations used water to sluice away rock debris to concentrate gold dust. The Egyptians had an early start on this process (as their magnificent tombs would attest), and later the Romans mined gold (and other metals) on a truly industrial scale.

Throughout history, a discovery of gold has always meant big changes in the neighborhood. Starting around 550 BCE, fortunes changed for King Croesus of Lydia (in present-day Turkey) when he presided over the invention of a new method of refining, generating pure gold from the naturally occurring gold-silver alloy known as *electrum*. This became a big business very quickly. Archaeologists have discovered an extensive gold-refining site at ancient Sardis, where the small Pactolus River deposited "golden sands" from Mount Tmolus, and Lydian chemists industrialized a technique involving molten lead and common salt to produce gold and silver coinage metal. The fantastic wealth that these metallurgists produced has kept the phrase *rich as Croesus* in use ever since.

The chemistry used by the Lydians is being worked out from analysis of pieces of the ancient ovens, metal fragments in the cracks of the crucibles, and even the dirt floors of the sites. No written details of the process have ever been found (it was surely a state secret). From assaying their coins, though, it's clear that once this method was discovered, the Lydians used their pure silver supply to dilute the gold content of their existing electrum currency, degrading the coins' intrinsic value while maintaining their face value with a special stamp. And so the profits rolled in!

SEE ALSO Bronze (c. 3300 BCE), Iron Smelting (c. 1300 BCE), Alchemy (c. 900), Aqua Regia (c. 1280), *De Re Metallica* (1556), Electroplating (1805), Cyanide Gold Extraction (1887)

Electrum pieces from Lydia and the rest of the Greek world were stamped with images of kings, heroes, myths, and animals. Note the variations in color according to the amount of silver present.

The Four Elements

Empedocles (c. 490–430 BCE), **Plato** (c. 428–c. 347 BCE), **Aristotle** (384–322 BCE), **Abū Mūsā Jābir ibn Ḥayyān** (c. 721–c. 815)

Thanks to the Greek philosopher Empedocles and his mid-fifth-century BCE poem *On Nature*, for almost two thousand years people thought that there were only four fundamental elements—earth, air, fire, and water—and that the composition of everything in the world depended on the varying proportions of them. So, why is this a milestone in chemistry? In fact, Empedocles was quite correct in the idea that there are such fundamental substances (which he called *roots*), and compared to some other philosophical systems, his conclusions were quite a leap forward. The world wasn't made up of just one substance that somehow manifested itself to us in different ways, and it wasn't made up of uncountable different substances, either. Rather, the world contained a countable number of basic building blocks from which everything else was assembled. From that standpoint, the difference between the four classical elements and the modern **periodic table** is merely a matter of degrees.

It was Plato who introduced the term *element*, and his famous student Aristotle worked out a scheme by which the characteristics of everything else could be understood as mixtures of them, giving each element two of four sensible qualities (e.g., air: wet-hot, earth: dry-cold) and adding a superior fifth element he called *aether*. Later philosophers introduced more elements and complexities to try to explain more phenomena, and the entire scheme of thought led eventually into **alchemy** via the writings of the Persian polymath Abū Mūsā Jābir ibn Ḥayyān (a.k.a. "Geber") more than a thousand years later.

This is an early example of *reductionism*: the search for knowledge by breaking things down into smaller and smaller units in the hope that fundamental truths will eventually be revealed. (In effect, the scientist asks, "OK then, what's *that* made of?" over and over.) Reductionism doesn't always work—some very important effects only show up when you start going back up the scale. (A living cell is far more than the sum of the chemical elements inside it, for example.) But reductionism is still a powerful technique that has long helped to advance chemistry and the other sciences.

SEE ALSO The Philosopher's Stone (c. 800), *The Sceptical Chymist* (1661), The Periodic Table (1869)

Earth, air, fire, and water: the building blocks of the world for two thousand years.

Atomism

Democritus (c. 460–c. 370 BCE)

When, you look back on the ancient Greeks, sometimes it's as if the years between them and us have suddenly grown thin enough to see through. You come across a theory that seems to be so perfectly on target that you wish you could reach out and shake the originator's hand. British mathematician G. H. Hardy captured this feeling when he said that, to him, the Greek mathematicians were "fellows of another College."

The fifth-century Greek philosopher Leucippus seems to have been the originator of the idea of *atomism*, which takes reductionism all the way down, proposing that everything is made out of extremely small, indivisible particles. His more famous pupil, Democritus, developed the theory further: Atoms, he believed, came in a huge number of varieties, and the physical properties of materials had something to do with the microscopic properties of the atoms themselves. Some of them were very slippery and tumbled easily past each other, while others stuck together strongly to make hard and dense materials. The explanations for why the atoms acted this way were not (of course) very sophisticated, but the key points are absolutely correct, and the theory is rightly considered one of the great achievements of Greek thought.

Atomism was also noteworthy for being thoroughly *materialist*—that is, it didn't try to explain things through talk of purposes and desires. Rather, things were treated mechanistically; we see something happening, therefore something "material" must have happened earlier to cause it. To take an example, a rock is hard because of some physical reason that could be investigated, not just because it was somehow necessary for it to be hard in the grand scheme of things. The outlines of modern scientific thinking are clearly visible here.

SEE ALSO The Sceptical Chymist (1661), Dalton's Atomic Theory (1808), Maxwell-Boltzmann Distribution (1877)

Democritus, in this 1628 work by Dutch painter Hendrick ter Brugghen (1588–1629), comes out looking rather Dutch himself.

Mercury

Qin Shi Huang (260–210 BCE)

Mercury has been considered strange and valuable since prehistory. Not only is it one of the elements whose pure metallic form can be found in nature without any refining, but it's also the only metallic element that stays liquid at common environmental temperatures. This gave it a reputation for magical powers that lasted for thousands of years. The problem is, this bizarrely heavy, shiny liquid is also quite poisonous. Interestingly, the pure metal isn't as hazardous as many of its compounds, which are taken up much more easily into the body, but exposure to the fumes that mercury slowly releases is a very bad idea because the toxic effects are essentially irreversible: it reacts with important sulfur-containing groups on many proteins and biomolecules, and once that happens, it cannot be removed.

Qin Shi Huang is the famous "first emperor of China," and he is also the first major customer for mercury that we're aware of. His prodigious buried army of thousands of life-size terra-cotta soldiers was discovered in 1974, to the amazement of archaeologists. According to *Records of the Grand Historian*, a monumental history of ancient China completed in 109 BCE by Han dynasty official Sima Qian, the rest of the tomb was equally grandiose, with scaled-down replicas of landscapes and palaces, interlaced with dozens of rivers of flowing mercury. Probes of the soil around it show greatly elevated levels of mercury, suggesting that the ancient descriptions were not exaggerated.

The emperor may actually have poisoned himself with mercury-containing medicines meant to make him immortal. Mercury compounds have been used as medicines for centuries, usually with similarly poor results, although they sometimes cured syphilis. In the modern era, mercury has been used not only in thermometers but also in electrical switches and fluorescent lights. This has led, however, to some terrible industrial pollution. Mercury compounds have concentrated in the food chain, winding up in potentially dangerous levels in some otherwise edible fish.

SEE ALSO The Philosopher's Stone (c. 800), Toxicology (1538), Mirror Silvering (1856), Salvarsan (1909), Boranes and the Vacuum-Line Technique (1912), Thallium Poisoning (1952)

Some of the famous terra-cotta warriors. Nearby, in unexcavated areas, there may still be enough mercury to make future archaeologists step carefully.

Natural Products

Pedanius Dioscorides (c. 40–c. 90 CE)

Dioscorides was a first-century Greek physician and student of medicinal plants. His position as a military surgeon with the Roman legions allowed him to travel extensively through the ancient world, and he collected specimens and local knowledge wherever he went. His multivolume work *De materia medica* (c. 60 CE) compiled all this information into what was likely the most comprehensive guide to medicines produced in the world up until that point. Its usefulness (and a certain amount of good luck) kept it in circulation for the next 1,500 years, covering the lifespan of the Roman Empire, the rise of the Islamic world, and the start of the Renaissance.

Many drugs have their origins in the activity of medicinal plants. The natural world is full of biologically active compounds, both beneficial and poisonous, known as *natural products*. Plants, animals, bacteria, and fungi all spend some of their metabolic energy synthesizing them. Some are used by an organism's own systems, while others are produced for external use as signals or weapons, but all of them have had vast amounts of time for evolution to sharpen their effects, and we humans can now take advantage of what there is to be found.

Tracking these compounds down, isolating them, and working out how they affect the body has advanced chemistry and medicine tremendously over the centuries. Even today, natural-products chemistry is a growing field, with strange and potent molecules being isolated from marine organisms, rare plants, and other sources. Purifying and identifying these substances are challenges for analytical chemists, who now employ **NMR** (nuclear magnetic resonance) and **Electrospray LC/MS** (liquid chromatography/mass spectrometry) equipment to assist them in the process, and synthesizing them in the laboratory has advanced organic chemistry greatly.

SEE ALSO Toxicology (1538), Quinine (1631), Morphine (1804), Caffeine (1819), Indigo Synthesis (1878), Fougère Royale (1881), Asymmetric Induction (1894), Aspirin (1897), Steroid Chemistry (1942), LSD (1943), Streptomycin (1943), Penicillin (1945), Cortisone (1950), Luciferin (1957), NMR (1961), Rapamycin (1972), Unnatural Products (1982), Electrospray LC/MS (1984), Taxol (1989), Palytoxin (1994), Shikimic Acid Shortage (2005)

Dioscorides Describing the Mandrake (1909) by English painter Ernest Board (1877–1934). A variety of active (and rather poisonous) natural products have been isolated from plants of this genus.

Roman Concrete

Pliny the Elder (23–79)

Concrete is everywhere in our civilization; modern construction wouldn't be possible without it. But its chemistry is surprisingly complex, depending on two elements (aluminum and silicon) that form strong bonding networks with oxygen atoms. These species, which are abundant in Earth's crust, form the basis for a huge variety of minerals and man-made ceramics. Concrete also requires calcium ions and a reaction with water to help hold everything together, but the technical name, *hydrated calcium aluminosilicate*, although an accurate description of concrete's chemical composition, doesn't roll off the tongue very easily.

The Romans had the finest concrete of the ancient world, and some of it can still be seen today in such magnificent structures as the famous Pantheon—completed around the year 126 and still the largest unreinforced concrete dome in the world. Roman civilization, though, was actually "science deficient"; considering their power and longevity, surprisingly little basic research was done. They didn't have much patience for mathematics, blue-sky experimentation, or abstract theories, but practical improvements in civil and military engineering were always welcome. As such, the Romans developed a variety of concrete mixtures for different applications. Their water-resistant mix was of very high quality, and according to the natural philosopher Pliny the Elder, a key ingredient in the mortar was the ashy volcanic deposits (now known as *pozzolan*) from the area of Mount Vesuvius. Pliny knew the area well—too well, in the end, since he was killed in the famous 79 eruption that destroyed Pompeii.

Just in the last few years, analytical chemists have been able to work out how this recipe for Roman maritime concrete must have been made. The process requires quite a bit less energy than modern Portland cement, which was developed in nineteenth-century Britain. In terms of the fuel needed to bake the starting limestone mix, the time needed to cure the finished product, and its durability in salt water, the Roman recipe has many advantages. After almost two thousand years, it may be making a comeback.

SEE ALSO Porcelain (c. 200)

The two-thousand-year-old Pantheon in Rome still has the largest unreinforced concrete dome in the world—a solid testament to Roman engineering.

Porcelain

Ehrenfried Walther von Tschirnhaus (1651–1708), Johann Friedrich Böttger (1682–1719)

Proto-porcelain had been produced in China a couple thousand years prior, but true porcelain doesn't appear on the archaeological record until the late Han dynasty period, which ended around 220. During the Sui and Tang dynasties (581–907), porcelain was created on a much larger scale. Durable and beautiful ceramics became a valuable export, first to the Islamic world and then, after 1300, to Europe. The remarkable thing is that during this whole period, no one else in the world could produce it.

Ceramics are an extraordinarily ancient art in China, with early examples possibly going back twenty thousand years. Porcelain was probably discovered, gradually, by craftsmen pushing the techniques of pottery-making further in search of new wares to sell. Its composition varies, but a good source of kaolin clay—which takes its name from a village in southwest China—is needed. Other ingredients include ground glass and minerals such as feldspar or alabaster, quartz, and bone ash. Two key factors in porcelain production are the amount of water in the mixture, which has to be kept within narrow limits, and the high firing temperatures (over 1200°C, or 2100°F), which allow for the formation of glassy phases in the final ceramic that are mixed with fine needles of the aluminosilicate mineral mullite.

Countless attempts were made to reproduce the Chinese techniques, but the first success occurred in Saxony (now part of Germany). A self-styled alchemist named Johann Friedrich Böttger had brought enough attention to himself by 1704 that Augustus the Strong (elector of Saxony and king of Poland) imprisoned him in Dresden in hopes of forcing him to produce gold. The German physicist-physician-philosopher Ehrenfried Walther von Tschirnhaus, who had been trying to make porcelain as another revenue stream for Augustus, was put in charge of Böttger, and after they were shipped samples of kaolin clay and alabaster in 1708, the breakthrough occurred. Von Tschirnhaus died suddenly that year, and Böttger, now free, was put in charge of the new porcelain factory in Meissen in 1710. Just two years later, a Jesuit priest revealed the Chinese methods his order had witnessed, and porcelain manufacture spread rapidly throughout Europe.

SEE ALSO Roman Concrete (c. 126)

This eighteenth-century Chinese porcelain figurine of the Buddhist deity Guanyin, goddess of mercy, now resides in the Hallwyl Museum in Sweden.

Greek Fire

Theophanes the Confessor (c. 752–c. 818)

Chemistry, sadly, has also been used to wage war. The Eastern Roman (or Byzantine) Empire lasted for many centuries past the collapse of the Western Roman Empire, but the Byzantines didn't survive because they were surrounded by friends. In fact, by this date, the Byzantines were very hard pressed by Arab armies during the initial expansion of Islam, but in response they developed a secret weapon: Greek fire.

Greek fire was first described by Theophanes the Confessor in his *Chronographia* (c. 814), ascribing its invention to an architect from Heliopolis (present-day Baalbek, Lebanon) around the year 672. We have several descriptions of Greek fires use in battle; the most well-attested form of the weapon describes it as fired like a hybrid of a flamethrower and a cannon. However, there's no universally agreed-upon recipe. The preparation was enough of a state secret that it eventually was lost completely. In fact, it may never have been written down at all. The mixture almost certainly used petroleum as a base, likely from natural crude oil seepage sites around the Black Sea, and probably pine resin as well. Sulfur is a likely ingredient, but after this point scholars have argued in every direction about the recipe.

What we know is that the flaming liquid was dispensed with explosive force, generated huge amounts of smoke, would burn on top of water, and was extremely difficult to extinguish—just the kind of thing you do not want your wooden-hulled invasion fleet to encounter. The Byzantines had special ships with trained crews whose only job was its deployment, and they used it with great success against their enemies (and against each other in civil wars) for the next five hundred years. After this, Greek fire gradually disappears from all reports.

SEE ALSO Gunpowder (850), Nitroglycerine (1847), Chemical Warfare (1915), Nerve Gas (1936), Bari Raid (1943)

Greek fire, as described in the Sicilian twelfth-century Codex Skylitzes Matritensis, *the only illustrated Byzantine chronicle that has survived.*

ςαυτο μεν ναυπικον εν τω πολω σαν παλεως αμαρται· και πυ χωρις προς ορμίζε ται τω βυρυ
δορ· εκ τε ρπ κορτα και τριακοσιοπ τω μ ἀ μενον πλοίων· πολεμικοπ τε και σ παρφων
δε του βασιλικου δουλου καταρχοντες· τη τουτων μεν σκοπε ελθον· νυκτος δε επιτιθενται ραω
ο χιον τοις εναντιοις· και τας φιαλας κατα πληξα μενοι· πολλας μεν ναυ ανδρος εχον
τω ν ρκων · Αμα δε και τας σκαφας προσ ολοι πυρι

ερω μου πυρπολ   τον των εναντιον·

ι χων ταυτ πλαϊος εξω γενομένου τουπλοιου · και προς τον κολπον τουβ λαχρω κατα ραι σε

By around 800, progress in the sciences was taking place almost entirely in the Islamic and Chinese cultures. Abū Mūsā Jābir ibn Ḥayyān (known in the West as "Geber") lived in what is now Iraq, practicing medicine, **alchemy**, astrology, and numerology, which at the time were considered equally worthwhile topics (and not really separate fields of inquiry at all). His writings attracted a great number of followers, many of whom seem to have written manuscripts of their own and attached ibn Ḥayyān's name to them, setting the stage for the hopelessly tangled alchemical literature to come. It does not help that these works are often written in an elaborate symbolic style that can be impossible to decipher—works from which we derive the word *gibberish*. For example, the detailed recipe for producing live scorpions from alchemical mixtures is surely not meant to be taken at face value, but what it was really meant to signify is impossible to say.

The more comprehensible works under ibn Ḥayyān's name, however, indicate that he was a dedicated experimentalist. At one point he warned readers that the only way to attain any sort of competence was to conduct practical work in the laboratory, and most modern chemists would agree. He believed that metals were fundamentally different from nonmetals, that they were composed of **mercury** and various forms of sulfur, and that metals could be changed into each another if the secrets of these mixtures could be worked out. Some powerful reagent or elixir (later called the *philosopher's stone*), he believed, could unlock this process, and this was a compelling enough idea to set up an alchemical research program for the next few centuries. A manuscript in Latin under the name *Geber* shows up in the thirteenth century with much more theorizing in this vein. Whoever its real author is, he must have decided that ibn Ḥayyān's reputation was still potent enough to be worth borrowing—even after four hundred years.

SEE ALSO Mercury (210 BCE), Alchemy (c. 900)

Viking Steel

There are a number of Viking swords that are head and shoulders above everything else that was available to the discriminating customer in the ninth century—in fact, no better steel was to be found in Europe until the Industrial Revolution. The Vikings may have come across superior ironworking technology in their trade with Asia, based on the wootz steels of India and Sri Lanka, whose furnaces took advantage of the monsoon winds. These blades had high carbon content, with far fewer impurities than the usual processes produced. The result was a metal alloy that remained at once tough, sharp, and flexible—a great advantage if you were trying to get one's sword unstuck from an enemy's shield, or perhaps from an enemy.

These superior swords all bear the name *Ulfberht* on the lower part of the blade, indicating some sort of workshop, trade name, or Viking metalsmith. The earliest blades are attached to handles that have been carbon-dated to c. 800, and they were made for about two hundred years, with no examples dating later than about the year 1000. Like so many other technological advances for which we lack written records, the exact technique used to make them has been lost, although some educated guesses have produced similar alloys in experiments. As with modern techniques, attention to detail and quality control are critical. Metals can form a wide variety of crystalline structures while being forged, and control of these factors has never been easy to attain.

Ulfberht swords were clearly rare and highly prized, and the workshop that made them was paid a high compliment by the many other swords made from inferior steels but bearing variant spellings of the name. The inescapable conclusion is that these were copies, trading on the fame of the originals. Knockoffs have clearly been a problem for a long time.

SEE ALSO Iron Smelting (c. 1300 BCE)

A Viking sword from Hedeby, which was an important trading settlement near the current German-Danish border.

Gunpowder

Gunpowder probably was discovered by alchemists trying to transmute metals or extend life rather than by weapons engineers seeking an explosive. A Chinese military compendium from 1044 listed a number of different recipes for gunpowder, showing that it had been the subject of a lot of action-packed research and development by the middle of the Song dynasty, but the first known mention comes from a mid-ninth-century Taoist text, which stressed its dangerously flammable nature. Sulfur was of great importance in **alchemy**, and any lab of the time would have had charcoal around for fuel. The third key ingredient—the oxidizer potassium nitrate—was available as the naturally occurring mineral niter (also known as *saltpeter*) or as crystals around deposits of bat guano in caves. Whoever first combined these powders and exposed the resulting mixture to a flame must have immediately realized that they were onto something big. Extending human life, though, turned out not to be gunpowder's strong point.

Knowledge of the new weapon diffused through China and past its borders, and the Mongol invasions of the thirteenth century spread the news even farther, from India to Europe. The Chinese kept raising the amount of potassium nitrate in their gunpowder as time went on, producing bigger explosions all the time. Early artillery shells, exploding arrows, and a variety of alarming bomb designs show up in several Chinese military manuscripts. In his *Treatise on Horsemanship and Stratagems of War* (c. 1280), detailing 107 different explosive mixtures, Syrian chemist Hasan al-Rammah referred to potassium nitrate as "Chinese snow." European militaries adopted gunpowder quickly: the first illustration of a firearm—a primitive metal cannon known as a *pot-de-fer* (French for "iron pot") with an enormous arrow emerging from its barrel—appeared in a 1326 manuscript by the English scholar Walter de Milemete. For better or worse, it has been with us ever since.

SEE ALSO Greek Fire (c. 672), Alchemy (c. 900), Nitroglycerine (1847), PEPCON Explosion (1988)

An explosion of gunpowder shells during the 1274 Mongol invasion of Japan, illustrated in a scroll commissioned some twenty years after the battle.

Alchemy

Abu Bakr Muhammad ibn Zakariya' al-Razi (865–925)

The Persian polymath al-Razi is one of the most famous of the alchemists, and his writings show how much modern chemistry was mixed into the beliefs of his time. As would be expected, he worked tirelessly to understand what made metals different from one another and from the nonmetals. Although he never claimed to have produced gold from base metals, he described how to make metals *look* like gold. He then went on to classify other substances into such categories as vitriols, salts, stones, spirits, etc., and to catalog their properties (e.g., how easily they melted or caught fire). Along the way, he rejected the classical **four elements** in favor of his more complex arrangement. Paying close attention to such similarities and differences is absolutely central to chemistry.

Al-Razi is also notable for rejecting magical explanations and the idea of manipulating the physical world through symbols rather than through physical causes. His greatest contribution in the history of chemistry comes from his book *Kitab al-asrar* (*The Book Secrets*), which revealed how dedicated he was to experimentation with real substances. With lengthy, detailed description of his apparatus, the book explained the uses of crucibles, tongs, bellows, flasks, funnels, mortars, heating baths, and more, offering his contemporaries a look at his state-of-the-art equipment and giving modern-day chemists an intimate view of a working lab from over a thousand years ago. Some of the experimenters who came after al-Razi adopted his classification scheme, while others disagreed or made their own. But for centuries to come, *Kitab al-asrar* was the closest thing that alchemy had to a standard laboratory equipment manual.

SEE ALSO Gold Refining (c. 550 BCE), The Four Elements (c. 450 BCE), The Philosopher's Stone (c. 800), Gunpowder (c. 850), *The Sceptical Chymist* (1661)

As fires rage in buildings and mines in the background, a goddess warns an alchemist of the element's dangers.

There aren't all that many mixtures that are still referred to by the same term used in the Middle Ages, but aqua regia (Latin for "royal water") is a memorable liquid. The corrosive mixture—one of the few reagents that can dissolve gold—was first mentioned by an anonymous European alchemist known as Pseudo-Geber (and identified by some scholars as the Franciscan Paul of Taranto) in his late-thirteenth-century discussions on transmutation of the elements and related subjects.

You can't order aqua regia from a chemical supply house; it has to be prepared fresh because it decomposes on standing. The classic recipe is one part concentrated nitric acid to three parts concentrated hydrochloric acid. The two acids react to form a vicious brew whose least toxic component is probably plenty of dissolved chlorine gas, which will be noticeable immediately if you are so unwise as to prepare this solution outside of a good **fume hood**. Gold is attacked fairly readily, and platinum more slowly, but some related metals aren't touched at all. In fact, the elements iridium and osmium—which occur in the same ores as platinum—were first discovered by the British chemist Smithson Tennant in 1803 after he found dark residues in the bottom of a flask after the aqua regia had eaten away the platinum.

Aqua regia is still used in some high-level gold refining processes, but its most famous use came early in World War II. Germany had invaded Denmark in 1940, and at the Niels Bohr Institute in Copenhagen, the Hungarian radiochemist George Charles de Hevesy dissolved the gold Nobel Prize medals of his colleagues, the German-Jewish physicists Max von Laue and James Franck, to prevent their confiscation by the Nazis. He left the acid solution in a storeroom and returned after the war to find it undisturbed. He then precipitated the gold back out through a reduction reaction and sent the powder to Stockholm, where it was recast into medals and presented again to the Nobel laureates.

SEE ALSO Gold Refining (c. 550 BCE), Acids and Bases (1923), Radioactive Tracers (1923), The Fume Hood (1934)

Fractional Distillation

Taddeo Alderotti (c. 1210–1295)

Distillation was originally used to liberate some liquid fraction from solids, but the next step in this process was to separate liquids with different boiling points from one another. To achieve this, the mixture had to be heated slowly and passed through a longer distilling apparatus, or column, making sure that the more volatile parts were brought over first. Heat things too vigorously, though, and the lower-boiling molecules end up mixing with the higher-boiling molecules as they distill over, and everything has to be done again. Naturally, the closer the boiling points, the more patience is needed.

The thirteenth-century Florentine alchemist Taddeo Alderotti seems to be the first person to describe fractional distillation in detail in the final section of his *Consilia medicinalia* (c. 1280). Famous for his medical knowledge, he used a three-foot-long apparatus to distill alcohol to a purity of 90 percent for strictly medical purposes. His work led to a profusion of different still types and a great deal of experimentation with other substances to see what could be purified from them.

In time, the theory of distillation became better understood, with advances in alcohol production often helping to move things along. The design of the so-called *still head* showed the greatest improvement, as fractionating columns provided greater surface areas for the vapor and condensate to mix, leading to better separations. Distillation is still extremely important both industrially and on a laboratory scale. It is one of the basic purification techniques for water and for industrial solvents, it is the fundamental technology behind the fractionation of oil in a refinery, and it remains the key to the making of high-proof alcoholic beverages. Some things never change.

SEE ALSO Purification (c. 1200 BCE), Thermal Cracking (1891), Liquid Air (1895), Deuterium (1931), Rotary Evaporator (1950)

A decorative (but not very practical) look at fractional distillation of "spirits of wine" by German surgeon and alchemist Hieronymous Brunschwig, 1512. In the middle, "a tube of cold water" condenses the vapors.

Toxicology

Paracelsus (1493–1541)

Swiss alchemist and natural philosopher Paracelsus spent his life practicing medicine, metallurgy, astrology, and whatever else the market would bear. He lived when alchemy was disappearing from the world and no one was quite sure what would replace it. The centuries-old quest to make gold and silver from base metals was, by that time, a long list of cold trails and dead ends. Paracelsus pointed the way forward when he said, "Many have said of Alchemy, that it is for the making of gold and silver. For me such is not the aim, but to consider only what virtue and power may lie in medicines."

By all accounts, he was not an easy person to get along with, and he did not mellow with age (which probably helps account for the long list of destinations he turned up in during his forty-eight years). One part of his personality served him very well, though: he became well known as someone who refused to believe that something was true just because orthodoxy deemed it so. Unfortunately, he demonstrated this by publicly burning copies of ancient medical texts while he proclaimed his contempt for them.

Paracelsus is most remembered today for asserting that illness was often caused by outside agents. His studies of the (all too numerous) diseases of miners furnished him with plenty of prescient observations, such as the idea that slow-developing lung problems were caused by toxic vapors (rather than by evil mountain spirits bent on revenge, the more popular theory at the time). These gave force to his immortal adage "the dose makes the poison," which appeared in his 1538 treatise *Septem defensiones*. Everything is toxic in a high enough dose, but some things never stop being toxic, even at miniscule doses.

SEE ALSO Mercury (210 BCE), Natural Products (c. 60 CE), Diethyl Ether (1540), Hydrogen Sulfide (1700), Hydrogen Cyanide (1752), Paris Green (1814), Beryllium (1828), Aspirin (1897), Salvarsan (1909), Boranes and the Vacuum-Line Technique (1912), Radithor (1918), Tetraethyl Lead (1921), Nerve Gas (1936), DDT (1939), Bari Raid (1943), Antifolates (1947), Thallium Poisoning (1952), Thalidomide (1960), Cisplatin (1965), Lead Contamination (1965), Glyphosate (1970), MPTP (1982), Bhopal Disaster (1984), Taxol (1989), Palytoxin (1994)

This watercolor portrait of Paracelsus at age forty-seven is loosely based on a fifteenth-century engraving of the vitriolic toxicologist.

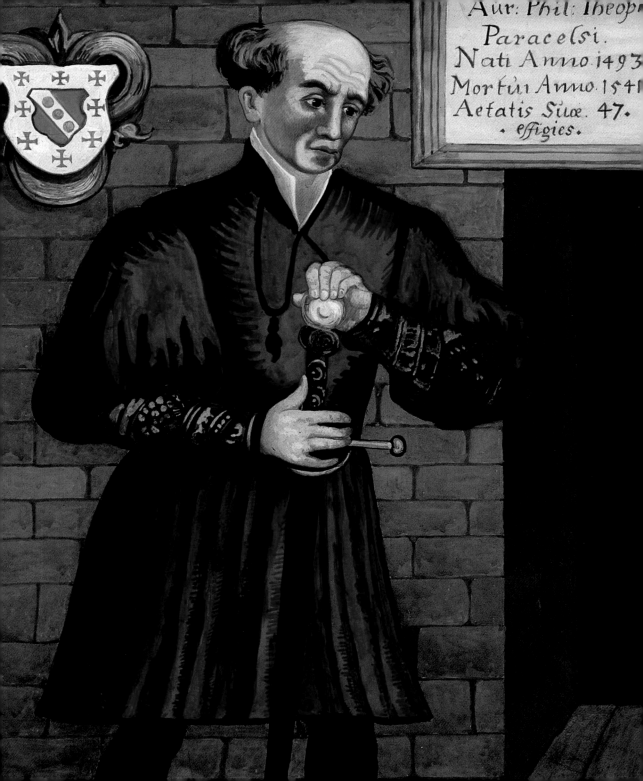

Diethyl Ether

Paracelsus (1493–1541), Valerius Cordus (1515–1544), Crawford W. Long (1815–1878)

Organic chemists divide compounds into classes based on the arrangements of their atoms. Where the oxygens, sulfurs, nitrogens, and other noncarbon heavy atoms go determines what "functional groups" the compound has, which allows us to infer a good deal about a compound's properties. One of the simplest classes is the ethers, which have an oxygen atom that looks as if it's been dropped into the middle of a chain of ordinary carbon-carbon single bonds. You can't make them by that method (yet), but the German physician and botanist Valerius Cordus discovered in 1540 that when ethyl alcohol is heated up with **sulfuric acid** (vitriol), diethyl ether is produced. (It's quite possible that earlier alchemists might have synthesized it, but no solid proof has surfaced.) The chemists of the time had no way of knowing quite what had happened—such details wouldn't be cleared up for another three centuries or so—but this "sweet oil of vitriol" was clearly a new compound.

Diethyl ether (known familiarly as simply *ether* today) is a light, low-boiling liquid with a powerful, instantly recognizable smell. The low boiling point lets it produce plenty of vapors even at room temperature, but this property also makes it wildly dangerous around flames, sparks, or even hot surfaces. Ether catches fire like few other substances can, and its heavier-than-air vapors can flow unpredictably along the floor of a room.

In his treatise *De naturalibus rebus* (c. 1540), Paracelsus noted that sufficient exposure to ether fumes could make chickens unconscious and unresponsive, and it surely wasn't long before the same effect was observed in humans. By the 1840s, diethyl ether had become the drug of choice at "ether frolics" thrown by medical students, and after the American surgeon Crawford W. Long used it to painlessly remove tumors from a patient's neck in 1842, it became the first surgical anesthetic—although it was soon replaced by less toxic (and less flammable) compounds.

SEE ALSO Toxicology (1538), Sulfuric Acid (1746), Functional Groups (1832)

This eighteenth-century engraving of ether production appeared in the British periodical Universal Magazine. *Making or using diethyl ether around an open flame would now be very strongly discouraged!*

natural philosophers of his day, he took a Latin pen name and wrote in Latin as well. At age twenty he was appointed rector of Greek at the Great School of Zwickau and went on to study physics, chemistry, and medicine. From 1527 on, all of his medical positions were held in important German mining towns, which seems to have been no accident: minerals and geology were his real passions.

De re metallica (*On the Nature of Metals*)—completed in 1550 and published in 1556—was the summing-up of all Agricola had learned, and it is one of the foundation stones of modern geology, as well as chemistry. Agricola spends a good part of the book describing the assaying of ores, the smelting and purification of metals, and the production of other reagents that were needed in the mining technology of his day. Latin, however, gave him difficulties in these sections because he had to describe things for which no classical terms existed.

Looking over the book's detailed diagrams of practical machinery and reading the painstaking procedures for the handling of various ores and liquid metals, you can almost see the tide of **alchemy** beginning to retreat. *De re metallica* relentlessly avoided any talk of the **philosopher's stone** and related concepts. (It does mention dowsing as a method for locating precious metals, only to reject it.) Instead, the book provides long descriptions of the best way to wash crushed ore before smelting it and woodcuts that show how to make a better furnace, adding words like *fluorspar*, *basalt*, and *bismuth* to the modern chemistry vernacular.

Agricola emphasized the integrity of his work in the introductory pages: "I have omitted all those things which I have not myself seen, or have not read or heard of from persons upon whom I can rely." The term *scientific method* had not been invented yet, but that way of looking at the world was clearly on its way up in the mid-sixteenth century.

SEE ALSO Bronze (c. 3300 BCE), Iron Smelting (c. 1300 BCE), Gold Refining (c. 550 BCE), Mercury (210 BCE), The Philosopher's Stone (c. 800), Viking Steel (c. 800), Alchemy (c. 900), Titanium (1791), Ytterby (1792), Beryllium (1828)

1605

The Advancement of Learning

Francis Bacon (1561–1626)

The polymathic English philosopher-statesman-scientist-jurist Francis Bacon—commonly referred to as the father of empiricism—would have been an odd fit for the modern age. But the world as we know it would not exist had he not produced such grand intellectual frameworks that encompassed science, literature, history, and religion simultaneously. In 1605, he wrote *Of the Proficience and Advancement of Learning, Divine and Human* in the form of a (long) letter to King James of England. Presenting an arrangement of all the fields of his day (generally divided into divine, natural, and human components), Bacon went beyond illuminating how the sciences relate to other fields of study and spoke about what (in his view) science was *for*. He also detailed how the goals of science were being realized and where they were falling short—themes he expanded on in his 1620 book, *Novum organum scientiarum* (*The New Instrument of Science*, commonly translated in English as *The New Method*).

Across his writings, the prevailing idea is that scientific discovery is something that will help humanity and is thus worth putting time and effort into. Earlier philosophies often emphasized some sort of spiritual duty to understand the works of God or the collection of knowledge for its own sake, but Bacon called for "a progeny of inventions" to "subdue our needs and miseries." In his utopian novel *The New Atlantis* (1627), he proclaimed science as the "effecting of all things possible"—a perspective that took hold in Western civilization as the seventeenth century progressed. When the Royal Society was founded in 1660—becoming officially "royal" with King Charles II's 1662 charter—its leaders often referred to "Baconian ideals."

We live in the world that such thinking has brought us. It seems normal to us that "science marches on," but it's worth remembering that for long periods of human history it did nothing of the kind. Francis Bacon was one of the key figures who gave science its marching orders.

SEE ALSO *The Sceptical Chymist* (1661)

Francis Bacon, the world's first highly placed science advisor.

Yorkshire Alum

Thomas Chaloner (1559–1615), Louis Le Chatelier (1815–1873)

Since Roman times, alum (a general name for a range of aluminum sulfate salts) has served a number of useful industrial and medical purposes as a water purifier, blood coagulant, pickling agent, antiperspirant, and flame retardant (among other things). It was also vital to the textile industry for fixing dyed cloth so that colors would not bleed. The British Isles imported almost all of their alum from papal territory in Italy, however, and those supplies were abruptly cut off in 1533 when King Henry VIII, seeking an annulment of his marriage to Catherine of Aragon, decided to start his own church rather than submit to Pope Clement VII.

The hunt began for native supplies, but several attempts failed before the English naturalist Sir Thomas Chaloner succeeded in Yorkshire sometime around 1600. In 1607 a company was formed to work the process on an industrial scale. That process—a brutal trek through inorganic chemistry—involved mining out gray shale, which had a high content of aluminum silicate minerals. Heaps of this rock were dumped onto woodpiles, which were set on fire and allowed to smolder for months (slower cooking meant higher yields). This caused the shale's iron sulfide to be oxidized into iron sulfate, which reacted with the aluminum-bearing minerals to create a crumbly, pink material. After this rock was soaked in large watery pits to remove the soluble sulfates, the resulting liquid was boiled down to concentrate it and treated with a potassium-bearing base. This was not an appealing step: ashes from burned seaweed were effective, but large quantities of human urine were often used instead. Finally, the desired potassium aluminum sulfate crystallized out much more readily when chilled, and this accident of solubility was used to separate it from the other (less valuable) salts.

The mining of the Yorkshire coast (and the collection of its inhabitants' urine) went on until the middle of the nineteenth century. A better route to synthetic alum was discovered in 1855 by the French chemist Louis Le Chatelier, and not long after that, the aniline-derived dyes (beginning with **Perkin's mauve**) made such fixatives obsolete.

SEE ALSO Paris Green (1814), Perkin's Mauve (1856), Indigo Synthesis (1878)

Remains of Peak Alum Works (established 1650) near Ravenscar in North Yorkshire. Sites of this kind can still be found throughout the region, providing excellent (and early) examples of "industrial ruins."

Quinine

Paul Rabe (1869–1952), **Robert Burns Woodward** (1917–1979), **William von Eggers Doering** (1917–2011), **Gilbert Stork** (b. 1921)

In 1631, a rare, expensive, extraordinary medication known as *Jesuit's bark* was brought from the New World and used in Rome. The city's local swamps were infested with malaria-carrying mosquitoes that infected countless people every year, but the Romans had not yet made the connection between the mosquitoes and the disease. In fact, the name *mal-aria* reflected the widely held belief that the disease arose from "bad vapors."

Quinine, as the active compound is now called, comes from the bark of the cinchona tree of South America. The Quechua people of Peru and Bolivia seem to have been the first to discover its medicinal properties, preparing a drink from the bark and using it to treat shivering in cold weather (quinine, as it turns out, is also a muscle relaxant). Since malaria also causes shivering and chills with its cycles of fever, it was natural for the Quechua to try the same treatment. It worked again, although not for the same reasons. Even now, although we know the bark has some kind of direct effect on the malaria parasites, our knowledge of its mechanism of action is still full of gaps.

Sometime between 1620 and 1630, Jesuit missionaries from Spain learned of the bark-based malaria treatment from the Quechua, and for the next three hundred years, quinine was the only substance known that could cure or prevent the disease. This discovery had a profound effect on human history, allowing the colonial powers of Europe to enter areas of the world that otherwise would have been deadly to them, and it also significantly advanced organic chemistry, as efforts to isolate, purify, and synthesize quinine went on for centuries. In 1944, the American chemists Robert Burns Woodward and William von Eggers Doering accomplished the first total synthesis of the molecule, based partly on German chemist Paul Rabe's 1918 partial synthesis. In 2001, Gilbert Stork synthesized quinine by a new route and started a controversy by suggesting the 1944 synthesis did not yield the correct molecule because Rabe's chemistry was not valid. Seven years later, however, both earlier routes were found to be correct.

SEE ALSO Natural Products (c. 60 CE), Caffeine (1819), Fluorescence (1852), Perkin's Mauve (1856)

A liquid quinine preparation from a London pharmacy, early 1800s. As one of the few medicines of the time with real and specific effects, it had a prominent place on pharmacy shelves.

The Sceptical Chymist

Robert Boyle (1627–1691), Robert Hooke (1635–1703)

By 1644, Robert Boyle, the fourteenth child(!) of Richard Boyle, 1st Earl of Cork (Ireland), was left enough land and resources by his father to devote his life to scientific research, which had been his great interest ever since visiting Florence, Italy (home to the then-elderly Galileo), at the age of fourteen. England was full of natural philosophers and nascent scientists at the time, and Boyle soon fell in with a London group calling itself the Invisible College. In 1654, he moved to Oxford, where he and Robert Hooke (the "English Leonardo") conducted experiments with his own version of the recently invented air pump. These experiments led to what is now known as Boyle's Law, i.e., the pressure and volume of a gas are inversely related.

Boyle went on to do an extraordinary variety of work in chemistry and physics. There was so much to be learned about sound, light, gases and liquids, crystallization, electricity, combustion, and more, that the Invisible College had its hands full, and in 1660 this scholarly circle formed the Royal Society. A year later, Boyle published *The Sceptical Chymist*, where he rejected the classical **four elements** in favor of a wider variety of indivisible substances. He believed that all matter was composed of atoms of these elements—alone and in more complicated clusters—and that physical and chemical phenomena were the result of their movement and reactions. He was, it hardly needs saying, completely correct about all this, and is thus often considered the father of modern chemistry.

Interestingly, although he believed that the alchemists (including Paracelsus) were wrong, Boyle also believed that alchemical transmutation of metals might be possible. One of his most prescient works, however, was a "wish list" of twenty-four inventions he hoped would be realized someday, including flying machines, diving equipment, drugs for pain, better lighting, the engineering of living species, transplantation, and more. If only he were able to see how many of his dreams came true!

SEE ALSO The Four Elements (c. 450 BCE), Atomism (c. 400 BCE), Alchemy (c. 900), The Advancement of Learning (1605), Conservation of Mass (1789), Dalton's Atomic Theory (1808), Ideal Gas Law (1834)

A painted portrait of Robert Boyle with what may be his seminal book.

Phlogiston

Robert Boyle (1627–1691), Johann Joachim Becher (1635–1682)

Phlogiston does not exist. It was supposed to be the essence of fire—an obsolete concept that might seem strange to us now. Things that burned easily were believed to have lots of phlogiston in them, and what we call *fire* was just the phlogiston being released. This theory was first put forward by the German alchemist and adventurer Johann Joachim Becher in his book *Physica subterranea* (1667), and by the early 1700s it was widely believed.

By the 1600s, it was well known that after things burned, they had lost a good deal of weight, and Becher attributed this to the loss of their phlogiston. The more they burned, the more phlogiston they had to lose in the first place. Air could only absorb a certain amount of it because, if you tried to enclose a fire, it soon went out—apparently choked by excess phlogiston gathering around it and keeping the rest from exiting. This "phlogisticated" air was known not to support life, so it was also believed necessary for animals to exhale phlogiston, which was not possible in air that was already saturated with it.

Becher's theory was an honest and reasonable attempt to explain the facts of combustion, which no one understood, and his description of phlogiston fit the facts as well as or better than any other theory. It also related combustion to respiration, which was quite correct in principle. There were only a few problems with the idea, but as the years went on, it became clear that they would not go away. One of the biggest difficulties was that some metals apparently gained weight when they burned—a fact much emphasized by chemist Robert Boyle. There was much arguing over this point, but over time the evidence became incontrovertible. Some advocates of Becher's theory suggested that phlogiston had negative weight, but that idea created more problems than it solved. The final blow to phlogiston came around 1774 with the discovery of **oxygen**, which flipped over the understanding of fire completely: it wasn't a loss of phlogiston but rather a gain of oxygen.

SEE ALSO Carbon Dioxide (1754), Oxygen (1774)

Phlogiston being released—if there were such a thing as phlogiston. The theory made more sense than it is given credit for, but it collapsed under the weight of too much contradictory evidence.

Phosphorus

Hennig Brand (c. 1630–c. 1692)

Phosphorus occupies a special place in the list of the elements because it was the first one discovered since ancient times. Hennig Brand, a German alchemist, was making a late attempt at the **philosopher's stone** in 1669, and this quest had gone far enough afield that he was distilling and concentrating his own urine. Heating the resulting solids to a high temperature and cooling the vapors by bubbling them through water produced a white solid that glowed in the dark. Brand himself probably thought he had found the philosopher's stone indeed, but what he had actually discovered was a crude preparation of what we now call *white phosphorus*, from the Greek word for "light bearer."

Like several other elements, phosphorus can exist as a pure solid in several forms, known as *allotropes*. The white variety is quite toxic and the most reactive by far—Brand had a lucky break when he formed it under water, since that's one of the few easy ways to keep it from igniting spontaneously. White phosphorus consists of clusters of four phosphorus atoms bonded to one another, and that compact form explains why it's comparatively easy to vaporize. Red phosphorus—a powdery material made up of jumbled networks of randomly bonded phosphorus atoms—is a more common form. High temperatures can convert red phosphorus to a more orderly purple form, while heating white phosphorus under extreme pressure converts it to black phosphorus, a flaky solid whose structure is repeating rings of six phosphorus atoms each, (and whose individual layers were found in 2013 to separate into the phosphorus structural analog of **graphene**).

Given all these possibilities, it might come as no surprise to find that the chemistry of phosphorus is complex, with a wide variety of compound classes known. The fully oxidized phosphates are absolutely crucial in biochemistry, as they are a key part of the structure of DNA and of the common chemical-energy molecule of every living cell: adenosine triphosphate, or ATP. Phosphates are also important regulators of proteins when they're added to and removed from their **amino acid** side-chains.

SEE ALSO The Philosopher's Stone (c. 800), Amino Acids (1806), Phosphate Fertilizer (1842), Cellular Respiration (1937), Graphene (2004)

A nineteenth-century etching of Hennig Brand's discovery of phosphorus. Brand almost certainly did not produce enough to light up the whole laboratory, but it still must have been a dramatic moment.

Hydrogen Sulfide

Bernardino Ramazzini (1633–1714), Carl Wilhelm Scheele (1742–1786)

Most people have caught a whiff of the "rotten egg" smell of hydrogen sulfide (H_2S), but no living person has ever breathed in a high concentration of it. The human nose can detect the disgusting odor at extremely low concentrations, and that's a very good thing, since it's actually more poisonous than **hydrogen cyanide** (whose smell is, unfortunately, rather faint). As with many other poisonous gases, it kills by damaging the lining of the lungs.

The first person to realize that H_2S was a separate gaseous substance was Italian physician Bernardino Ramazzini. His *De morbis artificum diatriba* (*Diseases of Workers*), first published in 1700, was a landmark book in the history of medicine, but it contributed to chemistry as well. Ramazzini noticed that people cleaning out cesspits often had irritated eyes and lungs and that copper or silver coins in their pockets turned black. He hypothesized that all these effects were caused by some irritating gas, formed in the decaying organic matter and released into the air by the workers. The same gas was found around some hot springs and volcanic areas, and its effects on silver and other metals had been noticed in those regions. In 1777, Swedish chemist Carl Wilhelm Scheele produced pure hydrogen sulfide chemically by treating iron pyrite (fool's gold) with **sulfuric acid**. He called it *Schwefelluft* (sulfur air) and described it as "stinking," as well he might.

Hydrogen sulfide is worth thinking about as a structural analog of water. Sulfur is the next heavier element down the **periodic table** from oxygen, but H_2S boils away at −76°F (−60°C)—nearly 300°F (150°C) below the boiling point of H_2O. This is a powerful illustration of **hydrogen bonding** in that water can form far stronger hydrogen bonds (O to H) than hydrogen sulfide (S to H), which makes water much stickier, stranger, and higher-boiling than any other substance made out of such small molecules.

SEE ALSO Toxicology (1538), Sulfuric Acid (1746), Hydrogen Cyanide (1752), The Periodic Table (1869), Claus Process (1883), Hydrogen Bonding (1920), Catalytic Reforming (1949)

Iceland's Holuhraun lava field erupts in 2014 in this false-color infrared image. The plume of gas is a foul and poisonous mixture containing carbon dioxide, sulfur dioxide, and hydrogen sulfide.

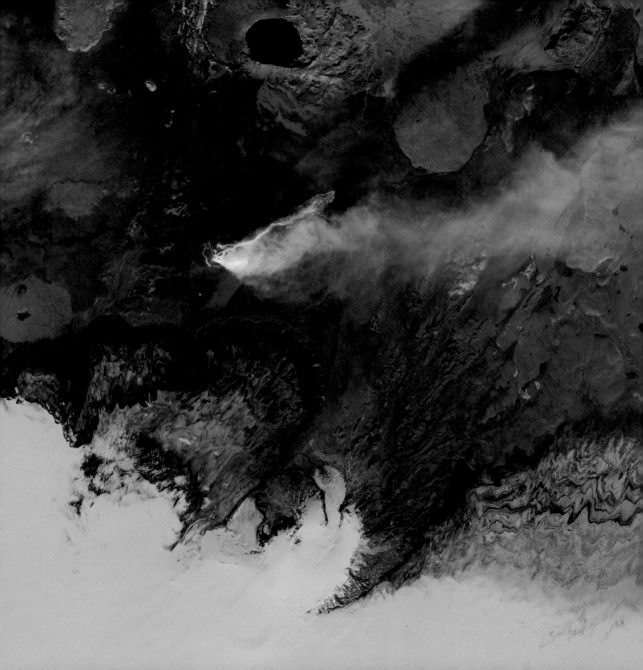

Prussian Blue

Georg Ernst Stahl (1660–1734), Caspar Neumann (1683–1737)

Here's something you may not have noticed: European paintings from before 1700 rarely feature much blue in them, and the color almost always accentuates the most exalted person in the frame. This is because the only durable blue pigment available for oil painting was ultramarine (from Latin, "beyond the sea"), made of wildly expensive lapis lazuli stones from Afghanistan. Smalt (ground cobalt glass) was available for blue ceramics, but oils caused it to discolor. Thus, lapis was the only nonfading blue since "Egyptian blue," whose recipe had been lost when the Roman Empire disintegrated.

But a chance discovery changed all that. Sources disagree on some details, but it appears that a German dye-maker named Johann Jacob Diesbach was trying to make a red pigment from cochineal (obtained from crushed beetles) around the year 1706 when, to his surprise, he obtained a blue substance instead. His reagents were contaminated, as it turned out, and within two years the synthetic blue paint was on the market as *Prussian blue*, *Berliner blue*, and similar names. By 1724 the German-Polish chemist Caspar Neumann had leaked the recipe to the Royal Society of London, which published it. Apparently, the mixture of cochineal, alum, iron sulfate, and animal oil–contaminated potash (potassium carbonate) was what yielded the brilliant blue color.

No one was going to reverse-engineer Prussian blue from its chemistry, though. The idealized version has three iron atoms in the +2 oxidation state, each with six cyanides around it, and around these complexes are four irons in the +3 state. Older preparations had all sorts of impurities, complicating the structure further, so it was not until the 1970s that the full chemical details were worked out. Attempts to understand Prussian blue helped advance inorganic chemistry for over 250 years—long after its importance as a pigment had, well, faded. Further, the substance lent its name to prussic acid (**hydrogen cyanide**), and its complexing properties (the cyanides are arranged in space and point inward) make it a valuable drug in cases of acute metal poisoning, since it exchanges some of its iron atoms for thallium or other toxic metal ions, which can then be passed safely from the body.

SEE ALSO Hydrogen Cyanide (1752), Titanium (1791), Coordination Compounds (1893), Thallium Poisoning (1952)

Japanese artist Hokusai's famous woodblock print The Great Wave off Kanagawa *(c. 1830). For some time, such prints were thought by European collectors to use some strange Japanese blue pigment, but it turned out to be Prussian blue imported from Europe!*

Sulfuric Acid

John Roebuck (1718–1794), **Peregrine Phillips** (1800–1888)

Sulfuric acid is a workhorse indeed. As a key component in many industrial processes, it has been a valuable commodity for centuries, but it's not something you'll find springing up out of the ground (fortunately); it has to be made on an industrial scale. The techniques required to create sulfuric acid—known to medieval alchemists as *oil of vitriol*—have always had the same last step: dissolving sulfur trioxide in water. But reaching the sulfur trioxide stage has been the tricky part . . . even though it involves nothing more than burning sulfur.

In earlier days, just producing equipment that could handle being soaked in sulfuric acid was a challenge. Five hundred years ago, sulfur was burned while suspended over water in glass jars, but that set limits on how much acid could be made per batch. Large glass vessels were not easy to come by and were unpredictably fragile—not a desirable property for vessels full of corrosive acid—but in 1746, English industrialist John Roebuck invented a better manufacturing process. He realized that lead was resistant to sulfuric acid, so he introduced large lead vessels, which produced ten times more acid than the glass ones. On any scale, though, this method needed several repeated doses of burning sulfur, and the resulting sulfuric acid had to be concentrated via boiling. Both of these processes are as hard to be downwind of as they sound, but the demand for the acid was so strong that Roebuck-style plants were opened throughout the industrialized world.

The lead-chamber technique was used until 1831, when English vinegar merchant Peregrine Phillips discovered a way to turn the more easily available sulfur dioxide gas into the trioxide by flowing it over a heated metal catalyst. This route, known as the *contact process*, is still used today, as sulfuric acid is more in demand than it ever was. Vast amounts are used for making fertilizers, and it shows up as a reagent in almost every part of the chemical industry.

SEE ALSO Hydrogen Cyanide (1752), Claus Process (1883), Acids and Bases (1923)

Sulfates (sulfuric acid salts) show up in many fertilizer mixtures. Sulfuric acid itself is not recommended for the garden, however!

Hydrogen Cyanide

Pierre Macquer (1718–1784), **Carl Wilhelm Scheele** (1742–1786), **Claude-Louis Berthollet** (1748–1822), **Joseph-Louis Gay-Lussac** (1778–1850)

Prussian blue was still at the cutting edge of chemistry throughout the eighteenth century, as its precise composition was a mystery. Then, in 1752, French chemist Pierre Macquer found that it could be broken down into iron salts and some sort of volatile gas, and that the process could be run in reverse to generate the pigment again. But what was that volatile material? Swedish chemist Carl Wilhelm Scheele had the good (or bad) fortune of working that one out: He reacted Prussian blue with **sulfuric acid**, noting that it produced a "strong, peculiar, and not unpleasant odor." A modern chemist would have told him to run for the door (or dive out the window), but he went on to actually *taste* the stuff (it was slightly sweet and gave a sensation of heat on the tongue). Scheele was describing famously poisonous hydrogen cyanide, and he was lucky to live to tell anyone about it.

The cyanides in Prussian blue are complexed (arranged in space and pointing inward) very tightly to iron atoms, and given the chance, they will do the same to the iron atoms found in the hemoglobin molecules of red blood cells—a reaction that makes them useless for carrying oxygen. Despite its dangers, hydrogen cyanide helped advance chemistry quite a bit. It is a weak acid, since in water it partly dissociates into H+ and CN–, and the presence of those H+ ions makes it an acid. At the time, though, it was believed that all acids had to contain oxygen in their formulae, the way sulfuric and nitric acids do. However, in 1787, prussic acid (named for its source of Prussian blue) was shown by French chemist Claude-Louis Berthollet to be oxygen-free, and in 1815 his fellow Frenchman Joseph-Louis Gay-Lussac worked out its precise chemical formula: HCN. Once cyanide ions were recognized as a separate species, the name for them suggested itself immediately, as *cyan* is the Greek word for "blue."

SEE ALSO Toxicology (1538), Prussian Blue (c. 1706), Sulfuric Acid (1746), Cyanide Gold Extraction (1887), Coordination Compounds (1893), pH and Indicators (1909), Acids and Bases (1923), Molecular Disease (1949), Miller-Urey Experiment (1952)

A statue of Scheele in Stockholm, erected in 1892. The tradition of decorating parks with statues of chemists should perhaps be revived.

Carbon Dioxide

Jan Baptist van Helmont (1580–1644), Joseph Black (1728–1799), Henry Cavendish (1731–1810), Joseph Priestley (1733–1804), Humphry Davy (1778–1829), Michael Faraday (1791–1867)

Carbon dioxide is now famous as a greenhouse gas, but it has a long history in the field of chemistry. Flemish chemist Jan Baptist van Helmont discovered, in 1625, that when charcoal was burned, the leftover material weighed less than it did before the fire—even if the smoke was trapped in a closed container and weighed with it. He suspected that the rest of the material had been turned into some invisible substance he called *gas sylvestre* (wood gas).

Over a century later, in 1754, Scottish physician and chemist Joseph Black found that he could produce a new gas from heated limestone (calcium carbonate), and this "fixed air" was strange stuff. It was heavier than regular air, so it could be "poured" like a thin liquid; it immediately put out flames; and it very quickly killed any animal enclosed with it. Perhaps his biggest discovery was that, when bubbled into a solution of lime (calcium hydroxide), the gas produced calcium carbonate again (and fell out of solution as a white powder). This discovery gave him a way to detect his invisible gas, which he then used to show that animals actually exhaled it. Not long afterward, English theologian-chemist-political theorist Joseph Priestley suspended a bowl of water over a beer vat in a Leeds pub, effectively infusing the water with carbon dioxide, which generated bubbles as it escaped and made the resulting "soda water" pleasant to drink.

Carbon dioxide was later found to show an interesting property as it was cooled. At the ordinary air pressure of a room, it skips the liquid state as it cools and "snows out" directly as a white solid once it gets to −109°F (−78.5°C). This "dry ice" in turn evaporates straight back into the gas phase as it warms—a process called *sublimation*. In the 1820s, English scientists Humphry Davy and Michael Faraday managed to produce liquid carbon dioxide by increasing the pressure significantly. Soon after, additional heat and pressure was found to transform it into a **supercritical fluid**, a strange state that is neither a liquid nor a gas.

SEE ALSO Phlogiston (1667), Oxygen (1774), Supercritical Fluids (1822), Greenhouse Effect (1896), Zymase Fermentation (1897), Carbonic Anhydrase (1932), Cellular Respiration (1937), Photosynthesis (1947), Carbon Dioxide Scrubbing (1970), Artificial Photosynthesis (2030)

Solid carbon dioxide (dry ice) bubbles away as it warms up under water, producing a distinctive thick, white fog on the surface.

Cadet's Fuming Liquid

Louis Claude Cadet de Gassicourt (1731–1799), **Robert Bunsen** (1811–1899)

Interesting things often happen at the border of two fields of science. Organic chemistry (the study of compounds with carbon-atom backbones) and inorganic chemistry (everything else!) meet at the organometallic compounds, which turn out to have a wide variety of useful properties. Oil refining, the production of plastics, many antipollution devices, and the synthesis of most drugs would not be possible without organometallic reagents and catalysts. The field is one of the most active in all of chemical research today.

But it did not get off to an impressive start. The first organometallic compound was also one of the more useless ones. In 1758, French chemist Louis Claude Cadet de Gassicourt used arsenic trioxide to produce a foul substance known afterward as *Cadet's fuming liquid*—and it was well named. We now know that Cadet's liquid was a mixture of tetramethyl diarsenic and its oxide, but before their compositions were worked out, they were known as *cacodyl* and *cacodyl oxide*, from the Greek word *kakodes* (awful-smelling). Organoarsenic compounds tend to smell like a reeking, less appetizing cousin of garlic, and they also tend to be quite poisonous.

In the case of cacodyl, its poisonous and foul nature is not balanced by much utility. Comparatively few organoarsenic compounds find any uses in modern chemistry, as opposed to workhorse metals like palladium, lithium, and magnesium. However, cacodyl did turn out to be a key piece of evidence later on when German chemist Robert Bunsen (inventor of the Bunsen burner) worked out the idea of interchangeable chemical groups or "radicals," since its methyl groups could be shown to transfer to new substances. But Bunsen himself remarked on the compound's overpowering stench, as well as the vapor's unnerving property of turning the surface of one's tongue black after exposure—and this was the high point of cacodyl's fame. A cacodyl compound was considered as a chemical weapon during the Crimean War, but the British commanders rejected the idea as inhumane. Very few living chemists have ever seen Cadet's discovery in person, and very few have any wish to.

SEE ALSO Grignard Reaction (1900), Salvarsan (1909), Chemical Warfare (1915), Ferrocene (1951), Metal-Catalyzed Couplings (2010)

A medicine bottle containing arsenic from the 1800s. The deep collar and ridged glass indicate the poisonous nature of the contents.

Hydrogen

Henry Cavendish (1731–1810), Antoine Lavoisier (1743–1794)

Hydrogen is the simplest of all the elements, and by far the most common substance in the universe, but much time passed before it was recognized as a discrete element. There's very little free hydrogen in Earth's atmosphere, partly because our gravity can't keep it from drifting off into space. On our planet's surface, most hydrogen has already been burnt, forming the clear liquid known as water.

We owe the discovery of hydrogen to British philosopher, chemist, and physicist Henry Cavendish, who (like many chemists and physicists of the time) was able to discover numerous basic principles by studying the behavior of gases. Robert Boyle and others had noticed that many metals gave off some sort of gas when they encountered strong acids, but Cavendish was the first to recognize it as an element in itself, and he offered the first description of hydrogen's properties in his 1766 paper, "On Factitious Airs." He reported its lightness and willingness to burn, but experiments in 1783 with what he called "flammable air" and Joseph Priestley's "dephlogisticated air" (**oxygen**) soon gave Cavendish a surprising and important result: the explosively flammable mixture turned out to produce water as it burned, proving that water was not an element in itself, but rather a simple compound of oxygen and hydrogen. That same year, French chemist Antoine Lavoisier replicated Cavendish's experiment, naming the substance *hydrogen* (from the Greek for "water creator").

Cavendish—whose unusually timid nature has led modern scholars to speculate that he suffered from Asperger's syndrome—went on to publish a number of important results on the nature of heat and combustion, which helped inspire Lavoisier's theory about the true nature of chemical reactions and the importance of oxygen. Tearing apart the simplest substances (air and water) and recombining them was putting chemistry on a better foundation than it had ever had before.

SEE ALSO Oxygen (1774), Avogadro's Hypothesis (1811), Hydrogenation (1897), Deuterium (1931), The Hottest Flame (1956), Hydrogen Storage (2025)

French engineers Jacques Charles and Marie-Noël Robert took to the skies over Paris in 1783 in the first manned flight of a hydrogen-filled craft. Almost 150 years later, the explosion of the Hindenburg would bring the airship era to a violent close.

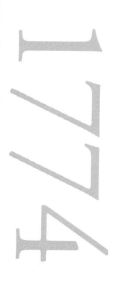

Oxygen

Henry Cavendish (1731–1810), Joseph Priestley (1733–1804), Antoine Lavoisier (1743–1794)

The history of oxygen and its chemistry is a tangled one, but its discovery demonstrated a very simple and very important point: air is not a single substance, but rather is divisible into different components. Chemists had produced several sorts of gases by this point—including **carbon dioxide, hydrogen, hydrogen sulfide,** and **hydrogen cyanide**—but their relevance to the air everyone breathed was an open question. For one thing, it was obvious that there couldn't be much hydrogen sulfide around, since no one could have missed that one.

English polymath Joseph Priestley was the man who discovered oxygen. He conducted experiments with the various "airs" known to him and made a key observation: "Fixed air" (carbon dioxide) was a known poison to animals, but it did not kill green plants. In fact, in a sealed container of fixed air, the plants somehow managed to "detoxify" the air in the container. Since it was known that animals breathed out carbon dioxide, Priestley speculated that plants somehow balanced things out in the atmosphere by removing the gas and (in turn) giving off something else. In 1774, he conducted another experiment, in which mercuric oxide (HgO) heated with a magnifying glass and sunlight began to decompose and give off what he thought must be the same substance, because instead of killing mice, this gas kept them alive, and instead of putting flames out, it made them burn brighter and hotter. Breathing it, Priestley found, was pleasant.

Priestley believed that this new air had had its **phlogiston** removed from it (and that Henry Cavendish's **hydrogen** might be phlogiston itself), but Frenchman Antoine Lavoisier offered a more accurate theory of combustion in a 1777 paper, "Reflections on Phlogiston." Priestley's new gas—which Lavoisier dubbed *oxygen* in 1778—was what burned things, what combined with metals and other elements when they burned, and what animals needed in order to breathe. There was no such thing as "phlogiston."

SEE ALSO Phlogiston (1667), Hydrogen Sulfide (1700), Hydrogen Cyanide (1752), Carbon Dioxide (1754), Hydrogen (1766), Avogadro's Hypothesis (1811), Ozone (1840), Cannizzaro at Karlsruhe (1860), Liquid Air (1895), Stainless Steel (1912), Superoxide (1934), Cellular Respiration (1937), Photosynthesis (1947), Molecular Disease (1949), The Hottest Flame (1956).

Quietly and invisibly, green plants around the world strip carbon from the air and release oxygen into it.

Conservation of Mass

Joseph-Louis Lagrange (1736–1813), Antoine Lavoisier (1743–1794)

French chemist Antoine Lavoisier was central to turning chemistry into a science. The discoveries being made with gases had advanced knowledge tremendously, but it was clear that the theories of the time couldn't accommodate them. A key example was **phlogiston**, and Lavoisier's work with **oxygen** helped to demolish that theory. But there were many more developments to come.

As he himself put it, Lavoisier wanted "to rid chemistry of every kind of impediment that delays its advance." One of these impediments was the naming of compounds. If compounds really were formed by the combination of elements, as English chemist Robert Boyle had proposed over a hundred years before, shouldn't the names reflect this in a systematic way? Lavoisier thought so, and he developed a system of chemical nomenclature that is still used today. For example, if iron combined with oxygen, the resulting compound was *iron oxide*. In *Elements of Chemistry* (1789)—considered the first modern chemistry textbook—Lavoisier laid out his system of nomenclature, along with a table of all the elements discovered to date, the compositions of compounds that had been determined, and his thoughts on the influence of temperature on reactions and on the formation of salts from acids and bases. He also stated a principle that no one had ever put into so few words: conservation of mass—the idea that a chemical reaction finished with the same amount of material it started with.

Unfortunately, Lavoisier's public profile and involvement in the *Ferme générale*—an organization that collected taxes on behalf of the deposed French monarchy—made him a target of the French Revolution during the Reign of Terror. Along with twenty-seven codefendants, he was summarily convicted of treason and beheaded on May 8, 1794. The next day, the Italian mathematician and astronomer Joseph-Louis Lagrange reflected, "It took them only an instant to cut off that head, and a hundred years may not produce another like it."

SEE ALSO *The Sceptical Chymist* (1661), Phlogiston (1667), Oxygen (1774), Dalton's Atomic Theory (1808), Chemical Notation (1813)

In this oil painting by British artist Ernest Board, Lavoisier explains the results of his experiments on air to his wife, who helped him with much of his research.

Titanium

Martin Heinrich Klaproth (1743–1817), William Gregor (1761–1817)

Titanium was a close race. Early in 1791, William Gregor (one of the many English clergymen who made contributions to the sciences) was analyzing a mineral sample from his parish in Cornwall and obtained an unknown metal oxide that he called *manaccanite* after Manaccan, the village near where it was obtained. Later that same year, German chemist Martin Heinrich Klaproth (who also discovered uranium and zirconium), discovered a metal he named *titanium* in the mineral known as rutile. Both metals were the same substance, but Gregor got credit for the discovery, while Klaproth got credit for the name.

Titanium is famously tough, light, and heat-resistant—an aerospace metal if ever there was one. But those qualities make it expensive to work with, so it only shows up in large quantities in "money is no object" items like advanced fighter planes and the inner hulls of some wildly expensive Russian submarines built during the Cold War. In smaller quantities, it is used for crucial high-performance machine parts and (less crucially) in high-end golf clubs.

Unless you build submarines, you've seen much more of the oxide than you've seen of the metal. Powdered titanium dioxide (TiO_2) is blindingly white, and it never, ever fades or breaks down. Paint formulators call it *the perfect white*, and most of the brilliant white paint, plastic, cardboard, lotion, or toothpaste you've ever seen has been made with it. When a new Earth-orbiting object was spotted by amateur astronomer Bill Yeung in 2002, its infrared spectrum turned out to be not that of asteroid rock but of titanium dioxide paint—it was a long-lost part from *Apollo 12*.

Like all materials, titanium is full of surprises for those who look closely. Titanium dioxide has at least eight polymorphs, and in 1967 some of these crystal forms were found to be catalysts for **photochemistry**. Because of this, titanium dioxide paint coatings may actually break down some air pollutants, and it's being studied for use in wastewater treatment, solar cells, and more. Over four million tons of titanium are mined each year, but the element still holds many secrets.

SEE ALSO *De Re Metallica* (1556), Prussian Blue (c. 1706), Photochemistry (1834), Artificial Photosynthesis (2030).

Architect Frank Gehry has famously used thin titanium sheet panels for the surface of many of his buildings, including the Guggenheim Museum in Bilbao, Spain.

Morphine

Friedrich Wilhelm Adam Sertürner (1783–1841)

Opium, the dried sap of the Oriental poppy, has been known as a medicinal substance since prehistoric times. All the civilizations of Asia and Europe made use of it, and rare is the ancient medical text that fails to mention it. Humanity had reason to be impressed: the morphine present in opium (up to 14 percent by weight) was, far and away, the most effective form of pain relief that had ever been found.

In about 1804, Friedrich Wilhelm Adam Sertürner, a pharmacy apprentice, extracted pure morphine from crude opium—a very long and tedious process with the tools of the time. He named the new compound after the Greek god of sleep and dreams, Morpheus, and began experimenting with it, first testing the compound on animals before trying it out on himself and a few local volunteers. From his records, it appears that they took many times the effective dose over a rather short period, which incapacitated the whole team for some time.

This work not only inaugurated the long history of opiate chemistry, but also the even wider history of alkaloids—nitrogen-containing natural products, produced by plants, that cover an extraordinary range of complex chemical structures. They're also well known for producing physiological effects. Many biochemical pathways have been discovered and elucidated by studying alkaloids and tracking down the proteins that they bind to. In the case of morphine, it binds tightly to what we now call the mu-opioid receptor proteins (among others) in the brain and spine. This suggested that the body produced its own compounds that also bind to these receptors, and after a long search, peptide neurotransmitters (the endorphins and enkephalins) were found to fill this role. Bizarrely, some recent research suggests that morphine itself is produced by many animal species, bringing the story around in a most unexpected loop. Tiny amounts of morphine can be detected in human cell cultures, and it was found that an oxygen **radioactive tracer** ended up in the morphine molecules, suggesting that the cells were making it themselves.

SEE ALSO Natural Products (c. 60 CE), Caffeine (1819), Radioactive Tracers (1923), LSD (1943)

In this 1811 painting by Pierre-Narcisse Guérin, Morpheus, the chief Greek god of dreams, is awakened by Iris, a messenger of still more powerful gods.

Electroplating

Alessandro Volta (1745–1827), Luigi Brugnatelli (1761–1818)

Electricity was at the frontier of technology in the late 1700s and early 1800s. Luigi Brugnatelli, an Italian professor and friend of electrical pioneer Alessandro Volta, used the new voltaic pile battery on various chemical solutions, and in 1805 he discovered that he could cause gold metal to deposit in a thin film on other metal objects if they were dipped into a solution of gold salts while they were wired to the negative terminal of the battery. (He was even able to electroplate nonmetallic things such as insects and flower petals if he covered them first with a thin layer of metal-based paint to conduct the electricity.)

Brugnatelli's discovery, now called **electrochemical reduction**, is another application of oxidation and reduction chemistry. **Iron smelting**, for example, involves both reactions: carbon is oxidized (combines with oxygen) when it burns, forming carbon monoxide (CO), which in turn strips the ore of oxygen, reducing the iron oxides down to the elemental metal. In the case of electroplating, the electrons at the negative end of the battery reduce the (positively charged) gold ions back down to elemental gold as they come near the surface of the metal object, causing a very thin layer of new gold to build up.

This is the sort of discovery that you might think would make someone wealthy and famous, but it was not to be. Brugnatelli presented his work to the French government (i.e., Napoléon Bonaparte), and there was some sort of catastrophic disagreement, so Brugnatelli's results were suppressed throughout Napoleonic Europe. Electroplating wasn't rediscovered for nearly twenty-five years. There were many variables to be worked out (voltage and current, the types of metal salts to use, and what sorts of materials would best take the new electroplated layer), but by the 1840s gold and silver plating was a commercial success, thanks partly to the discovery that cyanide solutions could dissolve large amounts of the metals. These and other metals are still electroplated on a large scale throughout the world, and electrochemistry has evolved into its own branch of science.

SEE ALSO Iron Smelting (c. 1300 BCE), Gold Refining (c. 550 BCE), Electrochemical Reduction (1807), Aluminum (1886), Cyanide Gold Extraction (1887), Chlor-Alkali Process (1892)

An early voltaic pile, made with alternating zinc and copper disks. This sort of battery provided the power for the first electroplating experiments and many other types of experiments as well.

Amino Acids

Nicolas-Louis Vauquelin (1763–1829), Pierre-Jean Robiquet (1780–1840), Franz Hofmeister (1850–1922), Emil Hermann Fischer (1852–1919)

The term *amino acid* is familiar. If you ask people for a definition, most of them might say something about nutrition or bodybuilding, but those who mention proteins would be the winners. Amino acids all have a central carbon atom with an amine (NH_2) group on it, as well as a carboxylic acid (CO_2H) group. The simplest amino acid compound is glycine. If the compound has a methyl group coming off the central carbon, it is alanine, and if that methyl group has a benzene ring attached to it, it is phenylalanine. The differences between amino acids are all in the *side chains* attached to the core structure.

The French pharmacist Nicolas-Louis Vauquelin and his student, Pierre-Jean Robiquet, isolated the first known amino acid in 1806 and named it *asparagine* after the asparagus they used as a source. Almost a century later, German chemists Emil Hermann Fischer and Franz Hofmeister independently discovered that proteins were polymers of these building blocks, stringing them together in long chains, which then fold up into structures like the **alpha-helix and beta-sheet**. Protein synthesis is carried out by an extraordinary bit of cellular machinery called the *ribosome*, acting on instructions from the cell's DNA sequence, in a process whose details are still generating Nobel prizes.

Joining two amino acids into a dipeptide requires a condensation reaction that results in the loss of a water molecule. The bond can then be broken by brute force or with the help of enzymes. If you've recently had anything to eat, your digestive system is dismantling proteins now. Digestive enzymes do this by holding them in just the right position and environment to be split, and thousands of other enzymes do the same for a wide variety of reactions in living cells. Our DNA codes for only twenty different amino acids, but that's enough: even a short ten-peptide chain has over ten trillion possible combinations.

SEE ALSO Polymers and Polymerization (1839), Spider Silk (1907), Maillard Reaction (1912), Carbonic Anhydrase (1932), Molecular Disease (1949), Sanger Sequencing (1951), Alpha-Helix and Beta-Sheet (1951), Miller-Urey Experiment (1952), Electrophoresis (1955), Green Fluorescent Protein (1962), Merrifield Synthesis (1963), Protein Crystallography (1965), Murchison Meteorite (1969), Glyphosate (1970), Enzyme Stereochemistry (1975), Engineered Enzymes (2010).

The asparagus plant is rich in asparagine, but it's not the only source of this amino acid. Vauquelin and Robiquet could have used potatoes or licorice to make their discovery, but then it would surely have a different name.

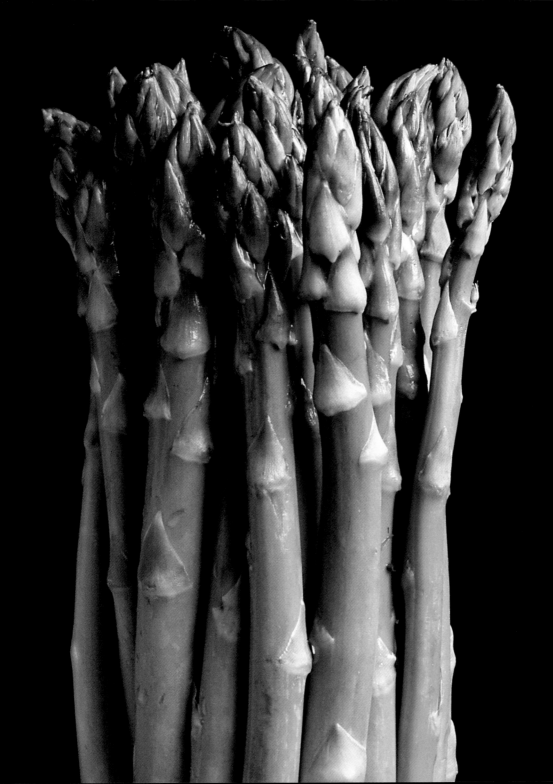

Electrochemical Reduction

Humphry Davy (1778–1829)

Cornish chemist, poet, and inventor Humphry Davy belongs to a heroic age of chemistry, as he conducted experiments with electricity and gases that put him at grave risk for his life. One effort to inhale what we now know was a mixture containing dangerous amounts of carbon monoxide was particularly foolhardy. On another occasion, he exposed himself to huge quantities of nitrous oxide (laughing gas) to see what would happen as he went past the customary doses. As he recounted, "I lost all connection with external things; trains of vivid visible images rapidly passed through my mind and were connected with words in such a manner as to produce perceptions perfectly novel. I existed in a world of newly connected and newly modified ideas. I theorized; I imagined that I made discoveries." He could say that because he knew what real discoveries *felt* like.

Davy followed the new electrical discoveries of the early 1800s with great interest, and he realized that batteries and electric piles were actually producing current due to chemical reactions. In 1807, trying the reverse experiment by seeing what electric current would do to salts of potassium and sodium, Davy became the first person to see these elements as pure metals. He went on to apply this technique of passing electric currents through salts (now called *electrochemical reduction*) to isolate magnesium, calcium, strontium, and barium.

In addition to being a founder of electrochemistry and making great contributions to the theory of acids and other substances, Davy is also remembered for his invention of the Davy lamp for miners, which used a wire gauze to dissipate the heat of the lamp's flame in such a way that it could not ignite the flammable gases in the mines. When he was being revived from his encounter with carbon monoxide, Davy told his assistants, "I do not think I shall die." And in a way, he hasn't. His impact on science has led to his name being attached to (among other things) a crater on the moon, a pub in his hometown, and to the Royal Society's prestigious Davy Medal, awarded every year for outstanding discoveries in chemistry.

SEE ALSO Electroplating (1805), Electrochemical Reduction (1807), Beryllium (1828), Oxidation States (1860), Aluminum (1886), Chlor-Alkali Process (1892), Hydrogen Storage (2025), Artificial Photosynthesis (2030).

Elemental sodium is a soft, silvery metal, but no one until Davy had ever seen it. That's because it explodes into flames when it touches water (note the bright yellow, as mentioned in the entry **Flame Spectroscopy***).*

Dalton's Atomic Theory

John Dalton (1766–1844)

John Dalton, one of the nineteenth century's impressive collection of polymaths-from-nowhere, took the ideas of **atomism** and updated them. He was a Quaker, and thus could not attend Oxford or Cambridge, but made himself extraordinarily well educated nonetheless. Outside of chemistry, he is remembered for being the first person to recognize and describe color blindness (his own!). His chemical investigations included countless experiments on every type of gas that he could isolate, and his study of their physical properties convinced him that they were fundamentally similar but had differences depending on the weights of their "ultimate particles."

In his *A New System of Chemical Philosophy*, published in 1808, Dalton gave a detailed account of his ideas. The ways that gases and liquids behaved, he proposed, meant that Democritus had been right about atomism and that Robert Boyle and Antoine Lavoisier had been on the right track to treat gases as if they were made of individual particles when explaining their behavior under changing pressures and temperatures. He also proposed that all matter is made out of atoms; atoms of the various chemical elements differ in weight and other properties (but are consistent among themselves); atoms combine in whole-number ratios to form new chemical compounds; and atoms cannot be created, destroyed, or otherwise broken up. Additionally, he introduced a table of atomic weights for six known elements, setting hydrogen to one and working up from there.

In retrospect, we can see that in many key details Dalton was right on target, and the effect on chemistry was far-reaching. A weakness in his theory was that he assumed nature to be a bit too reasonable—although he was far from the last chemist to make that mistake. He believed that the simplest possible combinations were the most likely to occur, so he formulated water as HO and ammonia as NH (water actually has two hydrogens, and ammonia has three). But his law of multiple proportions—the idea that atoms combine in whole-number ratios—was immediately successful, and his atomic theory was a fundamental part of chemistry from then on.

SEE ALSO Atomism (c. 400 BCE), *The Sceptical Chymist* (1661), Conservation of Mass (1789), Avogadro's Hypothesis (1811), Cannizzaro at Karlsruhe (1860).

John Dalton looking thoughtful in this 1823 engraving, as well he might.

Avogadro's Hypothesis

John Dalton (1766–1844), **Joseph-Louis Gay-Lussac** (1778–1850), **(Lorenzo Romano) Amedeo (Carlo) Avogadro (conte di Quaregna e Cerreto)** (1776–1856)

In 1811, Italian scientist Amedeo Avogadro published his hypothesis about molecular weights, but not enough people noticed. It would have saved a great deal of trouble if they had. Avogadro proposed that, if you took the exact same volumes of several different gases and compared their weights, these would correspond to the individual molecular weights of the gases themselves. This implies that identical volumes of different gases have the same number of molecules in them, so any differences in weight must come from the heavier or lighter molecules themselves. Avogadro concluded this after studying the work of John Dalton and Joseph-Louis Gay-Lussac, the latter of whom discovered in 1805 that when gaseous reagents reacted, the volumes of the reactants and their products were always in whole-number ratios to each other (first by showing that two volumes of hydrogen and one volume of oxygen were needed to produce water). Despite the fact that Gay-Lussac's results supported his own ideas, Dalton seems to have rejected them. Instead, it was Avogadro who cleared up a lot of the confusion about atoms and molecules.

Dalton always favored the simplest explanations, but Avogadro had the key insight that things were a bit more complicated: many common gases were actually composed of two identical atoms bonded together. Hydrogen gas, we now know, is H_2, oxygen gas is O_2, and nitrogen is N_2, but there was no reason to assume this in the early 1800s. It was known (from experiments by Gay-Lussac and others) that if you burned hydrogen gas with oxygen at a constant temperature and pressure, the volume of the water vapor was twice that of the oxygen you started with. This was a great problem for many theories, but Avogadro boldly proposed that this made sense if the oxygen started out as O_2 and then split to become part of two new water molecules. Despite Avogadro's sound logic, chemists of his day were stumped by their belief that chemical bonding resulted from positive and negative charges attracting each other. Why, then, should two identical atoms bond together? Decades later, the Italian chemist Stanislao Cannizzaro finally resolved this conundrum in a paper that gave Avogadro his due credit.

SEE ALSO Hydrogen (1766), Oxygen (1774), Dalton's Atomic Theory (1808), Chemical Notation (1813), Ideal Gas Law (1834), Cannizzaro at Karlsruhe (1860), The Mole (1894)

This photograph shows a long match igniting hydrogen soap bubbles, which results in an exothermic reaction between the hydrogen and the oxygen in the air.

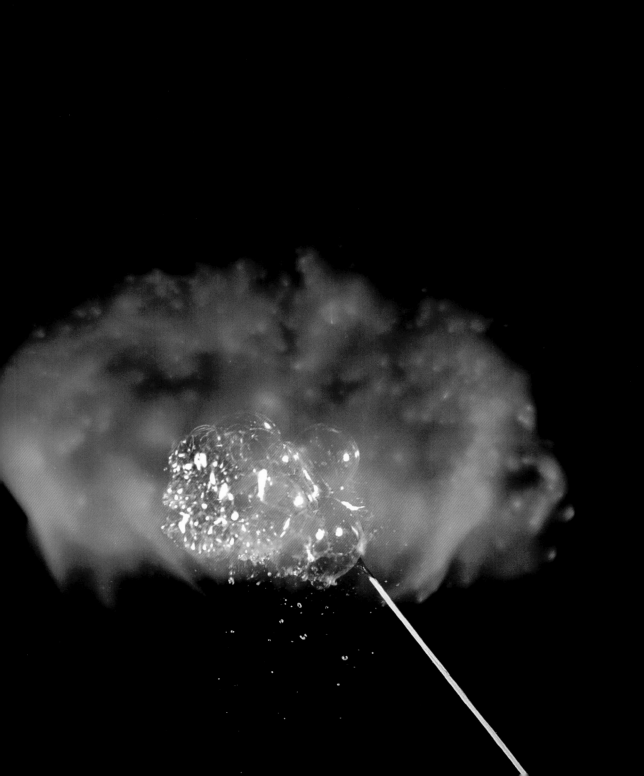

Chemical Notation

John Dalton (1766–1844), Jöns Jacob Berzelius (1779–1848)

Swedish chemist Jöns Jacob Berzelius began his career as a physician in 1802, but his impact on the field of chemistry is hard to overstate. By 1818, he was a professor at the prestigious Karolinska Institutet (which now awards the annual Nobel Prize in Physiology or Medicine), permanent secretary of the Royal Swedish Academy of Sciences, and the author of a highly influential chemistry textbook. He discovered the elements silicon, selenium, thorium, and cerium, and he was apparently the first chemist to realize that the entire field could be divided into the chemistry of carbon compounds (organic chemistry) and everything else (inorganic chemistry). He also coined words like *protein*, *polymer*, *isomer*, and *allotrope*, which can be found throughout this book.

Additionally, Berzelius helped to codify atomic and molecular weights. The former, he found, were not simply multiples of hydrogen, but the latter did seem to be combinations of whole-number ratios of the various elements, which was a big piece of evidence in favor of John **Dalton's atomic theory**, the cutting edge of chemical thought at the time. In 1813, this led Berzelius to start writing chemical formulas down in an element-plus-number style to keep track of them. He gave the elements simple one- or two-letter abbreviations and then wrote next to each element the number of that element a given compound seemed to be carrying, using superscripts rather than the subscripts we are familiar with today. Hence, table salt (sodium chloride) was written as NaCl, and baking soda (sodium bicarbonate) was identified as $NaHCO^3$. That is, each molecule of salt consists of one sodium and one chlorine, and each molecule of baking soda consists of one sodium, one hydrogen, one carbon, and three oxygens. The fact that there was another well-known molecule, now known as sodium carbonate (with the formula Na_2CO_3), immediately suggested something about the structures as well: CO_3 is one unit, and sodium, potassium, hydrogen, and other elements can be swapped in and out around it. Chemists were coming to terms with the idea that atoms combined in defined ways to make compounds with defined compositions. Berzelius's chemical notation illustrating this proved so useful that it is still employed today.

SEE ALSO Conservation of Mass (1789), Dalton's Atomic Theory (1808), Avogadro's Hypothesis (1811)

Chemical notation provides immediate information to those who have taken the time to learn the code. Shown here are the formulas for three corrosive acids.

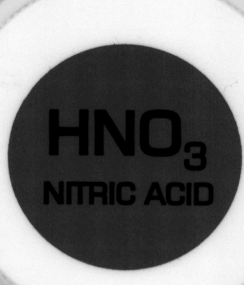

1814

Paris Green

There are several broad themes in the history of chemistry: stronger and better materials, life-saving new drugs, and life-destroying weapons and explosives, among others. Then there are the pigments and dyes that make paints and clothing colorful. This area of chemistry has had huge successes and a few huge failures. The story of Paris green falls into both categories. Developed in 1814 as a replacement for the less durable Scheele's green, this vivid crystalline powder was used to dye clothing, wallpaper, candles, and even food in the nineteenth century. It was relatively inexpensive and more colorful than the alternatives, but unfortunately it was also a compound of arsenic. As such, some very decorative William Morris wallpapers have turned out to be dangerously arsenic-laden, and Morris himself was even on the board of directors for one of the largest arsenic mining companies in the world. Coloring cakes and saturating one's fashionable walls with green arsenic compounds is exactly as bad an idea as it sounds, and toxicologists are still not sure what the worst aspect of those green walls might have been. The powdery flakes of pigment released over time were bad enough, but dampness (and the enzymatic actions of mold) also released volatile arsenic compounds into the air. Arsenic's poisonous effects are notoriously slow-moving, so although people understood that some forms of it were toxic, the realization that there were actually no safe forms took a painfully long time. Fortunately, other less insidious pigments were invented that displaced Paris green, which stayed on the market as an insecticide and rat poison.

Notably, the high levels of arsenic found in forensic samples of Napoléon Bonaparte's hair have led some researchers to speculate that the green wallpaper of his prison home on the island of Helena may have contributed to his demise. The era of poisonous Paris green left quite a mark in some areas of the world. Well into the twentieth century, dyed green candies sold poorly in Scotland, as older customers regarded them with suspicion. By then there was no cause for alarm, but one can understand the reaction.

SEE ALSO Toxicology (1538), Yorkshire Alum (1607), Perkin's Mauve (1856), Indigo Synthesis (1878), Salvarsan (1909), DDT (1939), Thallium Poisoning (1952)

Some of William Morris's favorite wallpaper is displayed in this nineteenth-century room interior. Who would have thought that interior decorating could be so deadly?

Cholesterol

François Poulletier de la Salle (1719–1788), **Michel-Eugène Chevreul** (1786–1889), **Otto Paul Hermann Diels** (1876–1954), **Adolf Otto Reinhold Windaus** (1876–1959), **Heinrich Otto Wieland** (1877–1957)

Cholesterol goes back a long way because it's relatively easy to isolate in a pure form. French chemist François Poulletier de la Salle studied a human gallstone in 1769 and found that it seemed to be a single waxy substance. In 1815, fellow Frenchman Michel-Eugène Chevreul realized that the same material was found in many other sources of animal fat and named it *cholesterine*, pairing the Greek word *chole* (bile) with *stereos* (solid). Gallstones are still the purest natural source of cholesterol—if not exactly a convenient one—but it gradually became clear that it was found throughout the body. It's a starting material for the steroid hormones and the main ingredient of bile from the gallbladder (which allows for the absorption of lipids and fat-soluble vitamins from food), but its most important role is as a key ingredient in every animal cell. It increases the flexibility of the lipid layers that make up the cell membrane, in sliding arrangements like **liquid crystals.** This allows cell-surface proteins to fit into the membrane with one face on the outside of the cell and one face on the inside, which lets them respond to signaling molecules from other parts of the body. Cholesterol has had bad press for its role in heart disease, but it is absolutely essential for human life.

Unraveling its structure took many years and strained the structure-determination methods of the time. It was not until 1932 that German chemist Adolf Windaus proposed the correct arrangement, common to all steroids: three six-membered rings and a five-membered one. He and Heinrich Wieland, among others, laid the foundations of **steroid chemistry**, took on much painstaking work in the pre-spectroscopy days. This meant observing what reactions an unknown compound could undergo, speculating on the product structures, and then trying to re-create those products through other routes to confirm (or complicate!) their hypotheses. Logic combined with knowledge and intuition was key. To a modern chemist, determining chemical structure by this method looks like an attempt to solve jigsaw puzzles in the dark, but instruments like **NMR** and **mass spectrometry** have turned the lights on for us.

SEE ALSO Soap (c. 2800 BCE), Liquid Crystals (1888), Mass Spectrometry (1913), Surface Chemistry (1917), Steroid Chemistry (1942), Cortisone (1950), The Pill (1951), NMR (1961), Enzyme Stereochemistry (1975), Isotopic Distribution (2006).

Polarized light shines through a thin layer of pure cholesterol crystals.

Caffeine

Pierre-Joseph Pelletier (1788–1842), **Friedlieb Ferdinand Runge** (1795–1867)

Caffeine is the world's most widely used consciousness-affecting drug. It had been clear for centuries that there was some sort of stimulant in coffee, tea, and other plant extracts, but it wasn't until 1819 that German chemist Friedlieb Ferdinand Runge isolated what he called *Kaffeebase* (coffee base) as a pure substance. It was independently discovered in France not long afterward (where the name *caffeine* was given to it) by researchers looking for **quinine** in coffee beans (no such luck). A few years later, a similar compound, theine, was reported to be isolated from tea. This turned out to be the exact same substance, however, and it gradually became clear that a wide variety of plants produced it.

But why would plants bother spending their metabolic energy making caffeine? Because it's a mild pesticide, for one thing, and it also inhibits the growth of some types of seeds nearby in the soil. Research suggests that it increases successful return visits by honeybees and perhaps other pollinators, too—an effect that coffee shops have found very profitable to exploit with their human customers.

Technically speaking, coffee shop customers are imbibing an adenosine receptor antagonist. An antagonist is a compound that blocks a receptor (a cell-surface signaling protein) from setting off the downstream signal inside the cell. The opposite sort of compound—one that binds and then causes the signal—is called an *agonist*. Adenosine (which is also found as a part of the structure of adenine, the A letter of the A, C, G, T genetic code in DNA) generally acts to suppress nervous system activity in the brain, so blocking its receptors with caffeine keeps that from happening. Too much blockade of the adenosine receptors can bring on the well-known side effects of nervousness, irregular heartbeat, and difficulty sleeping. Its lethal dose, fortunately, is too high to conveniently drink—at least seventy-five cups of coffee, all at the same time.

SEE ALSO Natural Products (c. 60 CE), Quinine (1631), Morphine (1804), LSD (1943)

The most widely used stimulant in the world is not seen as an optional substance for many of its users.

Supercritical Fluids

Charles Cagniard de la Tour (1777–1859)

French engineer and physicist Charles Cagniard de la Tour was interested in studying what happened when you heated liquids to (and above) their boiling points while preventing them from boiling away. He did this inside sealed cannon barrels, which was a wise decision considering the pressures being generated. Since the cannons allowed for no visual inspection of their contents, he added a flint ball inside each one to see if he could detect changes in the sounds it made, which would indicate a change in the liquid's behavior. (Don't try this experiment at home, unless you also own a cannon and are willing to ruin it.)

The results were surprising. Above some particular temperature, the contents of the cannon seemed to be no longer liquid—at least, there was no longer sloshing to be heard, and the sound of the flint ball rolling around changed. But there also wasn't room for the contents to expand into vapor, either. The temperature was peculiar to each liquid and didn't seem to be predictable. In fact, de la Tour had discovered what we now call *supercritical fluids*, and the temperature and pressure measurements at which the sloshing stopped marked what is now called the *critical point* for each liquid. Basically, the liquid phase gets less and less dense as it gets heated, while the gas phase next to it gets denser and denser as it pressurizes with more and more vapor. At the critical point, the two phases have the same density and mix into a new phase that's really neither gas nor liquid but that has new properties. Supercritical water, for example, is somewhat acidic compared to regular water, and nowhere near as polar a solvent.

Carbon dioxide turns out to have a supercritical phase that's relatively easy to reach and very useful. Supercritical CO_2, which dissolves many different substances and can be mixed with other solvents to dissolve even more, is now widely used for chromatography. And like other supercritical fluids, it spreads and penetrates very rapidly into many other substances—properties that have been exploited in materials science as well as dry cleaning and the extraction of **caffeine** from coffee beans.

SEE ALSO Carbon Dioxide (1754), Caffeine (1819), Resolution and Chiral Chromatography (1960)

Tube worms blanket the base of this "black smoker" (the Sully Vent) in the northeast Pacific Ocean, more than seven thousand feet down. Some of these deep-ocean vents are hot enough, and under enough pressure, to vent supercritical water directly into the open ocean.

Beryllium

Nicolas-Louis Vauquelin (1763–1829), **Antoine Bussy** (1794–1882), **Friedrich Wöhler** (1800–1882)

Beryllium is a strange element. With the atomic number of 4, it sits among all the well-known, lighter members of the periodic table, yet it's unfamiliar to most non-chemists. It has never been cheap or easy to obtain and is surprisingly toxic. So why bother with it? In fact, beryllium is extremely mechanically stable under heating and transparent to most useful wavelengths of X-rays (which is important when an X-ray source has to be sealed). It scatters and slows high-energy neutrons very efficiently, so it has long been used in nuclear physics and may well be a key shielding material for the walls of fusion power plants (which would give off dangerous amounts of neutrons). Copper-beryllium alloys are known for their strength and nonsparking behavior, making them valuable for wrenches and screwdrivers in rooms full of hydrogen tanks or other explosives.

Isolating the pure element took a lot of work, though. Beryllium has a very strong tendency to react with oxygen, especially when heated, which is exactly what you *don't* want when refining or working a metal. It's not a particularly common element, either. French chemist Nicolas Vauquelin first identified it as an unknown substance in 1798—naming it *glucine* because some of the element's salts have a sweet taste—after extracting its oxide from beryl (the gemstones emerald and aquamarine are both varieties of this mineral). It wasn't made in any semi-pure form until 1828, when Friedrich Wöhler and Antoine Bussy independently used another newly isolated element, potassium, to yield small grains of it. Pure beryllium took even longer to be produced, and it wasn't until the late 1950s that really high-grade material became available (**zone refining** helped).

Beryllium might have become an important industrial material during the 1930s and 1940s, as it was a component of early fluorescent light designs. However, it became clear that people working with it were coming down with a variety of health problems, especially from breathing the metallic dust. Its use has been severely restricted since then—with the exceptions of X-ray equipment and the high-performance alloys mentioned above. If fusion power is ever made practical, though, we might see more of the element.

SEE ALSO Toxicology (1538), *De Re Metallica* (1556), Electrochemical Reduction (1807), Zone Refining (1952).

Beryllium is found in a variety of gemstones, such as this aquamarine specimen from Nepal.

Wöhler's Urea Synthesis

Jöns Jacob Berzelius (1779–1848), **Friedrich Wöhler** (1800–1882)

You don't hear the word *vitalism* very often in conversation, but it represents a type of thinking that is still very much around. Vitalism refers to the idea that living things have an extra something—a spark, a spirit—that nonliving ones lack. That line of thought goes on to draw a distinction between the components of living creatures (organs, cells, blood, and eventually biomolecules and organic compounds) and the rest of the world, made up of inert objects, minerals, dead matter, and inorganic compounds. But the more we've learned about chemistry, the less room there seems to be for vitalism. The first big crack in that view of the world was the preparation of urea, a simple biomolecule, in 1828 by German chemist Friedrich Wöhler. Urea was well known as a substance that could be obtained from urine, whose only source was synthesis by living creatures, but Wöhler prepared it from starting materials that were absolutely, inarguably inorganic and dead—compounds like mercury cyanate, which no one associates with life. After his discovery, he wrote to his mentor, Swedish chemist Jöns Jacob Berzelius, declaring that he had found a way to make urea without a kidney.

While that wasn't a bad joke, many people found (and continue to find) the consequences to be no laughing matter. What made living creatures special, if not their ingredients? Many people today would still contend that there's a difference between, say, **vitamin C** extracted from an orange and vitamin C made in a laboratory. Some attribute the difference to other beneficial trace substances that might have been carried along from the fruit, but even if you insist on both compounds being absolutely pure, many people think there's something extra, or special, about the compound that came from a living plant. But there isn't. Mix them together or switch the labels, and they cannot be told apart (except perhaps by **isotopic distribution**, which is not what most people have in mind when they're willing to pay more for "natural" vitamins at the store). Compounds are compounds, and there is no vital essence in them.

SEE ALSO Asymmetric Induction (1894), Vitamin C (1932)

Looking like frost on a window, these crystals of urea are produced by evaporation of a concentrated solution.

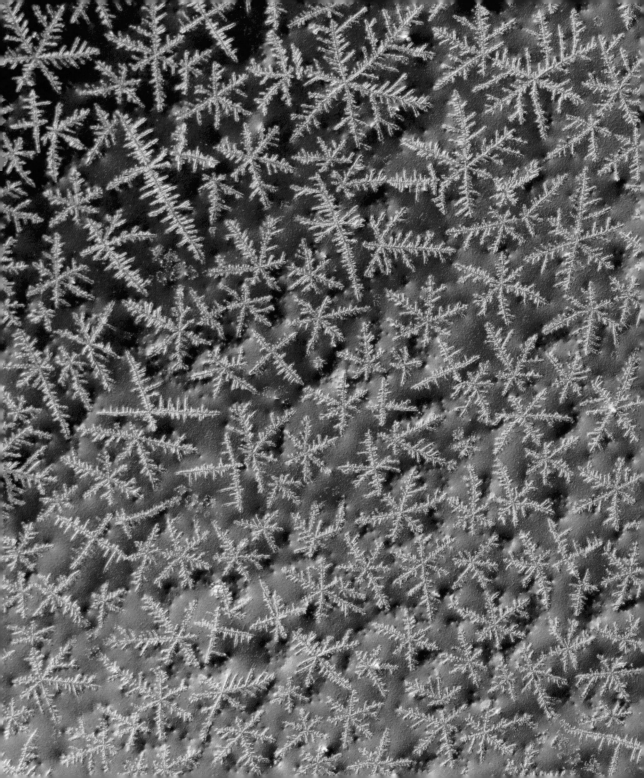

Functional Groups

Friedrich Wöhler (1800–1882), Justus von Liebig (1803–1873)

The publication of a paper in 1832 by German chemists Friedrich Wöhler and Justus von Liebig detailing the reactions of "oil of bitter almond," a compound we would now call benzaldehyde, was a real step forward in our understanding of organic chemistry. Benzaldehyde was known to undergo many other chemical reactions, and the formulas of the product molecules (the amounts of carbon, hydrogen, oxygen, and so on in each) could be determined. But no one really understood the relationship of those chemical formulas to the underlying compound structures, or how all those products were related to one another. In fact, no one really understood organic structures very well at all.

This was about to change. Wöhler and von Liebig showed that in these series of reactions, a consistent underlying structure seemed to follow from the starting material, benzaldehyde, all the way through to the last molecule formed from it. This structure corresponded to a formula of C_6H_7O, which they called the "benzoyl radical," with "radical" meaning a root structure that remained consistent. Other atoms or groups of atoms could be rearranged on this core—an extra oxygen atom made it into benzoic acid, and an extra chlorine turned it into a reactive compound that formed still more derivatives. But they all carried that benzoyl piece.

The idea that organic compounds reacted in parts like this illuminated many puzzling results, and as various compounds' structures were identified in later decades, it became clear how these piece-by-piece reactions could happen—there were "functional groups" on molecules that did most of the reacting, and they were attached to scaffolds that were less changeable but could still affect the molecules' reactivity.

Generations of organic chemistry students have learned reactions in this style, and they might agree with Wöhler's assessment: "At this time organic chemistry can drive one completely crazy. It seems to me like a primeval tropical jungle, full of the most remarkable things, an amazing thicket, without escape or end, into which one would not dare to enter."

SEE ALSO Diethyl Ether (1540), Mirror Silvering (1856), Infrared Spectroscopy (1905), Isoamyl Acetate and Esters (1962).

Justus von Liebig, c. 1846, by German painter Wilhelm Trautschold.

Ideal Gas Law

Robert Boyle (1627–1691), Jacques Charles (1746–1823), Joseph-Louis Gay-Lussac (1778–1850), Benoît Paul Émile Clapeyron (1799–1864)

Pressure, volume, and temperature are all related in the behavior of gases. Squeeze the volume down, and the pressure goes up (that's Boyle's law, named for English chemist Robert Boyle). Raise the temperature, and the volume goes up (Charles's law, named for French physicist Jacques Charles)—and if you don't let it, then the pressure goes up (Gay-Lussac's law, named for French chemist Joseph-Louis Gay-Lussac). In 1834, French engineer and physicist Benoît Clapeyron combined all these relationships into a single "equation of state" known as the ideal gas law. Every chemist should be able to recite it for you: $PV = nRT$.

Unpacking it isn't hard: P is pressure, V is volume, and T is temperature; n is the amount of substance in question (measured in moles), and R is the ideal gas constant. The value of R depends on the units that the other parts of the equation are being measured in, naturally, but no matter what the units, it's the way to convert the temperature of a given number of gas molecules into the amount of energy they contain. From this short bit of mathematics come refrigerators, air compressors, balloons, weather forecasting, and everything else that involves changes in the temperature and pressure of a gas.

As with any equation of this kind, the next question is "How well does it work with real gases?" For monoatomic gases, quite well, particularly at relatively high temperatures and low pressure—that is, under conditions where the atoms can behave like perfect little spheres. The ideal gas law makes no provisions, though, for differences in how real gas molecules might attract each other, or in how much they can be compressed—provisions that more detailed equations of state take into account. And it's those very deviations from the ideal, in any equation, that point out the real-world phenomena that need to be better understood. Chemistry and all the other sciences work this way: a new theory explains the facts better than the old one, but there are places where it breaks down or gives the wrong answer. The next theory brings those into line, and then the hunt starts for the conditions where it breaks down in turn.

SEE ALSO The Sceptical Chymist (1661), Avogadro's Hypothesis (1811), Maxwell-Boltzmann Distribution (1877), Gaseous Diffusion (1940), Methane Hydrate (1965)

One way to demonstrate the pressure/volume relationship: as a vacuum pump pulls the air out of the apparatus and lowers the pressure, the air pockets inside the marshmallows expand dramatically.

Photochemistry

Theodor Grotthuss (1785–1822), **John Draper** (1811–1882), **Hermann Trommsdorff** (1811–1884), **Giacomo Ciamician** (1857–1922)

Some chemical reactions begin as soon as the reactants are brought together, but many more need a push from an external energy source, with heat probably as the most common. But light can also drive chemistry: the fading of colors from exposure to sunlight was possibly the first photochemical reaction noticed by humanity.

Two photochemistry pioneers—German chemist Theodor Grotthuss in 1817 and English chemist John Draper in 1842—independently discovered that light must be absorbed by a chemical substance for a photochemical reaction to take place. Italian chemist Giacomo Ciamician, who in 1900 began the first systematic study of organic compounds exposed to light, helped to provide the groundwork for our understanding of photochemistry. But in 1834 twenty-three-year-old German pharmacist Hermann Trommsdorff had already noted that sunlight caused pure **crystals** of a plant-based product called *santonin* to turn yellow and burst apart.

We now know that different molecules can absorb different wavelengths of light, depending on their structures (which is what arranges their electron clouds, which really interact with the light). Absorption in the visible wavelengths gives materials different colors. The more energetic forms of light (shorter wavelengths up into the ultraviolet) can cause some types of bonds to break and others to become more reactive. A wide array of rearrangements and ring formations can take place, often in ways that no other sort of chemistry could easily produce. The bursting santonin crystals, whose full mechanism wasn't worked out until 2007, are a good example. The sizes and shapes of the molecules change so drastically that the crystal can't possibly hold together.

Photochemistry isn't just a lab curiosity, though. Molecular shape changes like these are the chemical signals that make our retinas sensitive to light, and vitamin D is produced by a sunlight-driven photochemical reaction in our skin. Where there's energy to be found, living systems will find a way to use it.

SEE ALSO Daguerreotype (1839), Free Radicals (1900), Infrared Spectroscopy (1905), DNA's Structure (1953), Woodward-Hoffman Rules (1965), CFCs and the Ozone Layer (1974), Tholin (1979), Unnatural Products (1982), Flow Chemistry (2006).

Photochemistry in a dentist's office: the ultraviolet light starts a polymerization reaction, which hardens the sealant for filling a tooth cavity.

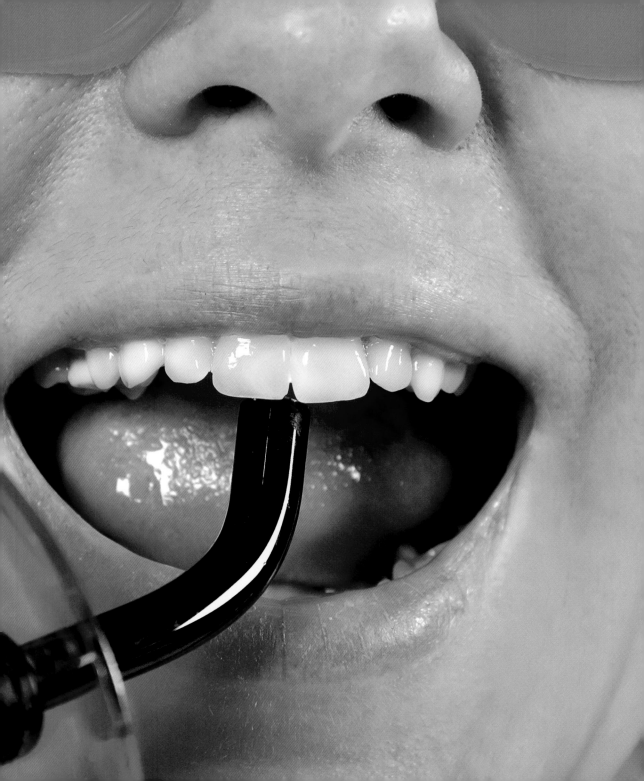

Polymers and Polymerization

Eduard Simon (1789–1856), Hermann Staudinger (1881–1965)

Building up a new substance (a polymer) by linking together small units into larger ones (through the process of polymerization) is a powerful idea. Polymers are essential to life, since proteins, starches, and DNA are all polymeric, and substances from silk to lobster shells are all built up in this way. There are many, many ways that polymers can form, so it's not surprising that examples have turned up unexpectedly.

Polystyrene definitely falls into that category. One of the most widely used plastics in the world (most of those hard, clear packaging shells are made of it), it was discovered in Berlin in 1839 by German pharmacist Eduard Simon, who had distilled the sweet gum tree resin to give a clear, oily substance with a strong odor. Later, though, he found that it had turned into a jellylike mass on standing. He decided (not unreasonably) that it had reacted with oxygen in the air, but later researchers showed that the same thing happened in the absence of oxygen. It took many more decades to appreciate what was really going on.

The original liquid was a simple molecule we now call styrene, which has a double bond that's activated and ready to react. When it does, it leaves behind a new reactive center that picks up another styrene, which picks up another, and so on. What results is a huge macromolecule, a long chain of single-bonded former styrenes. Depending on the mixing, solvents, and temperatures, you can produce a wide variety of final products. German chemist Hermann Staudinger figured out the general principles, and in the 1920s he proposed that substances as different as rubber, cornstarch, and proteins were formed by this repeated assembly of small units. He was absolutely right, and his prediction that chemists would soon figure out ways to use the same idea for their own purposes was right as well, as the huge variety of plastics and other man-made materials around us shows.

SEE ALSO Amino Acids (1806), Rubber (1839), Bakelite (1907), Polyethylene (1933), Nylon (1935), Teflon (1938), Cyanoacrylates (1942), Ziegler-Natta Catalysis (1963), Merrifield Synthesis (1963), Kevlar (1964), Gore-Tex (1969), Acetonitrile (2009)

Polystyrene, instantly recognizable to almost everyone, was discovered after Eduard Simon walked away from some tree resin for a while. Chemistry remains unpredictable.

Daguerreotype

Nicéphore (Joseph) Niépce (1765–1833), **Louis-Jacques-Mandé Daguerre (1787–1851)**

The most famous form of applied **photochemistry** is photography. Adapting the camera obscura—a mechanical method of projecting an image using light, lenses, and mirrors—French inventor Nicéphore Niépce attempted to use chemicals to record what formerly required an artist's eye and hand. In a process he called *heliography* (sun writing), Niépce coated a metal plate with the photosensitive agent bitumen (a tarry, naturally occurring petroleum fraction), positioned the plate in a camera obscura to receive the reflected image, and exposed it to sunlight for hours. Brightly lit areas hardened in the sun (probably via a **free radical**–induced polymerization), and unhardened, shaded areas were then washed away with a solvent, producing in 1826 the first permanent photographs. But because it required long hours of exposure, the technique was not a practical one.

French artist and photographer Louis-Jacques-Mandé Daguerre, who had been collaborating with Niépce, carried on after Niépce's death, using as a light-sensitive agent the more promising silver compounds. After much experimentation, Daguerre produced metal plates coated with silver iodide, so light sensitive that several minutes of exposure were enough to produce an image. The plate was developed by exposure to **mercury** vapors, and the resulting image was composed of dark silver-mercury alloy

(amalgam). But the plate was still light sensitive, and the unreacted silver iodide had to be removed to make the image permanent. Daguerre soon found that the final images could be tinted attractively (and made more durable) by a final exposure to gold salts.

The daguerreotype, revealed in 1839, was a sensation, especially when the process had improved enough for human portraiture. Still, the exposure times—between ten and sixty seconds—tended to produce rather a rather stiff appearance in the subjects, even under the best conditions. The whole procedure was difficult, expensive, and toxic. But it was the first, and it changed the world.

SEE ALSO Mercury (210 BCE), Photochemistry (1834), Free Radicals (1900)

LEFT: *A daguerreotype of Daguerre himself, 1844.* RIGHT: *Foremen of the Phoenix Fire Company and the Mechanic Fire Company in Charleston, South Carolina, c. 1855.*

Rubber

Thomas Hancock (1786–1865), **Charles Goodyear** (1800–1860)

Rubber, a well-known example of a natural polymer, is built from molecules of isoprene, a five-carbon compound found in a variety of plants that is believed to protect them from heat stress. When isoprene is polymerized, the first product created is the sticky latex sap given off by plants such as the South American rubber tree.

This sap can be processed further to natural rubber—as it has been for hundreds of years in Central and South America—but natural rubber has a lot of limitations, among them its relentless stickiness in hot weather and its propensity to crack in the cold. Many inventors tinkered with it, trying to turn it into something more useful, and after many impoverished years of experimentation, American chemist Charles Goodyear famously succeeded. With the addition of sulfur and heat, he discovered, whether by accident or by design (one version has him sticking a lump of rubber to a hot stove), the rubber was cured into an elastic, durable, nonsticky substance that looked as if it would have huge potential, if it could be made industrially. More years of experimentation followed, with Goodyear stretching the patience of his family and his creditors. By 1844 he had filed for a patent for what would come to be known as the vulcanization (after the Roman god of fire) of rubber and had built a factory to produce goods made from it. There were still many wild swings in his fortunes as he fought patent disputes in Europe, most notably with English manufacturing engineer Thomas Hancock, who was simultaneously experimenting with rubber and had received a British patent for the same process.

Chemically, the sulfur in vulcanized rubber crosslinks the polymer chains, altering the properties of the material by changing the ways that the molecules can move relative to each other. Serendipitous or not, the vulcanization of rubber was a significant industrial and commercial advance, and today is responsible for consumer goods as varied as tires, hoses, shoe soles, and hockey pucks, as well as many parts of the industrial machinery involved in making them.

SEE ALSO Polymers and Polymerization (1839), Claus Process (1883), Bakelite (1907), Polyethylene (1933), Nylon (1935), Teflon (1938), Cyanoacrylates (1942), Ziegler-Natta Catalysis (1963), Kevlar (1964), Gore-Tex (1969).

Rubber-tree sap, harvested the old-fashioned way.

Ozone

Christian Friedrich Schönbein (1799–1868), Jacques-Louis Soret (1827–1890), Carl Dietrich Harries (1866–1923), Rudolf Criegee (1902–1975)

While conducting experiments on the electrolysis of water (breaking it down by applying an electric current), German chemist Christian Friedrich Schönbein noticed a distinctive odor in his lab, evidence of a new substance. He named it *ozone* (after the Greek *ozein*, to smell) and described it in 1840. More than twenty years later, Swiss chemist Jacques-Louis Soret boldly proposed that this gas was in fact a new form of oxygen, making ozone the first substance to be recognized as an alternate form of a pure element—an *allotrope* (from the Greek for *other way*). Regular oxygen has the formula O_2, but ozone, also a gas, is O_3. It can be cooled down to an unpredictably explosive blue liquid, and cooling it even further can turn it into a deep violet–colored solid that very few chemists have ever seen.

Many people have encountered ozone, whether they've realized it or not. Produced by lightning bolts, ozone causes the faintly bleach-like "fresh air" smell sometimes noted around thunderstorms. (Its associations with cleansing storms and mountain air gave it a completely undeserved reputation for health—it's actually quite toxic.) But ozone's formation in the upper atmosphere forms a layer that absorbs ultraviolet light and protects living creatures below from its damaging effects.

Electric discharge is still the best way to make it in the laboratory. Ozone generators run pure oxygen past a high-voltage electric arc to provide the gas on demand. It has value because its structure lets it react with carbon-carbon double bonds in a **1,3-dipolar cycloaddition**, forming a five-membered ring containing three oxygen atoms in a row (a type of compound that can explode very easily!). This rearranges, fortunately, producing a species that can be split apart into two aldehydes, which (overall) gives a chemist a very clean, specific way to break an alkene into two reactive groups that can be used for many other reactions. German chemist Carl Dietrich Harries popularized this reaction in the early twentieth century, but it wasn't until the 1950s that another German chemist, Rudolf Criegee, worked out the detailed mechanism through **isotope** labeling studies.

SEE ALSO Oxygen (1774), Isotopes (1913), Dipolar Cycloadditions (1963), B_{12} Synthesis (1973), CFCs and the Ozone Layer (1974)

One reason for that fresh mountain air: ozone from lightning bolts.

Phosphate Fertilizer

Justus von Liebig (1803–1873), John Bennet Lawes (1814–1900), Erling Johnson (1893–1968)

For thousands of years, farmers have been managing the soil to grow better crops, beginning with the addition of manure and crop residue and eventually incorporating additives such as minerals, wood, and plant ash. In the nineteenth century, German chemist Justus von Liebig, continuing the quest to improve agricultural yields while reducing costs, pioneered the study of plant nutrition. He realized the importance of **phosphorus** and nitrogen (see **Haber-Bosch Process**), and made an unsuccessful attempt in 1845 at producing a synthetic phosphate fertilizer. A rival attempt by English agricultural scientist John Bennet Lawes was more successful, and he patented his process of treating phosphates with sulfuric acid in 1842. At the time, the standard way to enrich soil with phosphorus was the use of all-natural bird guano, but good guano did not come cheap.

In 1927, Norwegian chemist Erling Johnson developed a process to treat phosphate-containing rock with nitric acid, producing an effective nitrogen/phosphorus fertilizer. The best phosphate rock was found, inconveniently, on remote islands in the Pacific, but the tremendous importance of good fertilizer actually made shipping boatloads of rocks around the world a very profitable business, indeed. Too profitable, in a way. The island of Nauru, covered in guano phosphate layers, boasted some of the highest per capita income in the world while the boom lasted (through much of the twentieth century), but the exhaustion of its deposits left the economy in trouble and large parts of the island looking like the surface of another planet. Phosphate rock is still mined in other parts of the world, but instead of being turned directly into fertilizer, it's used primarily to make phosphoric acid, most of which is used to make ammonium phosphate fertilizer concentrate. (The rest goes mostly into detergent formulations or soft drinks.)

Unfortunately, many areas with suitable phosphate rocks also have noticeable levels of radioactive elements such as uranium, which get concentrated into the by-product of this process, phosphogypsum (calcium sulfate mixed with phosphoric acid), making it unusable. It piles up like the sulfur from the Claus process, waiting for someone to figure out what to do with it.

SEE ALSO Phosphorus (1669), Haber-Bosch Process (1909)

A view of what's left of the phosphate fields on the island of Nauru, 1990. No obvious natural resources remain.

Nitroglycerine

Christian Friedrich Schönbein (1799–1868), **Théophile-Jules Pelouze** (1807–1867), **Ascanio Sobrero** (1812–1888), **Alfred Nobel** (1833–1896)

Gunpowder reigned for centuries as the world's most powerful explosive, until Italian chemist Ascanio Sobrero's 1847 discovery of an explosive that today we know as nitroglycerine. Sobrero studied under French chemist Théophile-Jules Pelouze, who had worked with guncotton, made by treating cotton with nitric acid, in a process that we now know created nitrate esters on cotton's cellulose chains. This "nitrocellulose" had been discovered in a spectacular accident in 1832 by the German chemist who discovered **ozone**, Christian Friedrich Schönbein, when a cotton apron he used to wipe up a spill of nitric and **sulfuric acid** dried next to the fireplace and then burned up in an explosive flash. But guncotton was still too dangerous and unstable to replace gunpowder (as the explosion of some large-scale production runs unfortunately demonstrated).

Sobrero's breakthrough came when he nitrated simpler carbohydrate, the syrupy three-carbon glycerin. The resulting nitroglycerine was so explosive, and so hard to handle, that Sobrero didn't disclose his work for some time, warning correspondents in the strongest possible terms to avoid it. Undaunted, another student in Pelouze's lab, chemist Alfred Nobel, went home to Sweden in search of a way to make nitroglycerine stable enough to handle. Soaking it into an absorbing material worked, and dynamite was born (as were the beginnings of Nobel's fortune, which would eventually fund the prizes named after him). Nobel's hope that such explosives would make wars too terrible to wage, though, was a grievous misreading of human nature.

Several nitro compounds are well-known explosives, including TNT (trinitrotoluene) and the even more powerful RDX (Research Department explosive), used extensively starting in World War II. Their effectiveness comes from carrying **oxygen** molecules in their structures and their decomposition to nitrogen gas, a very stable substance that provides a large downhill energetic boost (like the formation of aluminum oxide in **thermite**). Chemists (well, chemists who would prefer to have reasonable life spans) know that any compound that can find an easy path to plain nitrogen should be viewed with suspicion and handled with care.

SEE ALSO Greek Fire (c. 672), Gunpowder (c. 850), Oxygen (1774), Ozone (1840), Gibbs Free Energy (1876), Thermite (1893), Haber-Bosch Process (1909), PEPCON Explosion (1988), Flow Chemistry (2006)

In its early days, dynamite was enough of a novelty to be the subject of advertisements, such as this one for the Aetna Dynamite Company of New York, c. 1895.

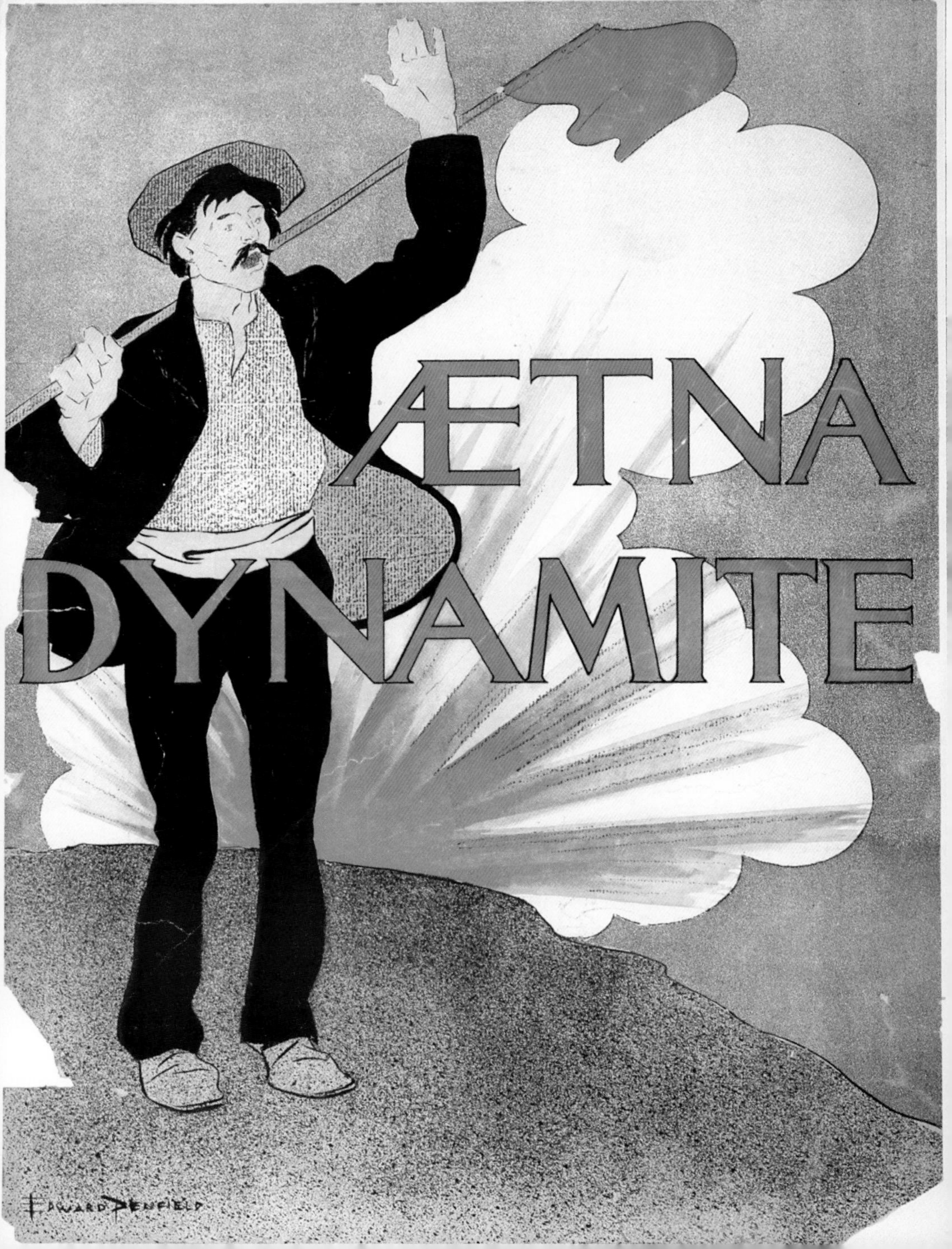
ÆTNA
DYNAMITE
EDWARD PENFIELD

Chirality

Louis Pasteur (1822–1895), **Joseph-Achille Le Bel**, (1847–1930), **Jacobus Henricus van 't Hoff** (1852–1911)

The discovery of polarized light in the early 1800s led to many experiments designed to determine its nature, and chemists observed various compounds under it to see how they behaved. One such compound was tartaric acid, found in grapes and capable of forming **crystals** in wine barrels. The French chemist and microbiologist Louis Pasteur found that a solution of tartaric acid crystals from wine could rotate the plane of the polarized light, as if it had been twisted while it passed through. But a solution of factory-made tartaric acid did not show the same behavior. Why did two batches of the same compound behave so differently?

This was a puzzle, because the two samples seemed otherwise identical. When Pasteur examined the crystals (as tartrate salts) under a microscope, he found that the nonrotating sample seemed to be a mixture of two sorts of crystals, mirror images of each other. He separated the two into pure samples and found to his delight that one variety rotated polarized light just like the wine-barrel tartaric acid, while the other rotated the same amount, but in the opposite direction. The original sample showed no rotation because it had been canceled out!

In 1848, the twenty-five-year-old Pasteur theorized that tartaric acid must have a structure that allowed it to exist in "right-handed" and "left-handed" forms, and that the same must be true for other compounds that could rotate polarized light. It wasn't until the 1870s, though, that Dutch chemist Jacobus Henricus van 't Hoff and French chemist Joseph-Achille Le Bel were able, independently, to explain how such "chiral" compounds (named later from the Greek word for *hand*) could exist.

Chirality is now known to be essential to life (proteins and sugars are both chiral), and many important drugs are chiral as well. We now also know that the sorts of mixed crystals Pasteur studied are relatively rare, and that Pasteur was not only very good, he was also very lucky. Still, as he famously said, "Fortune favors the prepared mind."

SEE ALSO Tetrahedral Carbon Atoms (1874), Fischer and Sugars (1884), Liquid Crystals (1888), Coordination Compounds (1893), Asymmetric Induction (1894), Thalidomide (1960), Resolution and Chiral Chromatography (1960), Murchison Meteorite (1969), B_{12} Synthesis (1973), Enzyme Stereochemistry (1975), Palytoxin (1994), Shikimic Acid Shortage (2005), Engineered Enzymes (2010)

Crystals of tartaric acid photographed with polarized light shining through them. The colors are produced by their varying orientations and thicknesses.

Fluorescence

George Gabriel Stokes (1819–1903)

The property of fluorescence was observed as early as the sixteenth century: some water extracts from exotic woods glowed blue around the edges of their containers when exposed to light and viewed from the right angle. Over the next few hundred years, many other substances were found with similar curious properties, including the mineral fluorspar and glass colored with uranium salts. With the discovery of ultraviolet light at the beginning of the 1800s, it became clear that many of these materials were somehow absorbing ultraviolet light's invisible rays and reemitting visible light instead.

Irish physicist Sir George Gabriel Stokes was the first to work this out (partly by using solutions of **quinine**, which has a fairly intense blue fluorescence), and he named the property after fluorspar. The mechanism for this glow was not immediately understood, though, because a full treatment needed an understanding of quantum mechanics. We now know that some materials can have some of their electrons excited by absorbing higher-energy radiation, and the excited electrons then fall back to their original ground state by emitting energy in the form of light (almost always light of a longer wavelength, and therefore more visible to us, than what was absorbed).

Fluorescent compounds abound, and many of them have been made deliberately. Fluorescent dyes are well known for their strangely glowing colors, a result of their producing more visible light than our eyes expect them to be able to. Although it's probably most common to think of fluorescence in relation to highlighters or orange safety vests, fluorescence is used in biomedical research as well: specific wavelengths are used for excitation and other specific ones are emitted, allowing substances (even inside living cells) to light up on command, completely distinct from their background. Fluorescent labels are even used to mark cancerous tissue in human patients for surgery. Spanning the visible spectrum, these fluorescent molecules have provided huge amounts of information that otherwise would have been hidden.

SEE ALSO Quinine (1631), Luciferin (1957), Green Fluorescent Protein (1962), Click Triazoles (2001)

A collection of fluorescent minerals under ultraviolet light. Some of these are also colorful under ordinary light, but not to this extent!

Separatory Funnel

1854

The separatory, or "sep," funnel is a common sight in any organic chemistry lab, and the principle behind it is used constantly in every part of the science. As anyone who has seen a mixture of oil and water has realized, not all liquids can mix with one another. When you dissolve a sample into an oil-and-water mixture, some of its components will be more soluble in the water, and some will be more soluble in the oil. This provides a quick and reliable way to separate a mixture into its "hydrophilic" (water-loving) and "hydrophobic" (water-fearing) parts. Underlying this are the same principles that are found in most kinds of **chromatography**.

A sep funnel allows the two liquid layers to be shaken together and mixed, but the key innovation is that when they've separated out again, the lower layer can be drained out of the funnel and into another flask. This is such a convenient method to clean up crude reaction mixtures of all kinds that extraction into two layers is second nature to most working organic chemists, and has been for hundreds of years.

The modern sep funnel has been around since roughly 1854 but versions were found earlier in the nineteenth century as well. In fact, even the alchemists used to use tall, thin variants of regular funnels to allow immiscible liquids to be drained into separate containers, and there are containers from much earlier in history that may well have been used for the same purpose.

The split between water-soluble and oil-soluble has been part of the science for a very long time. Most of the time in organic chemistry, the lower layer is the watery one, since water is denser than most organic solvents. But when you're using a chlorinated solvent (such as dichloromethane) it wins the density competition, and the water floats on top. Beginning chemistry students using that solvent often mistakenly pour away their desired products before they learn to take heed, but it's the sort of mistake you tend to only make once!

SEE ALSO Purification (c. 1200 BCE), Erlenmeyer Flask (1861), Soxhlet Extractor (1879), Borosilicate Glass (1893), Chromatography (1901), Dean-Stark Trap (1920), The Fume Hood (1934), Magnetic Stirring (1944), Glove Boxes (1945), Rotary Evaporator (1950), Reverse-Phase Chromatography (1971)

A row of separatory funnels. This classic design has hardly changed over the years, except for the plastic fittings (which used to be glass and sometimes became permanently stuck).

Perkin's Mauve

William Henry Perkin (1838–1907)

William Henry Perkin was a young Englishman in just the right place at just the right time. He was still in his teens when he began studying at London's Royal College of Chemistry under German chemist August von Hofmann, who challenged him to synthesize **quinine** (much in demand at the time for the treatment of malaria) from inexpensive starting materials. Given the state of organic chemistry at the time, neither Hofmann nor Perkin could have known what they were getting themselves into. Quinine was much more complicated than it looked: its structure wasn't worked out for another fifty years, and its synthesis did not come about for decades after that.

So it almost goes without saying that Perkin had little success synthesizing one of the most valuable medicines of his day from scratch. Experiments in his home laboratory, however, fared much better. In 1856, he found that he could produce a gorgeous purple compound from aniline, a component of coal tar, and he immediately saw its possibilities as a dye and pigment. While officially working on the doomed quinine synthesis at the college, Perkin began experimenting in his backyard with the help of his brother and a friend, looking for ways to increase the yield and purity of the world's first artificial dye, which he named *mauveine*, now also known as Perkin's mauve. He tested it on fabrics and other substances, found that it was colorfast (unlike many of the naturally occurring dyes then in use), and became convinced it had great commercial potential.

He was right. Perkin patented his discovery, set up his own factory with help from his family—although his professor advised against it—and was soon the only supplier of what became the most fashionable dyestuff in Europe. The Industrial Revolution had created both a huge textile sector and a lot of coal tar (from the coal-gas industry), and Perkin became rich by connecting the two with his chemical skills. He went on to discover many other dyes, and the chemistry behind them is still used in coloring agents today. In many ways, he helped create the modern chemical industry.

SEE ALSO Yorkshire Alum (1607), Quinine (1631), Paris Green (1814), Indigo Synthesis (1878), Sulfanilamide (1932)

A 1906 portrait of Perkin, holding a piece of the mauve-dyed cloth that made him famous.

Mirror Silvering

Justus von Liebig (1803–1873), Bernhard Tollens (1841–1918)

This reaction looks like a magic trick, but for nearly a hundred years it was a trick well worth knowing if you were in the mirror business. Mirrors used to be made by coating glass with a layer of tin foil and then exposing the foil to liquid **mercury**. The resulting tin/mercury amalgam, while reasonably reflective, could corrode over time (and give off drops of mercury, something to note if you own an antique mirror). The alternative was speculum (Latin for *mirror*) metal, a copper/tin mixture that could be polished to a not-all-that-reflective surface.

German chemist Justus von Liebig found a better way, and refined it enough that it gradually put the mercury method out of business. Liebig's invention depended on oxidation/reduction chemistry in which a silver/amine complex was formed and mixed with a sugar solution, and this was applied to a glass surface. The silver oxidized the sugar molecules to a soluble acid and was in turn reduced to elemental silver, which was deposited as a thin, highly reflective mirror layer on the glass.

Not only did this procedure provide high-quality mirrors—for personal use as well as for things like reflecting telescopes—but German chemist Bernhard Tollens also modified it to serve as a qualitative chemical test. Before modern analytical instruments, such tests were a major feature of chemistry, with a variety of reagent combinations that would give characteristic colors or precipitates with specific elements or functional groups. Tollens's test was used when a chemist wasn't sure whether the carbonyl group (a carbon double bonded to an oxygen) in a molecule had carbons on both sides of it (a ketone) or a hydrogen on one side (an aldehyde). Aldehydes are oxidized by the silver reagent to give a mirror coating on the inside of a test tube, but ketones don't react.

Tollens's test, an antique now, had to be performed with freshly made reagent. If silver/amine solutions are allowed to sit around, further reactions gradually form the unpredictable explosive silver nitride, which can detonate for any reason or none at all. The mirror-silvering people also had to learn this lesson, sometimes the hard way.

SEE ALSO Mercury (210 BCE), Functional Groups (1832)

Suspended in a beaker of warm water, Tollens's reagent quickly forms a thin layer of silver on the interior of this test tube. With careful technique and a very clean flask, an almost perfect mirror can be produced.

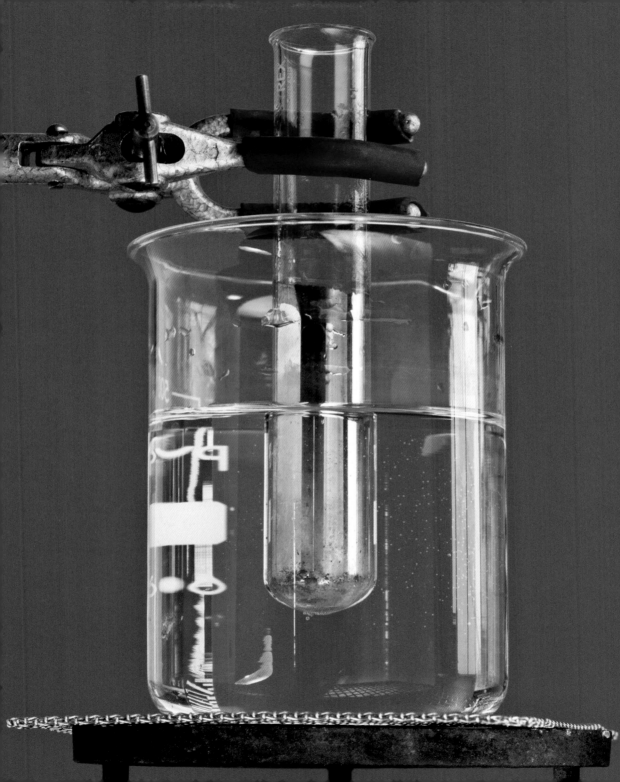

Flame Spectroscopy

William Hyde Wollaston (1766–1828), **Joseph von Fraunhofer** (1787–1826), **Robert Bunsen** (1811–1899), **Gustav Kirchhoff** (1824–1887), **Alan Walsh** (1916–1998)

Many non-chemists have heard of the Bunsen burner as a piece of lab equipment (although you don't see as many of them around these days), but not many people know why Robert Bunsen, a German chemist, invented it. He needed a better way to do flame tests, an old analytical technique that depends on the emission spectra of various elements—i.e., the light given off in the high-energy environment of a flame as electrons jump between energy levels. Sodium, for example, shows a vivid yellow-orange, strontium has bright red, and copper has blue-green. But a hot and colorless flame is needed to see all these colors, and Bunsen invented a better way to mix air and gas to create one.

His colleague, the German physicist Gustav Kirchhoff, then suggested that Bunsen use the new technique of splitting the light with a prism to make even finer distinctions. (Lithium, for example, also gives a red flame, which can be hard to distinguish from strontium with the naked eye.) In 1859, they built the first spectroscope, an extremely powerful tool for identifying elements—and for discovering them, too. A blue line in a sample of mineral water turned out to be the new element cesium, and Bunsen discovered another red-emitting element (rubidium) as well.

There was even more to be found with this technique. English chemist William Hyde Wollaston (followed by the German physicist Joseph von Fraunhofer) had noticed that the prismatic spectrum of sunlight had mysterious dark lines in it. Bunsen and Kirchhoff realized that many of them were in the same places in the spectrum as the bright emission lines they studied, which suggested that they were coming from elements in the sun itself that were soaking up light at these wavelengths. Suddenly it was possible to determine the chemical makeup of the sun (and other stars!) while sitting at a laboratory bench. Modern atomic spectroscopy instruments (invented by British physicist Sir Alan Walsh and others) use the same principles to analyze elements down to parts per billion to trace water contamination, among other uses.

SEE ALSO Helium (1868), Neon (1898), Deuterium (1931), Gas Chromatography (1952), Thallium Poisoning (1952).

The bright colors of fireworks are flame tests in the sky. Strontium and lithium are used for red, sodium for yellow, and barium for green. A good blue is said to be the hardest to formulate.

Cannizzaro at Karlsruhe

Stanislao Cannizzaro (1826–1910)

By the mid-nineteenth century, the scientific world agreed that atomic weights (and thus molecular weights) were extremely important. But what were the right weights? Weighing individual atoms was out of the question, and using the weights of starting materials and products of chemical reactions led to all sorts of arguments about how molecular formulas should be expressed. English chemist John Dalton, a pioneer in atomic theory, was sure that water had the formula HO.

The year 1860 saw the Karlsruhe Congress, the world's first international meeting of chemists, held to try to settle this and other chemistry questions. A paper from Italian chemist Stanislao Cannizzaro (actually published two years before but little known) made the biggest impression. Building on Avogadro's work, Cannizzaro's "Sketch of a Course of Chemical Philosophy" attempted to nail down the real atomic weights. Like Dalton, Cannizzaro started from hydrogen, assigning it a weight of 1, but like French chemist Joseph-Louis Gay-Lussac, he (rightly) took it to be a diatomic molecule, H_2, and he reminded the conference attendees of Avogadro's evidence that **oxygen** was diatomic as well. Avogadro had theorized that equal volumes of gases had the same number of molecules in them, and that they weighed different amounts only because the molecules of each gas had their own individual weights.

Based on Gay-Lussac's measurements, two volumes of hydrogen to one volume of oxygen were needed to create one volume of water vapor, indicating that the actual formula for water was not HO—as Dalton had suggested—but rather H_2O. Weighing the materials showed that oxygen was eight times as heavy as the hydrogen present in water vapor, which led Cannizzaro to conclude that its atomic weight was 16, given the 2:1 ratio of hydrogen to oxygen in water.

Cannizzaro's paper seems to have almost singlehandedly cleared up most of the confusion that had afflicted chemists about the elements' fundamental weights. It was so logical and so useful and fit the data so well that it almost *had* to be correct, and indeed it was.

SEE ALSO Oxygen (1774), Dalton's Atomic Theory (1808), Avogadro's Hypothesis (1811), The Mole (1894)

LEFT: *Stanislao Cannizzaro.* RIGHT: *Karlsruhe, Germany, c. 1900, where Cannizzaro put chemical formulas and weights on a solid footing at last.*

Oxidation States

Johann Rudolf Glauber (1604–1668), Henry Bollmann Condy (1826–1907)

Chemists often speak of the oxidation state of a given element. The convention is that the plain element, not part of any chemical compound, is in the zero oxidation state, and as electrons are removed or added (each electron being a single minus charge), the oxidation state moves up or down, respectively. The elements on the left-hand side of the periodic table lose electrons very easily and jump to a higher oxidation state at almost any opportunity, while the elements toward the right-hand side tend to pick up electrons. This is why sodium (all the way on the left-hand side of the periodic table) is usually in its +1 oxidation state and chlorine (nearly as far to the right as sodium is to the left) in its −1 state, which is how they balance out in table salt (sodium chloride).

Elemental sodium (in the zero oxidation state), on the other hand, is a soft, silvery metal that bursts into flame on contact with water, and elemental chlorine is a greenish gas that's fiercely reactive and poisonous—both of them are dangerously ready to move out of their original oxidation states. Exposing sodium metal to chlorine gas would give you salt, but you'd have trouble retrieving it from the salty debris of the resulting explosion.

High-oxidation state metals are often ready to move to a lower-energy lower oxidation state by oxidizing something else in turn. This process can decolorize pigments, kill bacteria, clean impurities from surfaces, and perform a lot of useful chemistry. German chemist Johann Rudolf Glauber first described potassium permanganate, a brilliant purple compound with its manganese metal in the high +7 oxidation state, in 1659, but it wasn't until 1860 that English chemist Henry Bollmann Condy, commercialized it as a disinfectant he called *Condy's fluid*. (As the manganese atoms drop in oxidation state, they turn brown or pink.)

Many other metals in the middle of the periodic table (the "transition metals") show similar color changes across their oxidation states, with chromium being another standout. These changes can be used to monitor their chemical reactions and also account for the vivid rainbow colors of the salts.

SEE ALSO Electrochemical Reduction (1807), Aluminum (1886), Chlor-Alkali Process (1892), BZ Reaction (1968), PEPCON Explosion (1988).

When dissolved in water, potassium permanganate has a vivid, unmistakable purple color that is instantly recognizable to any chemist who's ever worked with it.

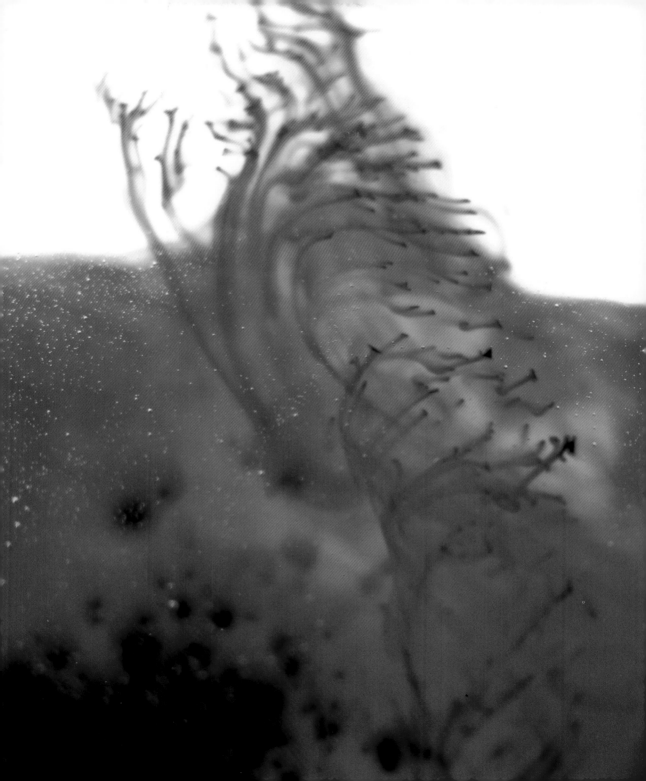

Erlenmeyer Flask

Richard August Carl Emil Erlenmeyer (1825–1909)

If you want an instant visual shorthand for chemistry, just show someone an Erlenmeyer flask. Unlike some of the glassware shown in popular depictions of the science, it's not a relic—chemistry labs around the world still have Erlenmeyers of all sizes on their shelves, thanks to German chemist Emil Erlenmeyer's ingenious design. Erlenmeyer was in his thirties in 1861 when he published a paper titled "Chemical and Pharmaceutical Technique" containing a description of the glassware, saying that he had displayed the flask at a conference in Heidelberg three years before. He noted that he had also asked some of the local glassblowers to offer the flasks for sale, and use of his design gradually spread throughout chemical practice over the next few decades.

What made the design so successful? For one thing, its conical shape makes it easy to mix the contents by swirling them around without spillage. In the days of colored indicators, this was a real advantage. Try hand-swirling with a beaker with its straight sides, and you're going to need at least a mop—and possibly a shower. The narrowed neck of the Erlenmeyer also keeps solvents from evaporating quickly and helps prevent splashing when the contents are blended with a magnetic stirrer.

Erlenmeyer, a well-known chemist in his time, was also the first to suggest carbon-carbon double and triple bonds in chemical structures. Today, he's mostly remembered for his glassware design, and though the use of Erlenmeyers has expanded to include locations outside the lab, such as in breweries and wineries (and though in Great Britain it's more likely to be called just a *conical flask*) its clever design is here to stay.

SEE ALSO Separatory Funnel (1854), pH and Indicators (1909), Soxhlet Extractor (1909), Borosilicate Glass (1893), Dean-Stark Trap (1920), The Fume Hood (1934), Magnetic Stirring (1944), Glove Boxes (1945), Rotary Evaporator (1950)

The symbol for chemistry the world over: the Erlenmeyer flask.

Structural Formula

Josef Loschmidt (1821–1895)

Chemists write down the structures of molecules in ways that many people find strange or even intimidating. For advertising directors and the like, nothing (well, except possibly an **Erlenmeyer flask**) says *chemistry* like a complicated structural formula.

But the rules for drawing molecular structures aren't hard. Each line represents a chemical bond, and if an atom isn't labeled, it's a carbon atom. Also, organic chemists tend to leave hydrogen atoms off their structural formulas, since they are everywhere. There can be single, double, and triple bonds to certain elements, and chemists often draw aromatic rings with a circle in them to show they're different (see **Benzene and Aromaticity**). This notation, though, assumes that everyone knows what the underlying structures really are, which makes Josef Loschmidt's work remarkable. An Austrian chemist whose work ranged over several fields, Loschmidt made his real mark on science with his 1861 book *Chemische Studien* (*Chemical Studies*), which presented a variety of his new structural notations for molecules. He drew the various atoms as different-sized circles, using shading to indicate different elements, and while many of them look strange to a modern chemist at first, they make reasonable sense after just a short inspection. Although he got several structures wrong, his notation for benzene and other aromatic rings is a strikingly modern circle with atoms or groups bonded to its edges, which is especially impressive given that this was four years before these cyclic structures were accepted as real.

Drawing molecular structures out this way is tremendously helpful to chemists; a structural formula conveys a lot of information in a small space and can tell an experienced chemist a great deal about how the molecule it represents will react, what its physical properties are probably like (even down, in some cases, to its smell), and how one might go about making or using it. Chemists take this sort of notation for granted, although we really shouldn't, as the rest of the nineteenth century was occupied with questions about how molecules were really put together.

SEE ALSO Benzene and Aromaticity (1865), Tetrahedral Carbon Atoms (1874)

These modern-day structural formulas for various common compounds bear only a passing resemblance to Loschmidt's versions.

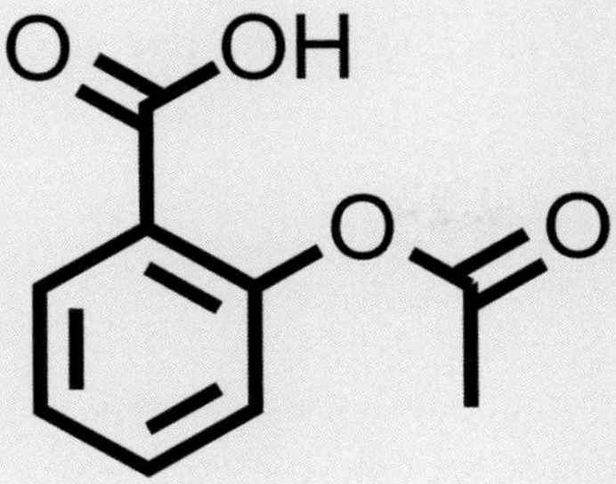

Aspirin

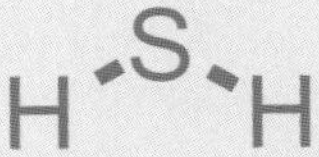

Hydrogen sulfide

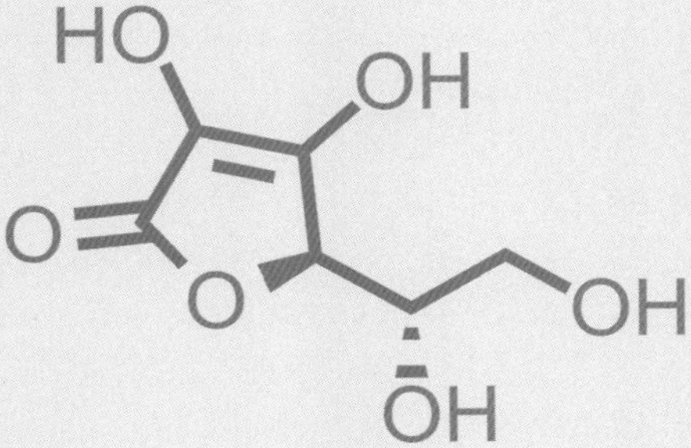

Vitamin C

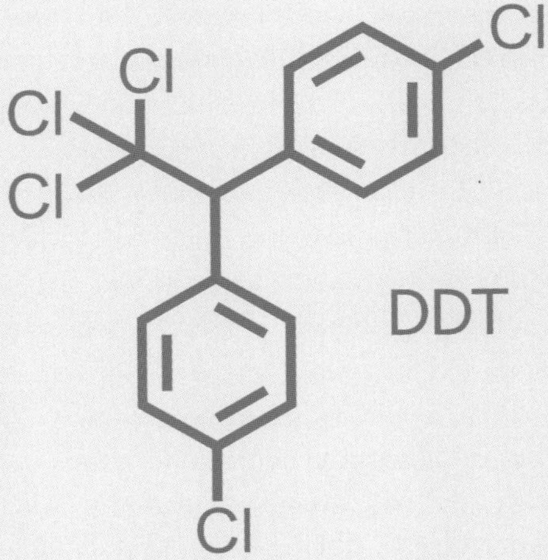

DDT

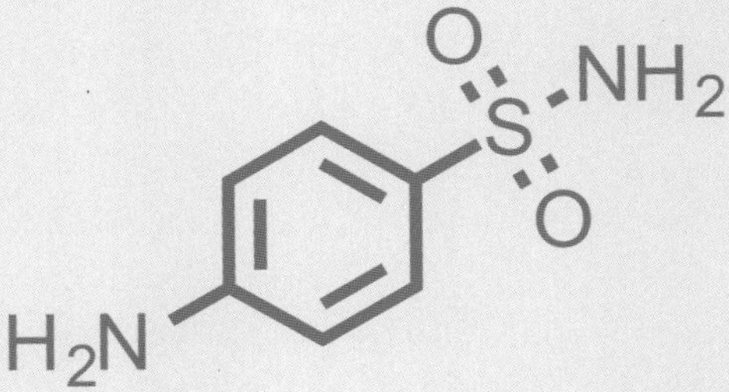

Sulfanilamide

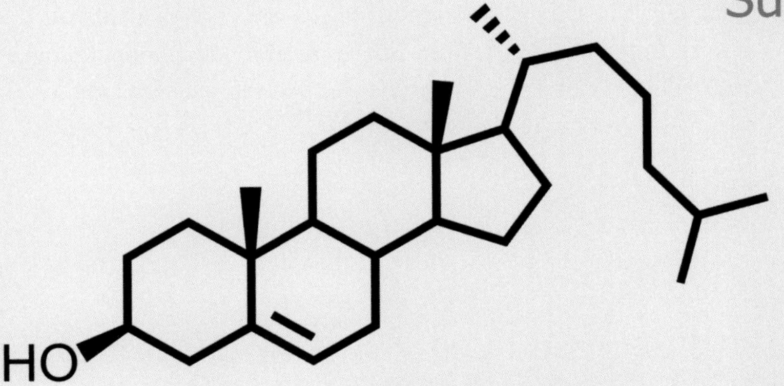

Cholesterol

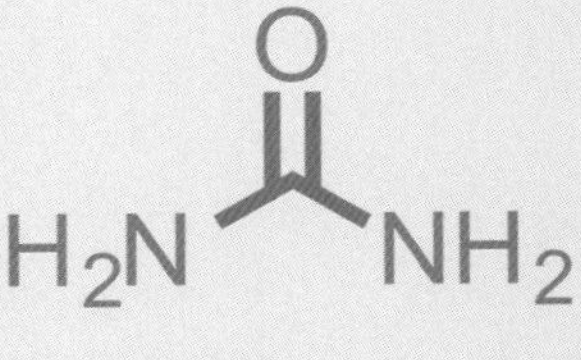

Urea

Solvay Process

Nicolas Leblanc (1742–1806), Ernest Solvay (1838–1922)

In eighteenth-century France, plant-based soda ash—a critical component in the manufacturing of glass, textiles, paper, **soap**, and other products—was becoming scarce, so a prize was offered to anyone who could develop an industrial method to manufacture it. One of the world's first industrial-chemistry breakthroughs was the 1789 synthesis of soda ash from sodium chloride (common salt) by French chemist Nicolas Leblanc. Leblanc's process wasn't easy (it required **sulfuric acid** and a great deal of heat), but it was a big success nonetheless, and it stayed relevant until Belgian chemist Ernest Solvay's 1864 invention of the Solvay process.

Like Leblanc's, Solvay's process also uses both salt, which can be obtained as concentrated brine from seawater, and calcium carbonate, which is ground limestone. The intermediate steps of the process use ammonia, but one of the handy things about the Solvay route is that it reclaims and recycles almost all of its ammonia, which was an expensive commodity in the early days of the industry. Sodium bicarbonate (more commonly known as baking soda) is an intermediate as well, but most of it is used to prepare the final sodium carbonate product. The final by-product is calcium chloride, commonly used to melt road ice. The overall process was a huge improvement on the Leblanc method, and Solvay became immensely wealthy as Leblanc's process was dropped and Solvay's technology was adopted around the industrialized world.

In North America, however, a huge discovery of naturally occurring sodium carbonate made the Solvay route uneconomical; it's difficult to compete with the simplicity of digging a product right out of the ground. Such mining operations, along with the increasing difficulties in finding a market for the calcium chloride by-product, which can be obtained more cheaply by other routes, have led to a gradual decline in the Solvay synthesis worldwide. Dozens of Solvay plants are still in operation, though, and the process developed 150 years ago is still in use (in certain locations) today.

SEE ALSO Soap (c. 2800 BCE), Sulfuric Acid (1746), Chlor-Alkali Process (1892), Borosilicate Glass (1893), Carbon Dioxide Scrubbing (1970)

A modern Solvay installation in Dombasle-sur-Meurthe, France. Even though such plants were once more common around the world, the process is too useful in the right situations for them to disappear completely.

Benzene and Aromaticity

Michael Faraday (1791–1867), **Friedrich August Kekulé** (1829–1896), **Kathleen Lonsdale** (1903–1971)

English chemist Michael Faraday discovered benzene in 1825. With a molecular formula of C_6H_6, it was considered for decades to be a hydrocarbon (a compound containing only hydrogen and carbon) like any other. The structure, however, was the tricky part. It had to be cyclic, because that ratio of carbons and hydrogens requires at least one ring and some carbon-carbon double bonds as well. Some strange features had to be explained as well. There were three—and only three—different dichlorobenzenes (benzenes with two hydrogens swapped out for chlorine atoms), for example, all with different melting points, and the same went for all the other disubstituted benzenes. No one understood why.

Several structures were proposed, but in 1865 German chemist Friedrich August Kekulé published one that seemed to fit the facts the best (years later, he said that the idea came to him in a daydream). His benzene was a flat, six-membered ring of carbons with alternating double and single bonds between them, with one hydrogen on each. All six carbons were identical, and this flat ring meant that the dichlorobenzenes could be easily drawn as different positions around the ring (two next to each other, or two separated by one carbon, or two across from each other).

Some mysteries remained. If Kekulé's structure was correct, two adjacent substituents could still be separated by either a double bond or a single bond in the ring, but no one ever saw those two types in reality. (In 1928, Irish crystallographer Kathleen Lonsdale's X-ray structure of benzene crystals proved that all the bonds were actually the same length, in between the lengths of a double bond and a single bond.) Benzene's double bonds were also much less reactive than they should be. This unusual behavior came to be called *aromaticity*—a name inspired by the distinctive smells many such compounds have—and was shared by several other types of cyclic compounds, all of which could be drawn with alternating double bonds. The specific characters of aromatic rings is part of the properties of everything from proteins to plastics to pharmaceuticals.

SEE ALSO Structural Formula (1861), Friedel-Crafts Reaction (1877), Reppe Chemistry (1928), Sigma and Pi Bonding (1931), Birch Reduction (1944), Graphene (2004)

This German stamp commemorates the benzene structure a century after its discovery. Kekulé's proposal has stood the test of time.

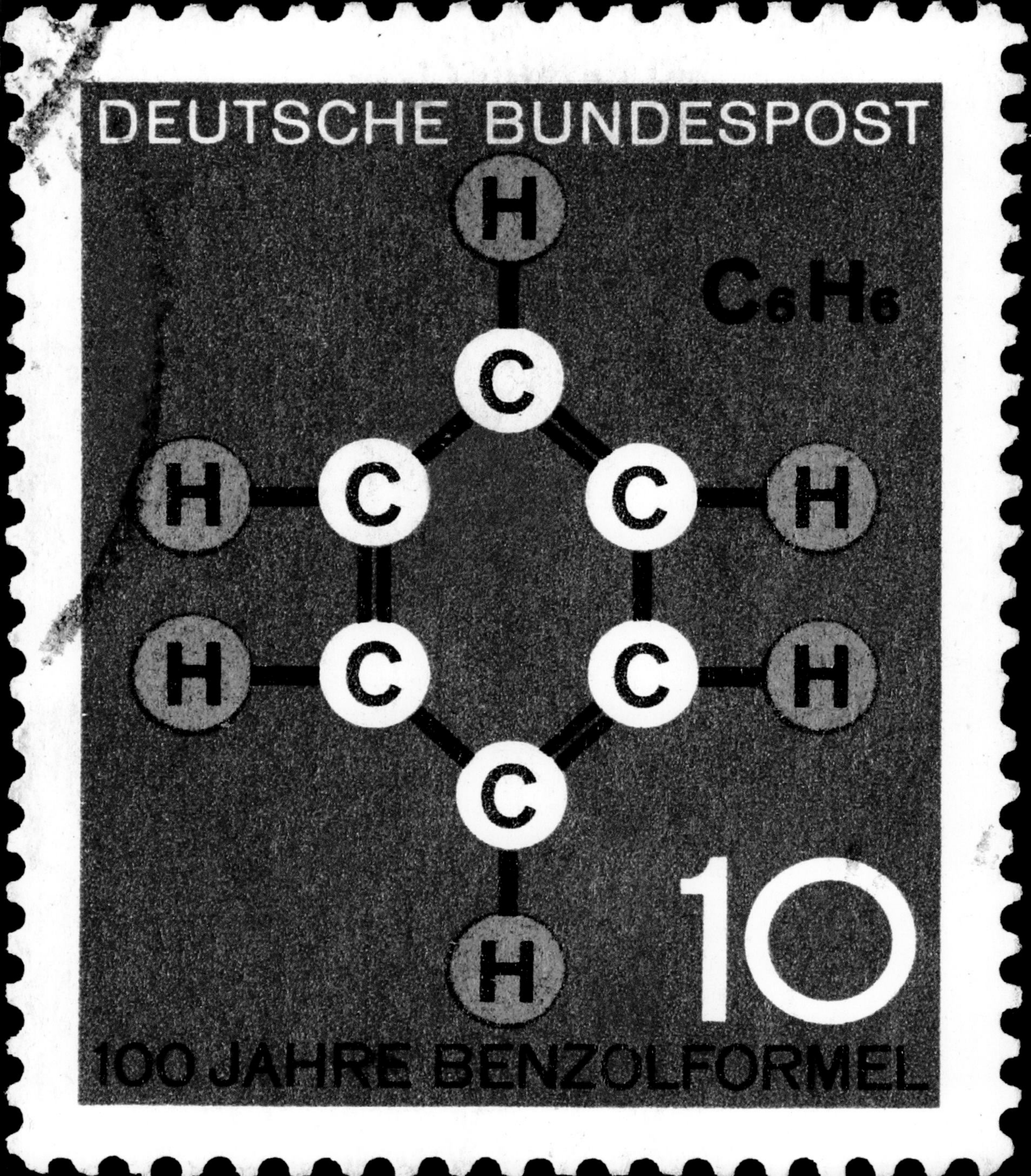

DEUTSCHE BUNDESPOST

C_6H_6

10

100 JAHRE BENZOLFORMEL

Helium

Luigi Palmieri (1807–1896), Pierre Jules César Janssen (1824–1907), Joseph Norman Lockyer (1836–1920), Hamilton Perkins Cady (1874–1943), David Ford McFarland (1878–1955)

German physicist Gustav Kirchhoff and German chemist Robert Bunsen's technique of analyzing sunlight with a spectroscope yielded an unexpected result in 1868. Astronomers Pierre Jules César Janssen, a Frenchman, and Sir Joseph Norman Lockyer, an Englishman, independently discovered a new yellow line in the light of solar prominences (masses of heated gas blown out from the sun) that belonged to no known element. Lockyer named this new element *helium* (from *helios*, the Greek word for *sun*), but the search for it on Earth was unsuccessful until 1882, when Italian physicist and meteorologist Luigi Palmieri found its spectrum in lava from Mount Vesuvius. The rare-earth ores later provided enough to actually isolate, but only barely.

Helium remained a rarity until a 1903 festival in Dexter, Kansas, went wrong. A drilling rig outside town had uncovered a huge deposit of natural gas, seeming to promise an economic boom. The townspeople planned to celebrate by lighting the plume into a huge flame before capping the well, but their flaming bales of hay were actually extinguished by the gas. This weird behavior caught the attention of the state geologist, who had a sample of the gas shipped to the University of Kansas. There, American chemists Hamilton Perkins Cady and David Ford McFarland found that the gas was mostly nitrogen, with only 15 percent valuable methane. They also realized the sample contained a wildly unexpected 2 percent helium. Other natural gas samples were later found to contain helium as well, meaning that the gas was now available in vast quantities.

World War I brought a demand for it, as a nonflammable way to inflate observation balloons. Soon helium was available for scientific study, where it led to a string of discoveries in fundamental physics. Today chemists use it as an inert carrier in **gas chromatography** (a method for separating volatile mixtures into their component compounds), and to cool massive **NMR** (nuclear magnetic resonance) magnets down to superconducting temperatures. In recent years, increased demand and fewer suppliers have caused its price to rise steeply—so in retrospect, the people of Dexter should never have called off the party.

SEE ALSO Flame Spectroscopy (1859), Gas Chromatography (1952), NMR (1961)

Beginning with the U.S.S. Shenandoah in 1923, the U.S. military experimented with helium-filled dirigibles for many years. They proved difficult to handle in stormy weather, however.

The Periodic Table

Lothar Meyer (1830–1895), **Dmitri Ivanovich Mendeleev** (1834–1907), **John Alexander Reina Newlands** (1837–1898), **Antonius van den Broek** (1870–1926), **Henry Gwyn Jeffreys Moseley** (1887–1915)

The periodic table is the undisputed centerpiece of chemistry. Built into its arrangement is a wealth of hard-earned knowledge about atomic structure, reactivity, bonding, and other crucial concepts. The building blocks of our world are all there, organized in a way that shows their deepest relations.

German chemist Lothar Meyer and English chemist John Alexander Reina Newlands were two of the first to realize (independently) that arranging the known elements by their atomic weights revealed underlying patterns. Elements with similar behavior tended to cluster together (such as sodium and potassium, both soft, highly reactive metals). In Russia, chemist Dmitri Ivanovich Mendeleev, unaware of the work of Meyer and Newlands, was thinking along the same lines, and in 1869 he presented his own arrangement based on atomic weights and the number of bonds the various elements tended to form. It not only had all the known elements, but it also boldly included gaps where new ones were predicted to exist. Their discovery (and the successful prediction of their properties) was powerful evidence that Mendeleev had it right.

Modern tables are ordered by increasing atomic number (the number of protons in the nucleus), as suggested by Dutch physicist Antonius van den Broek and by the work of English physicist Henry Gwyn Jeffreys Moseley. The columns (called *groups*) represent increasing numbers of electrons in each atom's outermost "shell" (called an *orbital*), from just one electron on the far left column (reactive sodium and its alkali metal group members) over to the unreactive noble gases on the far right, with their perfectly filled orbitals. Then a new row (called a period) starts, with a heavier alkali metal on one end, going all the way over to a heavier noble gas. The heavier elements demonstrate the greater number of electrons that go into the outer orbitals, and the table spreads out accordingly.

It's no exaggeration to say that many of the chemistry advances over thousands of years were leading up to this: Understanding how the elements differ and why has been one of the great works of the human race.

SEE ALSO The Four Elements (c. 450 BCE), Neon (1898), Silicones (1900), Technetium (1936), The Last Element in Nature (1939), Transuranic Elements (1951)

The current periodic table, where all of chemistry begins.

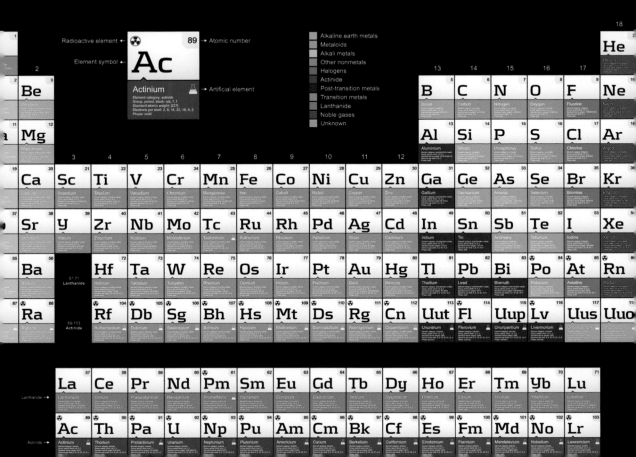

Tetrahedral Carbon Atoms

Jacobus Henricus van 't Hoff (1852–1911), Joseph-Achille Le Bel (1847–1930)

The idea that chemical compounds have defined three-dimensional shapes caused by the arrangement of their bonded atoms is now such a foundation stone of chemistry that it's hard to imagine a time when this property wasn't known. In the second half of the nineteenth century, though, chemists were still thinking two-dimensionally, grappling with the basic structures of molecules and trying to understand what gave rise to chiral compounds.

Dutch chemist Jacobus Henricus van 't Hoff attracted attention when he published a booklet in 1874 with his bold conjecture that single-bonded carbon atoms arrange their bonds in three dimensions, as if they were tiny tetrahedrons (triangular pyramids). French chemist Joseph-Achille Le Bel had independently come to the same conclusions in this same year, and for good reason. These ideas had a lot of explanatory power. For one thing, they gave an immediate reason for the origin of **chirality**. Van 't Hoff took time in his paper to explain just how these principles applied to French chemist Louis Pasteur's findings with the different forms of tartaric acid. Van 't Hoff suggested that a tetrahedral carbon with four different groups attached to it could exist as two mirror-image isomers. His theory also reinforced the idea that molecules had characteristic shapes, a factor that became critical when explaining their properties.

For a theory that fit the observed facts so well, van 't Hoff's proposal attracted some fierce criticism. Part of this was due to the pamphlet's unusually liberal use of illustrations, which in later versions even included cutouts for paper models of the tetrahedral carbons. The influential chemist Hermann Kolbe referred to van 't Hoff's work as "being dragged out by pseudo-scientists from the junk-room" and "devoid of any factual reality." But van 't Hoff was, in fact, the first winner of the Nobel Prize in Chemistry.

SEE ALSO Chirality (1848), Structural Formula (1861), Fischer and Sugars (1884), Asymmetric Induction (1894), Sigma and Pi Bonding (1931), *The Nature of the Chemical Bond* (1939), Conformational Analysis (1950), Resolution and Chiral Chromatography (1960), Enzyme Stereochemistry (1975).

This illustration shows the basic tetrahedron that a four-bonded carbon makes. If you colored each outer ball a different color, you'd find that you could have two mirror-image versions that can't be rotated to give the other one: a chiral carbon.

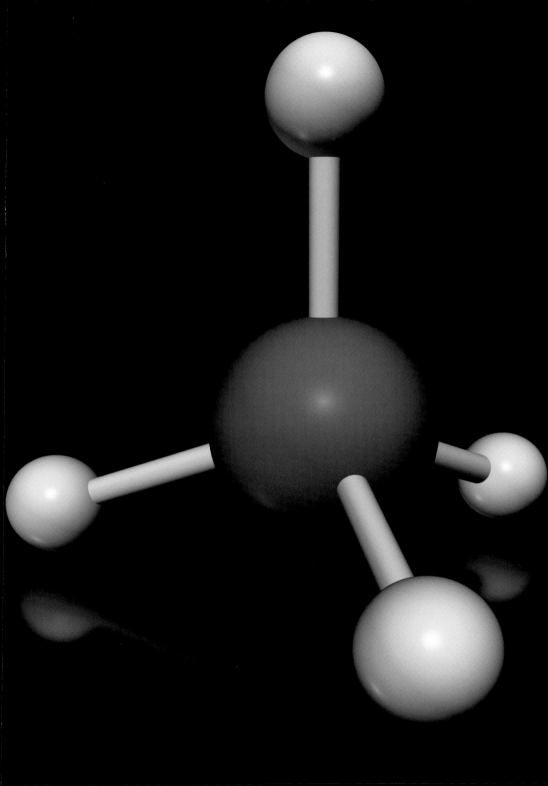

Gibbs Free Energy

Josiah Willard Gibbs (1839–1903)

If you want to look under the hood of chemistry and see what really makes things work, study thermodynamics. It measures changes in energy, the driving force for all chemical processes. Credit here goes to American scientist Josiah Willard Gibbs, whose theoretical insights and great mathematical ability turned thermodynamics into a precise scientific tool with applications in every possible area of chemistry, physics, and biology.

In 1876, he published his work on chemical systems and the "free energy" of reactions (now called *Gibbs free energy*, or G, in his honor). When a system changes from one state into another (chemically, as in a reaction, or physically, as in melting or boiling), the change in G (called ΔG or *delta*-G) is the work exchanged by the system with its surroundings (for example, the heat that is given off). Chemical reactions that can spontaneously give off energy show a negative ΔG. A fire is a perfect example. Reactions that have even larger negative ΔG values (such as the **thermite** reaction or the decomposition of **nitroglycerine**) can be dangerously energetic. In contrast, reactions with a positive ΔG—photosynthesis in plants, for example—require the addition of external energy, such as sunlight.

The other key thing to know about ΔG is that it's made up of two parts: enthalpy and entropy. Enthalpy (designated by the letter H) can be thought of as a pure measure of heat and energy, while entropy (S) is related to disorder and the reactants' "degrees of freedom" (i.e., how many different ways that they can move and vibrate). Chemists think in these terms constantly, gaining great insight into reactions by keeping all these factors in mind.

Some chemical reactions are spontaneous even though they actually get cold and soak up heat from their surroundings, like an instant "cold pack." This can happen because the entropy of the final state is so much higher (ΔS) than that of the starting materials, canceling out an unfavorable enthalpy change (ΔH), and giving an overall favorable ΔG. If both ΔH and ΔS are large and negative, though, you have an explosion in the making!

SEE ALSO Nitroglycerine (1847), Maxwell-Boltzmann Distribution (1877), Thermite (1893), Transition State Theory (1935), The Hottest Flame (1956), BZ Reaction (1968), Computational Chemistry (1970)

LEFT: *Josiah Willard Gibbs, 1903.* RIGHT: *This explosive thermite reaction has a large negative ΔG value.*

1877

Maxwell-Boltzmann Distribution

James Clerk Maxwell (1831–1879), **Ludwig Eduard Boltzmann** (1844–1906)

A chemical sample is really just a large collection of molecules. They're all moving around, but that doesn't mean that they're all moving in the same way and at the same speed. In the real world, in a container of nitrogen gas, for example, some of the nitrogen molecules will be zipping around much faster than the others, some of them much slower, with the rest spread out somewhere in between. This spread of different energies among the molecules is the Maxwell-Boltzmann distribution.

Scottish scientist James Clerk Maxwell (in 1860) and Austrian physicist Ludwig Eduard Boltzmann (in 1868) each modeled a container of gas as being full of tiny billiard balls, flying around and bouncing off of each other and off the walls. If you heat up the system (the collection of molecules), the balls fly around faster and bang harder against the inside of the container (and the pressure of the gas increases). From this simple picture, scientists began to understand the differences between the behavior of individual particles and the behavior of large groups of them, leading to further knowledge about bulk properties such as temperature, pressure, and many more. It's also very useful in describing reaction rates in chemistry, since it's often the case that only the most energetic molecules in a system are able to react. Learning how to deal with the behavior of large collections of particles like this is crucial to understanding the behavior of chemical systems (as well as for solving other problems in physics and even mathematics).

Though today we accept the presence of atoms without question, when these ideas were being developed, many physicists were not even sure if they believed in atoms as real things (rather than as useful little mental abstractions). Even after putting his work in its most complete form in 1877, Boltzmann in particular had great difficulty making headway with his proposals, which did not help his natural tendency toward depression (which eventually led to his suicide). Still, he and Maxwell are recognized (with American scientist Josiah Willard Gibbs, with his insights into **Gibbs free energy**) as discoverers of an essential way to deal mathematically with a world composed of small particles.

SEE ALSO Atomism (c. 400 BCE), Ideal Gas Law (1834), Gibbs Free Energy (1876), Gaseous Diffusion (1940).

Maxwell and Boltzmann used billiard balls to explain the movement of molecules in a gas.

Friedel-Crafts Reaction

Charles Friedel (1832–1899), James Mason Crafts (1839–1917)

Reactions involving aromatic ring compounds (like benzene) have marked a number of historic steps in the field of organic chemistry. One of the most famous, the Friedel-Crafts reaction, provided countless industrial uses—including the **catalytic cracking** of crude oil—along with insights into some deep principles of organic synthesis, such as the behavior of aromatic ring compounds.

Chemists Charles Friedel, a Frenchman, and James Mason Crafts, an American, were studying the reactions of organochlorine compounds and found that the addition of aluminum metal had a dramatic (and unexpected) effect. Very little happened at first, but when the reactions were heated up, they became so vigorous that they generated large amounts of heat on their own and produced a variety of products. This behavior suggested that some new species was being produced—slowly at first—that accelerated the reaction as time went on. It turned out, as Friedel and Crafts suspected, that aluminum chloride was being produced, and it was this compound that accelerated the reaction because, when it was added at the beginning, the reactions started immediately. In 1877, they published the results of their work, which quickly became known as the Friedel-Crafts reaction.

More study showed that positively charged carbons (carbocations) must be intermediates in the reaction, because the starting materials that formed these most easily were by far the most reactive. These carbocations readily attacked aromatic rings (see **Benzene and Aromaticity**), especially electron-rich ones, to produce new substituted ring systems. The positions where the new groups attached were mostly predictable, and we now know that the carbocations attack the positions with the most electron density (positive charges going toward negative ones).

A wide range of compound classes generate carbocation species for the reaction, and variations of the Friedel-Crafts reaction quickly became the preferred routes into many aromatic derivatives. The reaction is still widely used, from the research bench all the way up to the industrial scale. Aluminum chloride is still the classic recipe, but many other Lewis acids (see **Acids and Bases**) are also useful.

SEE ALSO Benzene and Aromaticity (1865), Acids and Bases (1923), Reaction Mechanisms (1937), Nonclassical Ion Controversy (1949)

LEFT: *Charles Friedel.* RIGHT: *Aluminum reacts with hydrochloric acid to form aluminum chloride.*

Indigo Synthesis

Johann Friedrich Wilhelm Adolf von Baeyer (1835–1917)

Adolf von Baeyer was one of the most well-known organic chemists of the nineteenth century, and the synthesis of indigo was one of his best-known accomplishments. Indigo was, and still is, a very important dye in the textile industry. (It's the color used for blue denim fabric, which should provide some idea of how much of it is produced and used every year.) At one time, indigo was isolated from the leaves of several different tropical plants in a process so ancient that no one knows who invented it, though India was a particular center of production. The plants themselves are not blue; a chemical precursor has to be hydrolyzed away from an attached sugar molecule and oxidized in air to produce the blue dye.

The year 1878 is the date of von Baeyer's first route to indigo, which was an outgrowth of his work on indole chemistry. The indole core (chemically, a benzene ring with pyrrole, a five-membered nitrogen ring, attached to one side) is found in a huge number of natural products, dyes, and pharmaceuticals, and von Baeyer was the first to synthesize it, in 1866. His initial route to indigo started from an indole derivative called *isatin*, but it was too expensive to make the synthesis commercially viable. His second route, from a benzene derivative, was financially unsustainable as well. Scientists in the German dyestuff industry, the world center of industrial chemistry at the time, continued their efforts to develop a feasible process, and by 1897 there was an economical route from aniline, a much cheaper starting material.

That spelled the end of the indigo plantations, which was not necessarily a bad thing. Conditions on an indigo farm could be brutal, and in the United States, before the Civil War, slaves did the work. This same story, a natural product–based industry remade by synthetic chemistry, would repeat itself many times over in the decades to come, as the owners of rubber plantations and wool mills could confirm.

SEE ALSO Natural Products (c. 60 CE), Yorkshire Alum (1607), Paris Green (1814), Perkin's Mauve (1856), Sulfanilamide (1932)

Vast amounts of indigo are used for dyeing blue jeans (and lots of other clothing).

Soxhlet Extractor

Franz Ritter von Soxhlet (1848–1926)

The Soxhlet extractor is an ingenious solution to a tedious problem: How do you wash a slightly soluble material (e.g., oil of peppermint) out of a mass of solid impurities (e.g., a handful of peppermint leaves)? Or if you're an organic chemist, how do you extract your reaction product from a mass of inorganic salts and by-products? One rinse is not enough, but rinsing over and over isn't a good use of a person's time (and wastes plenty of solvent, too). If only there was a way to recycle that solvent, sending it through fresh every time and leaving the extracted material behind . . .

Enter German chemist Franz Ritter von Soxhlet, who (sadly, or not) is remembered only for this piece of glassware, although he was the first to propose pasteurization of milk and other liquids. Soxhlet spent much of his career in agricultural chemistry, and his original goal for his invention was to extract lipids from milk. Its use has expanded, however, throughout the chemistry lab. Here's how it works: You put your solid in a porous pressed-paper "thimble" and put the whole extractor on top of a flask of solvent. The solvent is brought to a boil (or "to reflux," as chemists say), and the vapors are condensed back down into the sample. Once the thimble fills up, the function of the little curved tube on the side becomes clear. When the level of liquid goes higher than the top of the curved tube, the resulting siphoning action drains the chamber around the thimble, which starts filling up again with freshly distilled solvent for another cycle. Meanwhile, the extract concentrates in the flask below. As long as the material you're extracting is stable in the boiling solvent, the apparatus can be left running for days, while you go do something more useful.

Actually, what many chemists tend to do is watch the Soxhlet extractor itself. It's fun to try to guess how far it can get before siphoning off, and, in a large extractor, the siphoning action is fairly dramatic. Watching something else do your work for you somehow never gets old.

SEE ALSO Erlenmeyer Flask (1861), Borosilicate Glass (1893), Dean-Stark Trap (1920), The Fume Hood (1934), Magnetic Stirring (1944), Glove Boxes (1945), Rotary Evaporator (1950)

Modern Soxhlet extractors, like the ones pictured here, are a marvel to watch.

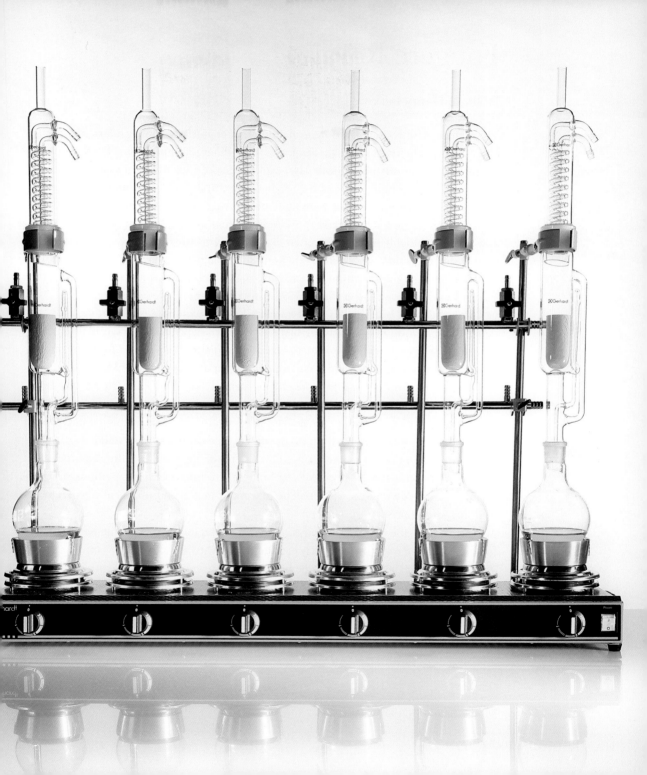

Fougère Royale

William Henry Perkin (1838–1907)

The human nose contains hundreds of different receptors for sensing odors, and subtle differences in chemical structure can make vastly different impressions. For thousands of years, the chemistry of perfumery has explored and exploited the human response to scent, but it was limited to techniques that concentrated and combined the scents of nature. Most perfume scents were (and are) derived from plants: flowers, naturally, but also aromatic seeds, bark, and roots. Animal secretions such as musk and ambergris also became valuable ingredients, often dissolved in mixtures of alcohol and water. The field advanced as new procedures were found to extract and preserve delicate natural essences, giving master perfumers new combinations to experiment with.

In 1881, however, the perfumer House of Houbigant marked a new era in perfumery when it launched its scent Fougère Royale (which translates rather pedestrianly as *royal fern*). It was the first perfume that depended on a synthetic chemical, the sweet-smelling coumarin, as an ingredient. Coumarin is found in many plants, but since English chemist William Henry Perkin (see **Perkin's Mauve**) had first synthesized it in 1868, it was far more easily obtained as a single synthetic compound than from any natural source. The perfume became a sensation, and it was produced—by its original makers and by copycats—for many years. The next synthetic ingredient to be added to a perfume was vanillin (the main constituent of natural vanilla extract), and soon perfumers were using compounds that had never appeared in natural sources at all. Instead of imitating flower scents (or combining them into bouquets), perfume designers found themselves with a far wider palette to work from and enthusiastically began to explore more impressionistic styles.

These days, the majority of perfumes use synthetic ingredients. Some natural fragrances (and flavors) are relatively easy to imitate with a few synthetic compounds, and their artificial versions are inexpensive. Others are very complex mixtures indeed, and they may also have key components that are difficult to synthesize, so perfumes made with natural extracts are correspondingly more costly.

SEE ALSO Purification (c. 1200 BCE), Natural Products (c. 60 CE), Perkin's Mauve (1856), Maillard Reaction (1912), Isoamyl Acetate and Esters (1962).

LEFT: *A c. 1884 bottle of Fougère Royale, the first scent to depend on a synthetic chemical.* RIGHT: *Perfumer Paul Parquet, the fragrance's creator.*

Claus Process

Carl Friedrich Claus (1827–1900)

Natural gas, as it comes out of the ground, is a messy substance. One of its most significant drawbacks is that it almost always contains **hydrogen sulfide**, which is very toxic and bears its highly objectionable rotten-egg smell. The same problem shows up as crude oil is refined—the crude deposits, characterized as "sour," have many sulfur-containing compounds. These are stripped out in a process that produces still more hydrogen sulfide, which has to be removed as well. This is done by the Claus process, invented and patented in 1883 by German-English chemist Carl Friedrich Claus. It was originally intended to recover sulfur from calcium sulfide, a waste product from manufacturing plants producing soda ash, but it was adapted to work for other sulfur-containing compounds as well. Refined in the 1930s, it is still in use worldwide.

That's because the chemistry involved is so direct that it's hard to improve on. First, hydrogen sulfide is burned to produce sulfur dioxide, which reacts with more unburned hydrogen sulfide to produce elemental sulfur and water. This takes place under both heating and catalytic conditions to maximize the removal of sulfur, which comes off as a hot gas and is then condensed to liquid sulfur so it can be pumped into tanks for storage. Any unreacted hydrogen sulfide gas that's dissolved in the liquid sulfur is usually removed by degassing and sent back to undergo the Claus process again. As the sulfur cools to a solid, it may be stored in huge piles—they look like bright yellow mountains—since a surplus of sulfur is produced by the Claus process in some oil-producing regions. Some of the sulfur produced this way is used in **sulfuric acid** manufacturing, in the **rubber** industry, as a fertilizer, and as a general chemical feedstock. And some of it just sits around. As long as the world needs to tap high-sulfur oil reserves, no one will ever get very rich selling it.

SEE ALSO Hydrogen Sulfide (1700), Sulfuric Acid (1746), Rubber (1839), Catalytic Reforming (1949)

Mountains of sulfur, extracted from natural gas and waiting to be put to use.

Liquid Nitrogen

Johann Gottlob Leidenfrost (1715–1794), **Zygmunt Wróblewski** (1845–1888), **Karol Olszewski** (1846–1915)

Nitrogen is the most abundant gas on earth, but it's a gas only because of the relative warmth of our planet. If you can cool it down to –321 degrees Fahrenheit (–196 degrees Celsius), it will condense into a thin, clear liquid. In 1883, Polish physicist Zygmunt Wróblewski and his colleague, the Polish chemist Karol Olszewski, first accomplished that feat, in Kraków, Poland. Using a laborious series of cooling and suddenly expanding of the gas, they were only able to produce small amounts, however.

Condensing nitrogen into the liquid state is a complex process (see **Liquid Air**), and its larger-scale production had to wait until the 1890s, with several more decades passing before liquid nitrogen became a common industrial chemical. Nontoxic, odorless, colorless, and nonflammable, this abundant gas is the most widely used cryogenic substance in the world, helping to cool down superconducting magnets for **NMR** (nuclear magnetic resonance) machines, cooling the traps of vacuum pumps (used in many scientific and industrial processes), and freezing tissue samples for medical research, as well as other applications. It's also crucial in many chemical reactions as well as food packaging, and its use has recently become popular among high-tech chefs.

In the lab, liquid nitrogen is often bubbling violently, simply because almost anything will warm it up enough to start it boiling away. It doesn't cool a warm object down as quickly as you might think, though, because a layer of vapor quickly forms and insulates it. This is known as the Leidenfrost effect, a phenomenon first described by German doctor Johann Gottlob Leidenfrost in 1756. Home cooks encounter it when flicking drops of water onto a hot griddle to test its temperature. In the same way, small amounts of liquid nitrogen will roll off the skin on contact—but beware, any more can cause painful and dangerous frostbite. Entertaining demonstrations of the effects of liquid nitrogen for students often include the quick freezing of common objects, leading to popular effects like the "shattering rose" and the "banana hammer," which more or less explain themselves.

SEE ALSO Liquid Air (1895), Neon (1898), NMR (1961)

In recent years, liquid nitrogen has become a favorite of avant-garde chefs and home-kitchen experimenters because it will produce unusual frozen concoctions in record time.

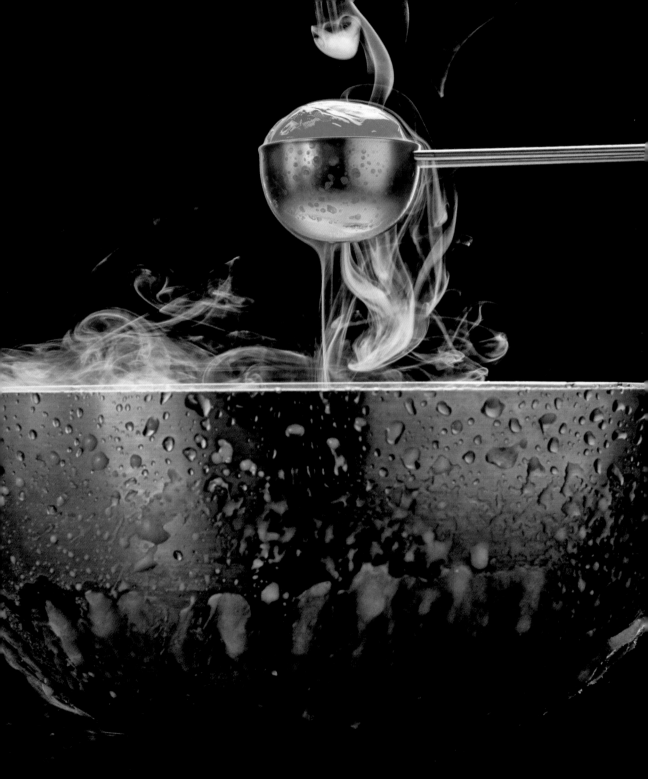

Fischer and Sugars

Emil Hermann Fischer (1852–1919)

Figuring out the structures and functions of proteins, carbohydrates, and lipids—the three major classes of biomolecules—has been, and still is, a major driving force in organic chemistry and biochemistry. It's impossible to look back to the beginnings of this great journey without paying respect to the pioneering German chemist Emil Hermann Fischer. He did fundamental work on proteins and lipids but is especially remembered for his sorting out of the sugars, a monumental task that began in 1884 and for which he was awarded the Nobel Prize in Chemistry in 1902.

When Fischer began his research, even the basic details of simple carbohydrate structures were debated. But Fischer showed, by painstaking work, that the sugars contained aldehyde groups that could react with hydroxyl (OH) elsewhere in their frameworks, giving rise to several different interconverting cyclic structures from each compound. (Glucose, for example, has several different structural forms, all with different properties—a situation that had understandably confused everyone.) The proposals from a decade earlier by Jacobus Henricus van 't Hoff and Joseph-Achille Le Bel for the stereochemistry of carbon atoms proved essential in understanding the relationships among the various sugars, as they often differed only in their three-dimensional arrangement around single carbon atoms.

Fischer worked out the general scheme relating all the simple sugars, a true tour de force. With this as his foundation, he was able to convert sugars to totally different forms of themselves by synthesis and build others up from simpler carbohydrate precursors. Everything that requires knowledge of sugar chemistry, from artificial sweeteners to the structure of DNA, goes back to Fischer. Achieving all this without modern analytical equipment, though, is a fearsome thought for a modern chemist. Fischer had to rely on forming chemical derivatives of his compounds, products with characteristic colors, melting points, and even, surprisingly, tastes. Tasting one's work was considered a perfectly reasonable measurement for chemists at the time, but no longer. Fischer died partly as a result of chronic exposure to his most important derivatizing reagent (the toxic phenylhydrazine) and by tasting all the compounds he'd made from it—a cautionary tale for any working chemist.

SEE ALSO Chirality (1848), Tetrahedral Carbon Atoms (1874), Asymmetric Induction (1894), Maillard Reaction (1912)

A 1904 photograph of Emil Hermann Fischer, father (and victim) of sugar chemistry.

Le Châtelier's Principle

Henry-Louis Le Châtelier (1850–1936)

Chemists everywhere are familiar with French chemist Henry-Louis Le Châtelier's principle: a chemical equilibrium changes to offset any disturbance that's made to it. Le Châtelier made the chemistry of cement his choice of study after being told by a senior chemist that the available fields of research were so numerous that he could pick a topic at random and find something interesting. Le Châtelier's cement studies led to the investigation of the laws of chemical equilibrium, and he first described his principle in an 1885 paper.

A good example of the principle is the **Dean-Stark trap**, an apparatus invented in 1920 by American chemists Ernest Dean and David Stark. In a condensation reaction, one equivalent of water is given off as the two reactants condense. Left undisturbed, a reaction like this will produce some product, but the water that's formed can react with the product and send the reaction back in the other direction, essentially undoing the work originally done. Eventually, the system reaches equilibrium, where the rates of product formation and product reversion cancel each other out. The Dean-Stark trap disturbs this situation by taking the newly formed water out of the system, so the reaction continues until the starting materials are consumed.

The same technique can be used to drive many reactions, either by loading more starting material onto one side of the equilibrium or by draining one of the products off from the other. The second method is usually preferred, so as not to waste material. You can distill out a more volatile product, for example, or run the reaction in a solvent that allows the product to crystallize out. As long as the product can't participate in the reaction system anymore, the equilibrium will shift to compensate. Changes in temperature, pressure, or other conditions can be applied as well.

Such a general idea has naturally diffused out to other fields. Its applications to biochemistry make it relevant to pharmacology and medicine, but it's also been used as far afield as economics. Le Châtelier's principle underlies every equilibrium reaction, and living creatures themselves are gigantic, complicated collections of equilibrating reactions.

SEE ALSO Haber-Bosch Process (1909), Dean-Stark Trap (1920), BZ Reaction (1968)

The equilibrium between two types of nitrogen oxide as they are being shifted from dark nitrogen dioxide (in a hot water bath) to lighter-colored dinitrogen tetroxide in ice.

Isolation of Fluorine

Joseph-Louis Gay-Lussac (1778–1850), **André-Marie Ampère** (1775–1836), **Humphry Davy** (1778–1829), **Louis-Jacques Thénard** (1777–1857), **Ferdinand-Frédéric-Henri Moissan** (1852–1907)

If any element in the periodic table can be said to have a personality, fluorine is the one. Unfortunately, that personality is like Lord Byron's, "mad, bad, and dangerous to know." Fluorine is by far the most electronegative element and is thus a spectacular oxidizing agent, wildly reactive with any material that can somehow donate electrons. The chemists who attempted to isolate it paid dearly for that knowledge.

In 1810, French physicist André-Marie Ampère was the first to suggest that hydrofluoric acid contained an unknown element, and English chemist Sir Humphry Davy proposed the name *fluorine* and set out to isolate it. The problem was that both fluorine and gaseous hydrogen fluoride (a constant presence in these experiments) are horribly dangerous, and Davy was among the first to be poisoned, in 1812, surviving with lung and eye damage. Despite his cautionary report, fluorine was enough of a prize that others risked their lives in experiments to isolate it. Inhalation of hydrogen fluoride gas injured French chemists Joseph-Louis Gay-Lussac and Louis-Jacques Thénard, and the Knox brothers, George and Thomas. Some years later, Belgian chemist Paulin Louyet and French chemist Jerome Nicklés were killed by fumes, and English chemist George Gore nearly died in a violent explosion.

French chemist Ferdinand-Frédéric-Henri Moissan finally isolated fluorine in 1886, without blowing anything up or being killed. For his apparatus he used electrolysis (as had Gore), with stoppers made out of the mineral fluorite (the Knox brothers' innovation), all inside a chemically resistant but ferociously expensive vessel made of platinum and iridium and cooled to −58 degrees Fahrenheit (−50 degrees Celsius). Moissan went on to perform the first controlled experiments on fluorine's reactivity—not surprisingly, he found that many things burst into flame or exploded outright.

Fluorine's unique properties make it valuable in everything from drugs to cookware, but even today, relatively few chemists encounter it in its elemental form. Special equipment and attention to detail are needed, as fluorine cannot be approached casually.

SEE ALSO Stainless Steel (1912), The Hottest Flame (1956), Noble Gas Compounds (1962), PET Imaging (1976)

This 1891 illustration shows Moissan isolating fluorine (explosions, fires, and deadly corrosive vapors not shown).

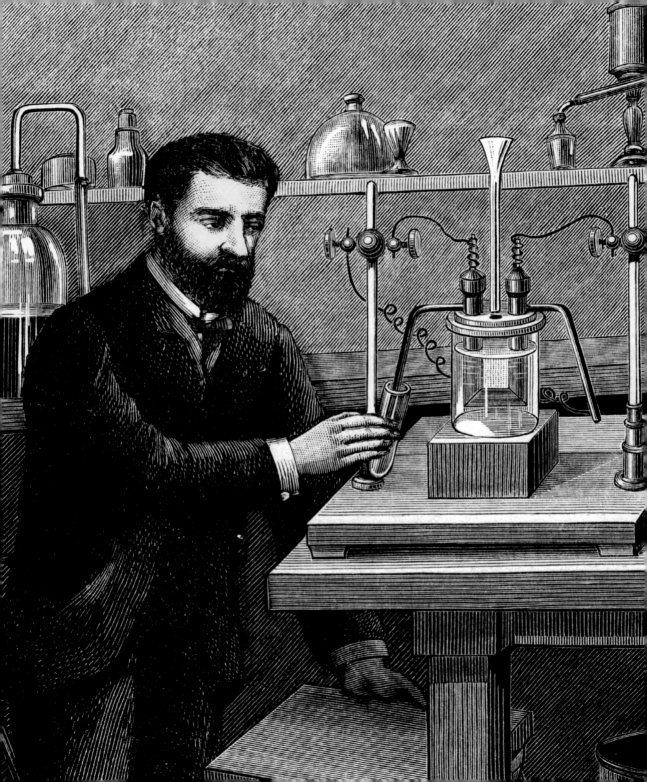

Aluminum

Frank Fanning Jewett (1844–1926), Charles Martin Hall (1863–1914), Paul-Louis-Toussaint Héroult (1863–1914)

Aluminum is now a familiar part of daily life, but it used to be a precious metal, which is one reason the capstone of the Washington Monument is made from it. When the monument was constructed, the only ways to refine aluminum from its ore were difficult and expensive, but it was clearly an extremely useful material—light, strong, and resistant to corrosion. Chemists and engineers all over the world sought an economical technique to make it available.

In America in 1886, chemist Charles Martin Hall (with the help of his professor, chemist Frank Fanning Jewett) found such a route, and in France chemist Paul-Louis-Toussaint Héroult was hit by virtually the same idea at the same time. Refining a metal chemically means taking its **oxidation state** down to zero, which can be accomplished in several ways. The existing process reacted aluminum salts with sodium or potassium (which were themselves made by **electrochemical reduction**), but the hope was to use electrochemistry to produce aluminum directly from its salts (and its ore, which is mostly aluminum oxide). But aluminum chloride, the most likely compound to use as a starting point, tended to absorb water from the air too quickly, spoiling the reaction. All the aluminum compounds that appeared to be feasible candidates also had very high melting points, so turning them into liquids fit for electrochemical work would require serious heat.

The key that Hall and Héroult hit upon independently was using the mineral cryolite (sodium aluminum fluoride), high melting point and all, to dissolve aluminum oxide. Not many other things would, and because cryolite's density is lower than aluminum's, the molten metal accumulates at the bottom of the vessel. The fierce conditions also called for special attention to that vessel—standard clay crucibles had their silicates dissolved right out of them, contaminating everything—but once these problems were solved, the Hall-Héroult process was complete. It's used in aluminum smelting plants today, and though it still uses significant amounts of electricity, and thus results in significant greenhouse gas emissions, no method has been found to improve upon it.

SEE ALSO Iron Smelting (c. 1300 BCE), Electroplating (1805), Electrochemical Reduction (1807), Oxidation States (1860), Acetylene (1892), Thermite (1893).

Workers assemble the aluminum framework of a B-17 bomber during World War II. Before the Hall-Héroult process, such a plane would have been perhaps the most wildly expensive item in the world.

Cyanide Gold Extraction

Carl Wilhelm Scheele (1742–1786), John Stewart MacArthur (1856–1920)

It seems like some sort of nasty joke: the cheapest, most reliable way to extract and purify gold turns out to use massive amounts of poisonous cyanide. Scottish chemist John Stewart MacArthur developed the method in 1887 in Glasgow, building on the much older discovery by Swedish chemist Carl Wilhelm Scheele that gold, well known to be unreactive, in fact dissolves in cyanide solutions. MacArthur partnered with two Glasgow doctors, Dr. Robert Forrest and Dr. William Forrest, and pioneered a way for lower-grade ore to be stripped of its gold in a manner that had been impossible before. The MacArthur-Forrest process swept through the gold-mining world, and it is still in use today.

Some care has to be taken along the way, though. Ground-up ore is first slurried with cyanide in water, but the mixture has to be kept basic (at a high pH) to prevent the formation of poisonous **hydrogen cyanide** gas. With that potential hazard addressed, as long as there is **oxygen** present (usually provided by bubbling air through the mixture), a soluble gold-cyanide complex can form, which is then adsorbed onto activated carbon for later recovery.

Regardless, a huge amount of cyanide-laced water is left at the end of the process, enough to kill off everything in its path were it to be released untreated. Various oxidation reactions are used to turn it into a different ion, cyanate, which is much less toxic, and the water thus treated is stored in holding ponds for residual decontamination. Despite these precautions, there have been numerous spectacular (and spectacularly awful) spills when containment walls have been breached. Although cyanide is cleared from the immediate environment relatively quickly (partly by microorganisms using it for food, if it's not too concentrated), it can leave a trail of destruction before it's gone.

These problems have led to the banning of the whole process in some jurisdictions, but the demand for gold remains high, and the majority of gold extracted each year—for jewelry, investments, and electronic connectors—still reaches the world this way.

SEE ALSO Gold Refining (c. 550 BCE), Hydrogen Cyanide (1752), Electroplating (1805)

A gold mine by night. The metal is still so valuable that all sorts of unappealing chemistry are tolerated in order to purify it.

1888

Liquid Crystals

Otto Lehmann (1855–1922), **Friedrich Reinitzer** (1857–1927), **Georges Friedel** (1865–1933), **Daniel Vorländer** (1867–1941), **George William Gray** (1926–2013)

Liquid crystals are the technology behind modern electronic displays, but they were first studied in 1888, when Austrian chemist Friedrich Reinitzer and German physicist Otto Lehmann began exchanging letters and samples of a cholesterol derivative that showed some odd behavior. It seemed to have two melting points: on heating the solid turned into a cloudy liquid, but at a higher temperature this changed into an ordinary clear liquid. All of these were reversible, and closer study showed that the cloudy phase had properties in between those of a solid and a liquid.

In fact, it was the first "orderly liquid": the shapes of its molecules allow them to stack together without forming a real solid phase. In 1922, French mineralogist Georges Friedel classified these in a scheme still used today: the smectic liquid crystals have layers of flat molecules slipping over each other like sheets of paper, and the nematic liquid crystals feature long, rodlike molecules that slide past each other like dry spaghetti noodles.

German chemist Daniel Vorländer synthesized most known liquid crystals, but for many years no real use was found for them. Then, in 1962, American chemist Richard Williams found a way to make nematic phases twist and untwist in an electric field. These molecules rotated polarized light as it passed through them, and the amount of light that transmitted depended on the voltage. A black number or letter could be thus made to appear or disappear in the display almost instantly.

The only problem was that the liquid crystals had to be heated up for this to work. But when Scottish chemist George William Gray discovered room temperature liquid-crystal mixtures in 1973, the new flat displays appeared in calculators and cheap digital watches. Liquid crystal displays (LCDs) kept improving, and what started as a strange-looking cloudy liquid took over every portable flat screen in the world.

SEE ALSO Cholesterol (1815), Chirality (1848), Surface Chemistry (1917), Kevlar (1964)

LEFT: *Otto Lehmann in his laboratory, 1907.* RIGHT: *A film of liquid crystals with polarized light shining through them. The orientation of the molecules in each region affects the rotation of the light, giving rise to both the colors and the dark bands.*

Thermal Cracking

Vladimir Shukhov (1853–1939), William Merriam Burton (1865–1954)

Petroleum is a messy substance—dark, thick, and smelly—as it comes from the ground. Chemically it's even messier. It contains a huge variety of large carbon compounds, many of which are not particularly useful as a mixture. Breaking down larger hydrocarbons into smaller, more useful ones is called *cracking,* and it has been an important field of research for more than a hundred years.

The first breakthrough in this area came from Russian inventor Vladimir Shukhov, a wide-ranging scientist who made contributions in engineering and architecture as well as petroleum chemistry. In 1891, he was issued a patent by the Russian Empire for his thermal cracking process, in which petroleum was heated under pressure to at least 700 degrees Fahrenheit (370 degrees Celsius) to break and re-form the carbon-carbon bonds. This process (we now know) takes place through **free radicals,** formed when carbon-carbon bonds break under harsh conditions, and the distribution of straight and branched hydrocarbons that result from it can be at least partially predicted by knowledge of the various bond strengths and the stabilities of the various radicals produced.

The more volatile products are distilled out of the process, which in its heyday was mostly used to produce gasoline. Conditions can be adjusted to produce different mixtures of gasolines and fuel oils, depending on demand, and in theory, more petroleum could be fed into the cracking equipment to run in efficient continuous flow conditions. But in practice, thermal cracking can produce a high-molecular-weight residue of tar, which gradually clogs the equipment. This "coking" process is a constant concern for industrial chemistry of many kinds—especially processes that run at high temperature where mixtures of products can form.

William Merriam Burton obtained a U.S. patent in 1913 for thermal cracking, and though competitors tried to invalidate it by reference to Shukhov's earlier work, Burton's route is still recognized as the first industrially successful process. Later variations of the process were conducted at even higher temperatures, in an effort to improve the extraction of useful products, but thermal cracking had its limits and was eventually surpassed by the invention of **catalytic cracking.**

SEE ALSO Fractional Distillation (c. 1280), Free Radicals (1900), Fischer-Tropsch Process (1925), Catalytic Cracking (1938), Catalytic Reforming (1949), Flow Chemistry (2006)

The early days of oil exploration were followed by the early days of oil refining, which meant energy-intensive, messy thermal cracking plants. Here, bystanders witness the filthy expulsion of an oil gusher at Port Arthur, Texas, in 1901.

Chlor-Alkali Process

Karl Kellner (1851–1905), Hamilton Young Castner (1858–1899)

When Belgian chemist Ernest Solvay displaced the older Leblanc process for sodium carbonate in 1864, there was a follow-on effect. Most of the hydrochloric acid in the world was a by-product of the Leblanc route, and most of the world's chlorine was made from that hydrochloric acid. Both of these were valuable in their own right, used for metal processing, bleaching, and other industries. As the supply of hydrochloric acid dried up, electrochemistry came to the rescue in the form of the chlor-alkali process, the first economical form of which was discovered (independently) and patented in 1892 by Austrian Karl Kellner and American Hamilton Castner.

This route involves the electrolysis of concentrated brine, which produces **hydrogen** and sodium hydroxide (alkali) from one electrode and chlorine gas (chlor) from the other. The electrodes are separated by a permeable membrane, and sodium ions from the brine side cross the membrane so that one side gradually becomes a sodium hydroxide solution (alkali). All of these products are used in the paper, textile, and fine chemical industries, but the chlorine is usually the product in least demand—making "What do we do with all this chlorine?" one of the persistent questions for any plant owner. For that reason, the hydrogen and chlorine gases are sometimes allowed to react (which they do vigorously) to produce hydrogen chloride gas, which is absorbed into water to produce high-quality concentrated hydrochloric acid. This method made up for the Solvay shortfall of hydrochloric acid (and also reversed the earlier industrial relationship between it and chlorine).

With different types of mixing, and at different temperatures, the process can also be run to form a solution of sodium hypochlorite (bleach), or the powerful oxidizer sodium chlorate, which is converted to chlorine dioxide, the biggest bleaching agent in the paper-making industry. The classic chlor-alkali cell (still used in some parts of the world) has a layer of liquid mercury as one electrode, and inevitably produces toxic mercurial waste. Even the less toxic permeable-membrane method (like the Hall-Héroult process for aluminum) uses huge amounts of electricity, but it has proven too useful to be replaced.

SEE ALSO Hydrogen (1766), Electroplating (1805), Electrochemical Reduction (1807), Oxidation States (1860), Solvay Process (1864), Chemical Warfare (1915), Artificial Photosynthesis (2030)

Chlorine oxidants from the chlor-alkali process are used on a large scale to bleach paper (among many other things).

Acetylene

Friedrich Wöhler (1800–1882), **James Turner Morehead (1840–1908)**, **Francis Preston Venable (1856–1934)**, **Thomas Leopold Willson (1860–1915)**

The Hall-Héroult process of 1886 was the winning entry in the contest to refine **aluminum** economically. But the many efforts that preceded it to find a workable method led to other discoveries, one of the great features in general of experimentation and discovery. Thomas Leopold Willson, a Canadian inventor working in Spray, North Carolina (now called Eden), held patents on electric arcs for smelting ores and wanted to apply this technology to aluminum. He established a water-powered electric plant with the backing of James Turner Morehead, a local textile manufacturer, and began trying to reduce aluminum oxide with carbon in his arc furnace.

It worked—barely. The process was terribly inefficient, so Willson tried to introduce calcium into the system as a more reactive metal. In 1892, when he used a carbon arc to reduce calcium oxide (lime), he obtained a gray-black product, which he tested by its reaction in water. Calcium metal should have bubbled and produced **hydrogen**, and his product did produce a flammable gas. But it had a sooty flame, and a hydrogen flame is simply incapable of producing soot (which comes from carbon; a hydrogen flame produces only water vapor). Willson brought in Francis Preston Venable, a chemist at the University of North Carolina, to try to understand this result.

Venable realized that a reaction discovered by German chemist Friedrich Wöhler back in 1862 had occurred. Willson had made calcium carbide (the calcium salt of acetylene), and the gas being produced was in fact acetylene, which until then had been a rare chemical. Over the next few months, it became apparent that the arc furnace was never going to produce aluminum at a profit, so Willson and Morehead decided to make a go of it in the acetylene business. An acetylene flame was extraordinarily bright and hot, and it was strong competition for the new electric light industry (especially for portable lanterns) as well as a new way to weld and cut metals. It turned out to be a valuable feedstock for many other processes as well (see **Reppe Chemistry**). Morehead eventually sold his patent rights to what became the Union Carbide Corporation, and Willson moved back to Canada, becoming one of its wealthiest industrialists.

SEE ALSO Hydrogen (1766), Aluminum (1886), Reppe Chemistry (1928), The Hottest Flame (1956), Single-Molecule Images (2013).

A welder demonstrates the correct welding technique in the Gary, Indiana, plant of the Tubular Alloy Steel Corporation, 1943. Acetylene rapidly became (and remains) the most important gas for the welding industry.

Thermite

Hans Goldschmidt (1861–1923)

As mentioned in the entries on the Hall-Héroult process and **stainless steel**, aluminum is an odd metal. Thermodynamically, it is far happier as an oxide. This (stated more scientifically) means that the energy state of a mixture of iron oxide (rust) and elemental aluminum, for example, is much higher than that of a mixture of aluminum oxide and iron. If you could just start a reaction between the elemental aluminum and the iron oxide, you'd liberate a lot of heat along the way.

And that's thermite. Nothing more, in its original recipe, than powdered iron oxide mixed with powdered aluminum, but they will not start reacting without the addition of an outside energy source. A fuse of burning magnesium metal is the standard, but any sufficiently hot flame (greater than 4,000 degrees Fahrenheit, 2,200 degrees Celsius) will do. The effect is dramatic and dangerous, and once the reaction starts, nothing will stop it until it goes to completion. Aluminum has a relatively low melting point (1,220 degrees Fahrenheit, 660 degrees Celsius), letting it react very quickly as it soaks into the iron oxide, but it has a relatively high boiling point for such a reactive metal, so it doesn't evaporate away. Enough heat is generated to produce nearly white-hot liquid iron at temperatures up to 4,500 degrees Fahrenheit (2,500 degrees Celsius), and the aluminum oxide product is so light that it tends to float on top of it. You can also run the thermite reaction with metal oxides other than iron (copper, manganese, chromium), because the energy payoff from aluminum going down to aluminum oxide is enough to drive a wide range of reactions.

German chemist Hans Goldschmidt discovered the thermite reaction in 1893 when he was trying to find a new way to refine metals. It can do that, albeit uneconomically and violently, but thermite's real uses have to do with all that heat. A portable source of molten iron (initiated on demand with added **oxygen**) turned out to be very useful for welding steel rails, for example (or cutting them, if you're in wartime sabotage mode). It's instant liquid fire.

SEE ALSO Oxygen (1774), Gibbs Free Energy (1876), Aluminum (1886), Stainless Steel (1912), The Hottest Flame (1956)

Welding railroad tracks with thermite.

Borosilicate Glass

Friedrich Otto Schott (1851–1935)

Glass is mostly silicon dioxide, but the number of different additives and recipes for it is beyond counting. The most common glass is the so-called "soda-lime" variety, which contains salts of calcium and sodium (slightly different recipes are used for window glass versus bottle glass). These additives make the glass easier to melt and manipulate, while keeping the product hard and colorless, but there are some disadvantages. One big problem is thermal stability: sudden heating and cooling cycles almost invariably cause an object made of soda-lime glass to crack.

That's not very good for laboratory work, where glass is otherwise an almost ideal material for storing and reacting many chemicals. There are reagents that will degrade it (hydrofluoric acid and strong hydroxide base solutions, for example), but many other nasty compounds can be stored safely in glass containers for years. If these containers are at risk of suddenly shattering when they're warmed up or cooled down, however, that's not a safe storage solution at all.

Enter borosilicate glass, which is made by adding boron oxide to the silica mixture. It was first sold in 1893 by Friedrich Otto Schott, a well-known German glass chemist of his time, and varieties of it quickly cropped up in other countries. It's even harder and more chemically unreactive than standard glass varieties, and it has tremendous thermal stability. This is the type of glass used in baking dishes—making them oven-, dishwasher-, and microwave-safe—and in labware. Most chemists have never seen an Erlenmeyer or round-bottom flask that wasn't borosilicate—virtually all laboratory glassware is made of it.

Although borosilicate is more difficult to soften and work than soda-lime glass, the benefits are worth the trouble. It's not impervious to thermal shock, but it takes some work to get it to break, and under normal lab usage it's very durable indeed. The only problem is that chemists get used to glass behaving so well and sometimes get surprised in the kitchen with other varieties!

SEE ALSO Erlenmeyer Flask (1861), Separatory Funnel (1854), Solvay Process (1864), Soxhlet Extractor (1879), Dean-Stark Trap (1920), The Fume Hood (1934), Magnetic Stirring (1944), Glove Boxes (1945), Rotary Evaporator (1950)

Laboratory glassware is almost always made from borosilicate mixtures; it has to go through too many temperature changes for most other kinds of glass to have a chance.

Coordination Compounds

Alfred Werner (1866–1919), **Victor L. King** (1886–1958)

The difficulty of understanding the structure of compounds like **Prussian blue** was proof that there was something different about the chemistry of metals. They formed complex molecules, which seemed to be made up of metal atoms interacting with other molecules, called *ligands*. Different elements seemed to pick up two, four, six, or eight ligands at a time, in patterns that were very hard to interpret. Swiss chemist Alfred Werner threw some much-needed light on the subject with his theories of coordination, first published in an 1893 paper. Metals, he found, attracted these ligands according to their oxidation state, and in specific patterns in space. The ligands (things like ammonia and other amines, cyanide, chloride, nitrate, and others) could be arranged in a square, a shape like two pyramids base to base, or toward what would be the six faces of a cube centered around the metal, among others. This accounted for why there were metal complexes known with the exact same formulas but different properties (color, solubility, reactivity, etc.): their ligands were arranged differently around the metal. We would now call the metal a Lewis acid, and each ligand a Lewis base, since it is an electron pair on each ligand that is coordinating with the metal. And we now know that coordination chemistry is a fundamental property of metallic elements, which shows up in everything from minerals in the Earth's crust to drug molecules (see **Cisplatin**).

These arrangements in space suggested that some metal complexes could have chirality, meaning they could exist in right-handed and left-handed forms. Werner's lab spent over a decade on the problem, which his American student Victor L. King finally solved by preparing chiral cobalt complexes in 1907. King had been greeted for months by fellow students in Zurich with the question "Does it rotate yet?" referring to the rotation of polarized light that a chiral compound produces. Werner himself was reportedly so overjoyed at the news that he stopped near strangers on the street to tell them about the discovery. (See **Noble Gas Compounds** for another example of the same problem!)

SEE ALSO Prussian Blue (c. 1706), Chirality (1848), Ferrocene (1951), Noble Gas Compounds (1962), Cisplatin (1965), Coordination Frameworks (1997)

These are various possible coordination compounds. The metal atom sits in the middle, with the ligands arranged around it in defined geometries.

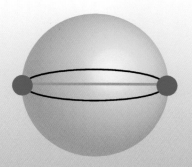

linear
2 ligands

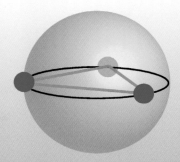

trigonal planar
3 ligands

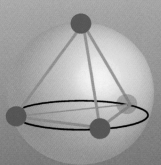

tetrahedral
4 ligands

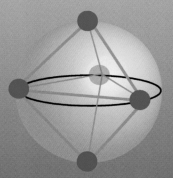

trigonal bipyramidal
5 ligands

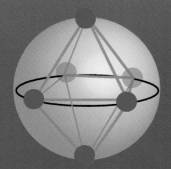

octahedral

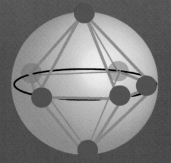

pentagonal bipyramidal
7 ligands

The Mole

(Lorenzo Romano) Amedeo (Carlo) Avogadro (conte di Quaregna e Cerreto) (1776–1856), Stanislao Cannizzaro (1826–1910), Friedrich Wilhelm Ostwald (1853–1932)

In 1894, German chemist Friedrich Wilhelm Ostwald introduced the term *mole* into chemistry. Based on the work of Italian chemists Amedeo Avogadro and Stanislao Cannizzaro, it's the amount (in grams) of any substance that matches its atomic or molecular weight. A mole of carbon (atomic weight 12) weighs 12 grams, and a mole of aspirin (molecular weight 180.16) weighs 180.16 grams. What makes it a useful concept is that a mole of any substance always has the same number of molecules in it, so if you have a chemical reaction in which Molecule A and Molecule B react together in a one-to-one ratio, you can simply add one mole of each. It can be a quarter of a mole each, or a thousandth of a mole each (one millimole), but as long as you work in molar ratios, everything will come out even.

For the many non-chemists who have encountered the idea of a mole and been confused by it, here's another way to think about it: It's a big dozen. A dozen eggs, a dozen bowling balls, a dozen elephants—they weigh different amounts, but there are twelve of each of them. A mole of oxygen, a mole of plutonium chloride (stand back), a mole of insulin—they weigh different amounts, but there's an Avogadro's number of molecules in each of them.

So, how many molecules does a mole of anything contain? 6.022141×10^{23}, to be exact, which is known as Avogadro's number, to honor his work with molecular weights and ratios. That's a huge number. One mole of carbon isn't much—the graphite from a dozen pencils, more or less—but there are over six hundred thousand million million million carbon atoms in there. Even reagent solutions in the lab are labeled in terms of how many moles of compound they have in them per liter. Thinking in terms of moles is second nature to chemists, allowing them to keep track of ratios and equivalents in chemical reactions without worrying about molecular weights.

SEE ALSO Avogadro's Hypothesis (1811), Cannizzaro at Karlsruhe (1860)

One mole of nitrogen (in the balloon), with a mole each of aluminum, iron, copper, sodium (as sodium chloride), and mercury in front of it.

Asymmetric Induction

Emil Hermann Fischer (1852–1919), Willy Marckwald (1864–1942)

Chirality, or the feature of left- and right-handedness of some molecules, was discovered by French chemist Louis Pasteur and given a sound basis by Dutch chemist Jacobus Henricus van 't Hoff and French chemist Joseph-Achille Le Bel. It raised a number of interesting issues, though. For example, many natural products were chiral, and few (if any) other substances seemed to be. This was one of the arguments in favor of vitalism, the idea that there was something fundamentally different about living chemistry that distinguished it from nonliving things. The idea of vitalism had been taking hits ever since **Wöhler's urea synthesis**, but it was hard to argue with the fact that alkaloids, sugars, and **amino acids**—all produced by living creatures—were so relentlessly chiral.

German chemist Emil Hermann Fischer, the world expert on sugars, determined that natural carbohydrates all fell into the same chiral series because they were all derived from the same chiral starting material. He showed, in an 1894 paper, that one chiral center (from an alkaloid called brucine) could influence the reactions next to it to help set the chirality at the next carbon, especially if you were using chiral reagents or catalysts as well. This was the concept of *asymmetric induction*—using one chiral center to help set another one—and it has been used by synthetic organic chemists ever since to synthesize complex structures. A few years later, German chemist Willy Marckwald used brucine again to run a reaction on an artificially made starting material that had no chiral center at all. His product was chiral, though, demonstrating that chiral reagents alone could be enough to make chiral products.

This rule still applies: to achieve chirality, you must start with a chiral material or use a chiral reagent. Questions remain, however. How did living systems select chiral carbohydrates and amino acids in the first place? Did the ultimate ancestor cell pick one series at random? Did a source of polarized sunlight shift the balance? Can the laws of physics explain why one "handedness" is favored over the other? These are still open issues and very important in research on the origin of life.

SEE ALSO Natural Products (c. 60 CE), Amino Acids (1806), Wöhler's Urea Synthesis (1828), Chirality (1848), Tetrahedral Carbon Atoms (1874), Fischer and Sugars (1884), Resolution and Chiral Chromatography (1960), Murchison Meteorite (1969), B_{12} Synthesis (1973), Enzyme Stereochemistry (1975), Shikimic Acid Shortage (2005), Engineered Enzymes (2010).

These tasty-looking fruits are actually full of alkaloids: deadly strychnine and the bitter (but less toxic) brucine. Both of these are chiral and have been used in asymmetric organic synthesis, but brucine is understandably more popular.

Diazomethane

Hans von Pechmann (1850–1902)

Diazomethane is a useful molecule, and nothing can take that away from it. Not even the way it explodes when exposed to sharp edges or ground-glass joints (which means that you have to use special glassware and flame-polished glass pipettes to handle it). It can also explode when exposed to strong sunlight, metal salts, heating, and probably a whole list of other things that you'd rather not explore in detail.

It's highly toxic as well, and its volatility doesn't make guarding against exposure any easier, nor does the fact that lethal exposure can occur without any evidence at the time, since the effects on the lungs take hours to develop. It can't be stored for any period of time, so it has to be made freshly using all that special glassware. Even for short periods of storage, it needs to be kept cold in dilute solution.

So why, you may be asking yourself by now, would anyone tolerate such a reagent? As German chemist Hans von Pechmann discovered and described in a paper he published in 1894, and as generations of chemists have found since, diazomethane is a compound with unique reactivity. (It's worth noting that von Pechmann also described the symptoms of diazomethane poisoning in the same paper.) It can do several reactions to form small rings, and it will turn acids into methyl esters, a common and useful chemical intermediate (see **Isoamyl Acetate and Esters**) so quickly and cleanly that you can hardly believe it. Add diazomethane, a bit of nitrogen gas bubbles away, and there you have it. For complex structures, there is no milder reaction possible. All sorts of complex, sensitive structures can pass safely through this step without danger of decomposition.

The diazo group is responsible for all this—its two nitrogens are bonded together in such a way that makes them quite close to turning back into nitrogen gas, and given a chance, that's just what they'll do. Compounds like this, balanced on a knife-edge of reactivity, can be powerfully useful and powerfully dangerous at the same time. Small wonder that such things are considered natural candidates for **flow chemistry**.

SEE ALSO Polyethylene (1933), Isoamyl Acetate and Esters (1962), B_{12} Synthesis (1973), Flow Chemistry (2006)

Winter semester at the Ludwig Maximilian University of Munich, 1877–1878. Pechmann is sixth from the right in back row. Seated in front of him with the large cap is Emil Fischer.

Liquid Air

Carl von Linde (1842–1934), James Prescott Joule (1818–1889), William Thomson (1824–1907)

German scientist Carl von Linde hit on what some people would call the perfect business. In 1895, he took air and turned it into a range of worthwhile products. He developed an ingenious refrigeration process that cooled gases until they liquefied, using the thermodynamic properties worked out by previous investigators. Compressing a gas heats it, and allowing it to expand cools it (the Joule-Thomson effect, after the English physicists James Prescott Joule and William Thomson, Baron Kelvin of Largs). Linde's device took compressed air and cooled it back to room temperature and below using groundwater-chilled pipes. It was then allowed to expand through a nozzle into a larger insulated space, which lowered its temperature sharply—the technique is still used in refrigerators and air conditioners today.

The key step, though, was that this chilled air was then pumped back around to cool down the air in the earlier stages, running through a jacket around the outside of the apparatus (an arrangement called a heat exchanger), which meant that the next batch was even colder when it hit the expansion chamber, which allowed it to cool down the batch after it even further, and so on. Eventually the air would start to condense into liquid in the collection chamber, and whatever didn't was sent back around to the heat exchanger until enough liquefied air had been produced.

Liquefied air was an industrial novelty as it was, but Linde did not rest on his business model. The next stage was **fractional distillation**, which separated out the nitrogen from the **oxygen**. Those two elements had very different uses indeed—oxygen could be used for breathing-gas mixtures and to make high-temperature furnaces—and they were worth paying for either in their chilled form or in pressurized bottles of the now-purified gases. Argon was next, and when the other noble gases were discovered a few years later, **neon** became still another valuable commodity.

SEE ALSO Fractional Distillation (c. 1280), Oxygen (1774), Liquid Nitrogen (1883), Neon (1898)

A factory producing liquefied air for separation into the various industrial gases.

Greenhouse Effect

Svante August Arrhenius (1859–1927)

Carbon dioxide's effect on Earth's temperature has been a huge topic in recent years, but most people will probably find it surprising that Swedish chemist Svante August Arrhenius first proposed the greenhouse effect in 1896, in a paper titled "On the Influence of Carbonic Acid in the Air upon the Temperature of the Ground." Arrhenius's work covered a wide range of topics and won him a Nobel in 1903. He was one of the first to recognize that salts form positive and negative ions when dissolved in water and that these are their actual reacting species, and he explained the behavior of acids and bases according to the same scheme. He was also one of the first to propose that life could have been carried from planet to planet in the form of spores or other robust microorganisms, an idea that is taken very seriously now that we know that fragments of the Martian crust (and other bodies) occasionally land on Earth as meteorites.

In 1896, Arrhenius's work on the infrared properties of the atmosphere showed that carbon dioxide and water vapor were responsible for strong absorbance at particular wavelengths, and he realized that this meant that they kept heat (from sunlight or other sources) trapped, like the walls of a greenhouse do. Arrhenius was attempting to explain the ice ages, and he believed that raising the levels of carbon dioxide in the atmosphere (through burning fossil fuels) was good insurance against another one. He also believed that global warming would make living in colder climates easier and would help to feed the world's growing population.

It was only decades later that evidence for his hypothesis was found by using **infrared spectroscopy**. Until then, many climate scientists thought it likely that the oceans would absorb carbon dioxide quickly enough to keep its levels in the atmosphere from increasing significantly. Most people were generally not as enthusiastic as Arrhenius about global warming, however. Arrhenius himself neglected to consider clouds, circulation of heat in the atmosphere, and many other factors, but his basic idea is still the key behind the entire global warming debate.

SEE ALSO Carbon Dioxide (1754), Infrared Spectroscopy (1905), Carbon Dioxide Scrubbing (1970), Artificial Photosynthesis (2030).

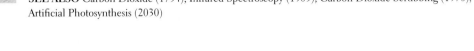

LEFT: *Svante Arrhenius, c. 1895.* RIGHT: *The carbon dioxide and water vapor in the Earth's atmosphere trap infrared radiation (and thus heat). This makes the planet's climate warm enough for us, but the big question is if we're now making it too warm by our own actions.*

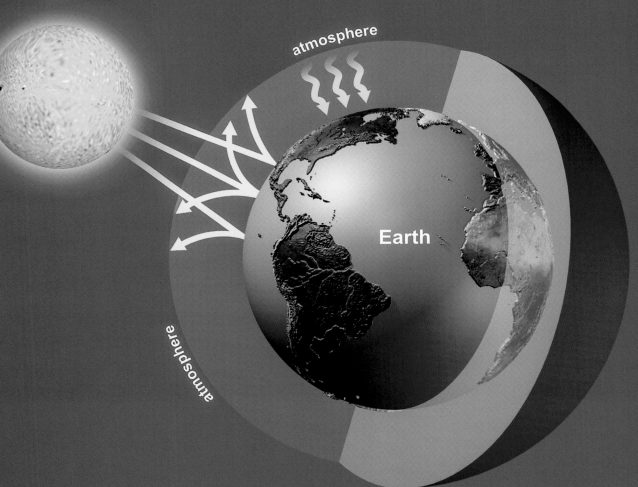

Aspirin

Edward Stone (1702–1768), **Johann Andreas Buchner** (1783–1852), **Pierre-Joseph Leroux** (1795–1870), **Charles-Frédéric Gerhardt** (1816–1856), **Arthur Eichengrün** (1867–1949), **Felix Hoffman** (1868–1946)

Willow bark was known by the Egyptians, the Chinese, some American Indian tribes, the Greeks, and the Romans to reduce aches and fever. But this ancient knowledge was lost in Europe for centuries after the fall of the Roman Empire. Edward Stone, an eighteenth-century English minister, experimented with willow bark because the tree's preference for damp conditions, thought to cause aching joints, led him to believe that the bark might be a remedy. By sheer luck, he rediscovered willow bark's medicinal use, reporting on his findings in 1763. More than sixty years later, German pharmacologist Johann Andreas Buchner and French pharmacist Pierre-Joseph Leroux independently isolated the very potent but very bitter active compound from willow, salicylic acid. It was effective but caused extreme irritation to the throat and stomach.

A more palatable form of the compound was sought in the 1890s. In 1897, German chemist Felix Hoffman at the pharmaceutical firm Bayer found that adding an acetyl ester to the molecule created a new substance that was equally effective but much better tolerated. (It converts back to salicylic acid once ingested.) French chemist Charles-Frédéric Gerhardt had reported this derivative in 1853 but did not realize it had medical uses. Hoffman's colleague Arthur Eichengrün also claimed to have made the discovery, and the name Bayer became synonymous with aspirin (the company's trade name for acetylsalicylic acid), even long after its patents expired.

Aspirin's simple structure is deceptive; its wide-ranging effects in the body were only established in the 1970s. It's an inhibitor of the cyclooxygenase family of enzymes, which are essential for producing prostaglandins, powerful signaling molecules that affect inflammation, blood clotting, and other processes. Pharmaceutical companies' attempts to make new selective cyclooxygenase inhibitors led to product safety recalls and several expensive failures. But aspirin is still with us, used for everything from treating headaches to preventing heart attacks.

SEE ALSO Natural Products (c. 60 CE), Toxicology (1538)

A pain-reliever formula containing aspirin, next to an aspirin bottle marketed under the trademark "Tabloid." Both medicines were imported to the U.K. by Burroughs, Wellcome & Co., a drug manufacturer and distributor founded in London in 1880 by two young American pharmacists.

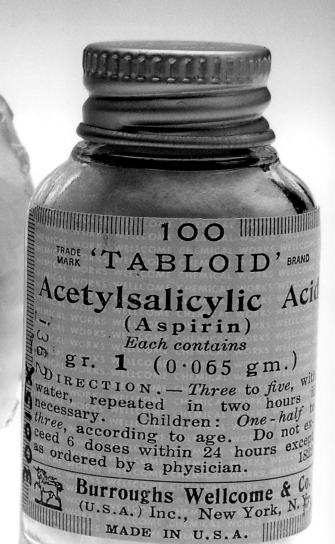

Zymase Fermentation

Hans Ernst August Buchner (1850–1902), Eduard Buchner (1860–1917)

Fermentation has been practiced since prehistory—bread, wine, yogurt, and pickles are older than any written language—although it wasn't until 1857 that French chemist Louis Pasteur determined that it was tied to the microorganisms identified as yeast and bacteria. Clearly, fermentation, the breaking down of sugars to small acids and **carbon dioxide**, happened only in living cells.

That explanation endured until 1897, when German chemist Eduard Buchner—with the help of his older brother, Hans—startled the scientific world by fermenting sugars without any living organisms at all. He broke up the cells of dry yeast to release their contents and pressed the results through fine filtering material, a suggestion of his brother's assistant, Martin Hahn. When Eduard added simple sugars to the pressed extract, the mixture consumed the sugars and evolved carbon dioxide. No yeast cells were present (they definitely would have been noticeable through a microscope, since the yeast would have grown rapidly). Not only had he discovered active proteins (enzymes, in particular the complex named *zymase*), but he demonstrated that they could be freed from their native environments and remain functional. It was another blow to vitalism. Rather than being full of some vital essence that disappeared upon death, the insides of living cells were apparently full of machinery that could be removed.

Others had actually tried this before, but with no success. One of the reasons for previous failures, it was found, was the use of ground glass to break up the cells, which apparently inactivated the cell contents through a reaction with the glass surface. The success of this process also varied greatly depending on the strain of yeast used. Hans Buchner was interested in preparing bacterial extracts, and he and Martin Hahn worked out the pressing technique that made the difference.

Cell-free enzyme chemistry is now a huge field, with applications in everything from detergents to therapy for rare genetic diseases. Even today, it can be a tricky business to keep things working correctly, so the amount of work the pioneers put into it is worth remembering.

SEE ALSO Carbon Dioxide (1754), Wöhler's Urea Synthesis (1828), Engineered Enzymes (2010)

Yeast of the genus Candida *growing on a laboratory plate. Brewing and baking yeasts are so vigorous that anyone working in a cell-culture facility must be extremely careful not to contaminate the lab cultures if they do any baking or home brewing as an after-work activity!*

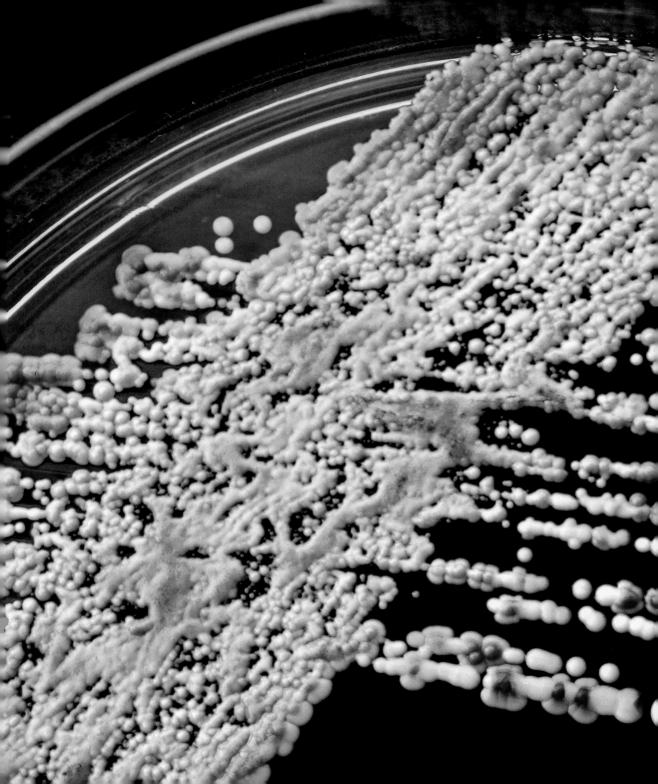

Hydrogenation

Paul Sabatier (1854–1941), **Roger Adams** (1889–1971)

If you could somehow add a molecule of hydrogen (H_2) across a carbon-carbon double bond, you'd take it back to the equivalent single bond—each of the two carbons would pick up a single hydrogen, with single bonds all around. But hydrogen alone doesn't do anything of the sort—it's quite unreactive with organic molecules. In 1897, though, French chemist Paul Sabatier discovered that finely divided metals catalyzed the reaction. This "hydrogenation" was to become one of the most important reactions in all of organic chemistry.

The best metals for the reaction are, alas, rather expensive: platinum and palladium. But only tiny amounts are needed (and they're usually recovered for sale to metal dealers—no one throws platinum away if they can help it). Nickel, rhodium, ruthenium, and other metals are sometimes used in special cases, such as hydrogenating aromatic rings (see **Benzene and Aromaticity**). The choice of catalyst, solvent, and pressure gives chemists many options, and there are many techniques to reduce one functional group without touching others. When it works, it can be one of the cleanest reactions in all of organic chemistry, but if you hydrogenate a bond that you didn't plan to, there's often no way to reverse it.

These reactions work much better under pressure, and in 1922 American chemist Roger Adams invented a device to do this easily. Just four years later, in 1926, the first commercial hydrogenating equipment was available, and these "Parr shakers" (named for the Parr Instrument Company) are still found in labs around the world. Generations of chemists have hooked up their thick-walled glass bottles, turned up the hydrogen pressure from a tank, and set the equipment moving in a familiar (and noisy) ratcheting shake. Newer machines can do hydrogenations via **flow chemistry**, but the Parr shaker is proving hard to dislodge from the labs.

Industrial hydrogenation produces modified oils for the food industry, which are easier to process and store. Unfortunately, this can also form isomers of some double-bond compounds (the "trans" form), and these trans fats have been shown to be harmful to human health (and are being banned in many countries).

SEE ALSO Hydrogen (1766), Flow Chemistry (2006), Engineered Enzymes (2010), Hydrogen Storage (2025)

Making partially hydrogenated oil can produce small amounts of so-called "trans fats," which have been detected in such foods as McDonald's french fries and are now known to raise human LDL cholesterol levels, increasing the risk of coronary heart disease.

Neon

John William Strutt (1842–1919), William Ramsay (1852–1916), Morris Travers (1872–1961)

The advent of techniques for liquefying air sent British chemist William Ramsay on a hunt for rare gaseous elements. In 1895, after English physicist John William Strutt, 3rd Baron Rayleigh of Terling Place, noticed that nitrogen from the air was denser than nitrogen made chemically, Ramsay set out to discover why. He found the nitrogen in air also contained an inert (chemically unreactive) gas he named *argon* (after the Greek word for lazy). The layout of **the periodic table** suggested that there were other "noble gases" filling in the gaps around it. In the summer of 1898, Ramsay and his research partner, English chemist Morris Travers, concentrated the residue from liquefied air and removed the nitrogen by **fractional distillation**, followed by the **oxygen** and the argon. (The small amount of **carbon dioxide** was a solid frost at these temperatures, and was easily managed.) They were glad to find, after this painstaking work, a small residue of unknown liquefied gas. Krypton was discovered first in this mixture, then neon, and finally xenon, all within about six weeks of one another.

Since these gases are colorless, unreactive, and present in low concentrations, the best way to characterize them was by their emission spectra. And that's when Ramsay and Travers got a surprise. Krypton had a weak but colorful glow under electric discharge, but when the first tube of neon was hooked up to the electric lines, the lab filled with a bright red-orange light that no human being had ever seen before. According to Travers, "The blaze of crimson light from the tube told its own story, and it was a sight to dwell upon and never to forget."

Neon went on to light up the twentieth century. The French company Air Liquide began isolating industrial quantities of neon. (It is rare on Earth, since our gravity doesn't hold it, but common across outer space). An attempt to use it for indoor home lighting failed (as can well be imagined), but in 1912, the first neon advertising sign was introduced in a barbershop in Paris.

SEE ALSO Fractional Distillation (c. 1280), Flame Spectroscopy (1859), The Periodic Table (1869), Liquid Nitrogen (1883), Liquid Air (1895), Noble Gas Compounds (1962)

Downtown Las Vegas, one of the traditional homes of neon signage. LED signs have cut into the classic neon business, but neon's warm glow is so distinctive that it will surely live on.

Grignard Reaction

Philippe Antoine Barbier (1848–1922), François-Auguste-Victor Grignard (1871–1935)

Every organic chemist knows the Grignard reaction, which produces some of the most useful organometallic reagents in the toolbox. They are generally easy to prepare, stable on storage, and they react in handy and predictable ways. Uses are still being discovered for them, and the number of variations and modifications that have been found must be nearly beyond counting by now.

In 1900, French chemist François-Auguste-Victor Grignard was a twenty-nine-year-old doctoral student under the supervision of French chemist Philippe Antoine Barbier when he made the discovery that would earn them the 1912 Nobel Prize in Chemistry. Organometallic pioneer Barbier had suggested that Grignard investigate magnesium compounds, and this suggestion led directly to Grignard's breakthrough: discovery of the preparation of magnesium alkyl halides (Grignard reagents).

Preparing Grignards usually requires just some magnesium metal, a solvent, and a starting compound with a carbon-bromine or carbon-chlorine bond. Many of these halide compounds will start to react with pieces of magnesium metal after only a bit of encouragement (a little heating, a speck of iodine, or some scratching of the metal surface to remove the oxide layer). Chemists learn quickly that if they start the reaction with heat, they'd better be prepared to cool it back down, because it can be vigorous after it gets going. Once prepared, the reagent acts largely as if there was a plain negative charge on the carbon that used to have the halide atom, and these species can react with a variety of partners to form new carbon-carbon bonds. Since those are the currency of organic synthesis, Grignards and other such reagents have been used constantly when assembling new molecular frameworks.

You can actually buy solutions of many common Grignard reagents, ready to be dispensed by syringe. An older generation of chemists, used to making their own as needed, rolled their eyes at such luxury, but these preparations are now widely used. The "handmade" Grignards are in no danger of disappearing from the world, though. Over a hundred years later, the reaction is going strong.

SEE ALSO Cadet's Fuming Liquid (1758), Silicones (1900), B_{12} Synthesis (1973), Nozaki Coupling (1977), Metal-Catalyzed Couplings (2010)

Victor Grignard, many years after his graduate school days, when he was a famous and honored professor in his own right.

Free Radicals

Moses Gomberg (1866–1947)

Most of the molecules described in this book have all their electrons accounted for. A typical single bond is the sharing of two electrons, one from each atom involved, with no odd ones left over. Or are there? In 1900, Russian-born chemist Moses Gomberg reported the first known exception. He was the first chemist to make tetraphenylmethane, the compound with a single carbon atom sprouting four aromatic rings. Along the way, he naturally had several triphenylmethyl intermediates, and he noticed some odd reactivity when he tried to couple them head-to-head to make his desired product. He had apparently formed some sort of triphenylmethyl species that was air sensitive, and it quickly reacted with chlorine, bromine, and iodine, among other compounds. This was not triphenylmethane—Gomberg had plenty of that around, and it was not air sensitive at all. The reactivity of his new product didn't match up with any known species, and Gomberg made the bold suggestion that he had produced and isolated the first organic free radical, whose unpaired electron made it highly reactive.

Some chemists were convinced, but many others were not. Over the next thirty years, though, evidence began to pile up from other labs that free radicals were intermediates in many reactions, and that they could be stabilized by such things as multiple phenyl groups. Gomberg is now recognized as the founder of free radical chemistry.

Free radicals often get a bad rap in health-related reporting, as they can contribute to illnesses related to aging, heart attacks, stroke, and inflammation. However, living cells produce many free radicals through their use of **oxygen**, among them **superoxide**, and free radicals play important positive roles in metabolism and in fighting disease. For example, white blood cells use free radicals to destroy cells infected with bacteria or viruses.

Gomberg is also remembered for being perhaps the last chemist to try to invoke a nineteenth-century-style "gentleman's agreement" for research priority. At the end of his first triphenylmethyl radical paper, he said, "This work will be continued and I wish to reserve the field for myself." It didn't happen.

SEE ALSO Oxygen (1774), Photochemistry (1834), Tetraethyl Lead (1921), Polyethylene (1933), Superoxide (1934), CFCs and the Ozone Layer (1974)

Microcrystals of **Vitamin C**, *an antioxidant that inhibits the production of free radicals, in polarized light.*

Silicones

Frederick Stanley Kipping (1863–1949)

Silicon is the element directly below carbon in **the periodic table**, and it shares some of its properties. But despite the presence of silicon-based life forms in a wide variety of science-fictional stories, it's a different enough element that the comparison doesn't hold up very well. For example, the silicon-**oxygen** bond is very strong, and oxygen atoms tend to single-bond to another silicon atom instead of double-bonding as they can do with carbon. So you have **carbon dioxide**, the familiar gas that we exhale, versus silicon dioxide, which is glass. Evidently, silicon has its own way of doing things.

Those differences created difficulties for early chemists, who tried to treat silicon in chemical reactions the same way they did carbon. Englishman Frederick Stanley Kipping, the chemist most associated with the dawn of organosilicon chemistry, seems to have often had a terrible time of it. In 1900, he was able to extend Grignard-like chemistry into the silicon field, using organometallic reagents to try to make new compounds, but the products from these reactions kept condensing to thick, gluey resins and jellylike substances that were difficult or impossible to analyze with the equipment of the time. Kipping worked in the field for forty years, and at his retirement in 1936 he told an audience that "the prospect of any immediate and important advance in this section of chemistry does not seem very hopeful."

But he lived long enough to see that he was wrong. World War II jump-started research in the field of organosilicon chemistry, as chemists at General Electric and Dow Corning had found that the organosilicon-oxygen compounds (known confusingly as sili*cones*, a word coined by Kipping) were excellent insulators and lubricants. Kipping had never quite used the right conditions to produce some of these compounds, but building on his pioneering work, entire companies turned their attention to the new substances, which ended up being very useful indeed, with excellent stability to heat and corrosive agents. They could be made clear or opaque, with varying degrees of flexibility—similar to new forms of **rubber**—and their first use was in protecting airplane ignition systems from water. The use of silicones in breast implants has been controversial, but they're also found in pacemakers and many other medical devices.

SEE ALSO Rubber (1839), The Periodic Table (1869), Grignard Reaction (1900).

Silicone bakeware, a fate for his compounds that Kipping could never have imagined. New uses continue to be found for silicones, in the kitchen and well outside it.

Chromatography

Mikhail Tsvet (1872–1919)

Russian-Italian botanist Mikhail Tsvet had been working with plant pigments (chlorophylls and the like), and had found a new way to separate and purify them. When he poured a small sample of liquefied plant extract onto a column of powdered chalk (calcium carbonate), and then washed that through with organic solvents, he found that the sample began to separate as it moved down the column. Colored bands of individual components became clear, as some of them interacted more with the powdered support, dragging behind. He reported his great discovery in 1901, but it would be years before the significance of his breakthrough was recognized.

Tsvet called his process *chromatography*, after the Greek word for color, and it has become the most versatile technique in all of analytical chemistry. A wide variety of "stationary phases" have taken the place of Tsvet's chalk, with powdered silica a common choice for organic compounds. Solvents used to be dripped through columns by gravity, as Tsvet described, but are now pumped through at higher pressures, allowing constantly varying mixtures of solvents to be used, increasing in strength to wash out slower-moving compounds. The columns themselves have been used in sizes ranging from less than an inch long to huge devices that require that holes be cut through ceilings and floors.

Since organic chemists can't rely on their compounds being as brightly colored as plant pigments, special detectors are used to monitor the stream of solvent coming off a column, such as changes in its ultraviolet spectrum to indicate the presence of a particular compound. More expensive systems are automated, mechanically injecting the samples, varying the solvents, and draining each new peak of material coming off the column into separate containers as they're detected.

Chromatography is essential to every kind of chemical analysis and is used constantly in medicine, forensic investigation, the food industry, and more. Tsvet would have been amazed to see what became of his work!

SEE ALSO Purification (c. 1200 BCE), Separatory Funnel (1854), Gas Chromatography (1952), Resolution and Chiral Chromatography (1960), HPLC (1967), Reverse-Phase Chromatography (1971), Tholin (1979), Electrospray LC/MS (1984), Acetonitrile (2009)

A demonstration of chromatography of dye mixtures, spotted at the bottom of a thick piece of filter paper and allowed to move as the solvent soaks up from the bottom. Some of the individual color components are already moving faster than others.

Polonium and Radium

Antoine-Henri Becquerel (1852–1908), **Pierre Curie** (1859–1906), **Marie Salomea Skłodowska Curie** (1867–1934)

The 1890s was a period of great tumult in physics and chemistry. The discovery of radio waves and then X-rays prompted new areas of research, among them French physicist Antoine-Henri Becquerel's attempt to determine whether the luminescence of some minerals could be connected to X-rays. In 1896, he found that uranium salts could expose photographic plates by themselves, without any energy source, which meant that uranium compounds were emitting another undiscovered form of radiation.

Polish-French physicist and chemist Marie Curie, who immigrated to Paris to study physics and mathematics, and her husband, French physicist Pierre Curie, the head of a research laboratory studying magnetism and crystals, were two of the most productive and fearless explorers of these new phenomena. Becquerel's reports caught Marie's attention, and she began a systematic search for more such substances. She found that other elements such as thorium gave off the same kind of radiation, and regardless of their chemical form, the amounts of uranium or thorium in a sample determined its radioactivity.

The mineral pitchblende, however, was much more radioactive than could be explained by its uranium content, which led Marie to examine it for unknown radioactive elements. Painstaking work led to the isolation of a new metal, for which the Curies proposed the name *polonium* (for Marie's native Poland). Another new element was isolated several months later, which they called *radium*. Proving the new elements' existence required isolating them from tons of pitchblende—a massive effort. The Curies worked for the next few years in an unheated (and uncooled) shed, finally isolating useful amounts of both elements in 1902. Marie's resulting doctoral dissertation, the most famous in chemical history, was the basis for two Nobel Prizes.

It all came at a cost. Both Curies were severely affected by radiation. Their lab notebooks are hazardous material, stored in lead-lined boxes and examined only while wearing protective clothing. They will be dangerous for centuries to come.

SEE ALSO Isotopes (1913), Radithor (1918), Radioactive Tracers (1923), Technetium (1936), The Last Element in Nature (1939), Transuranic Elements (1951).

Marie and Pierre Curie in the lab. It is safe to assume that everything in this photograph was at least mildly radioactive, including both Curies.

Infrared Spectroscopy

William Weber Coblentz (1873–1962)

The infrared (IR) region of the spectrum hits at just the right place to be absorbed by specific bonds and groups in molecules. The energies involved aren't enough to break the bonds (that would be up in the ultraviolet and beyond, where a good deal of **photochemistry** happens), but infrared light provides enough energy to set molecules stretching, bending, and wagging. The resulting absorption bands can tell you instantly if (for example) your molecule has a carbonyl (a carbon double-bonded to an oxygen), and the position of that carbonyl band in the spectrum can tell you if it's a ketone, an aldehyde, an ester, or an amide.

American physicist William Weber Coblentz put infrared spectroscopy, the analysis of infrared light's interactions with molecules, on its feet. While Coblentz was a postgraduate at Cornell University in the early 1900s, he built his own equipment and took thousands of measurements, working through the wavelengths manually, bit by bit, with all sorts of compounds. (These were published, with large folding charts, in 1905.) He was the first to show, systematically, that functional groups had specific IR signatures that could be relied on across various structures. He went on to apply his instruments to astronomy, measuring things like the daytime and nighttime temperatures of the surface of Mars, an impressive feat for the technology of the day.

Despite their obvious utility, his advances in analytical chemistry did not catch on quickly. Infrared developed a reputation as an esoteric tool of physics rather than a workhouse of the chemistry labs. This was partly due to instrumental difficulties. Coblentz's spectrometer was (as he himself freely admitted) very tedious to use, because it had to be reset by hand every step of the way through the whole range of infrared wavelengths.

However, infrared spectroscopy ruled the middle of the twentieth century as an analytical technique in chemistry, and it's still used for quality control in manufacturing and for monitoring **carbon dioxide** in greenhouses. Other, more powerful methods have displaced it in many applications, but in its day it provided insights that nothing else could.

SEE ALSO Carbon Dioxide (1754), Functional Groups (1832), Photochemistry (1834)

An emission Fourier-Transform Infrared (FTIR) spectroscopy instrument, which allows atomic bonds and their relative abundance to be detected by determining discrete infrared wavelengths from the complex mixture of light coming from various transitions within the excited atoms.

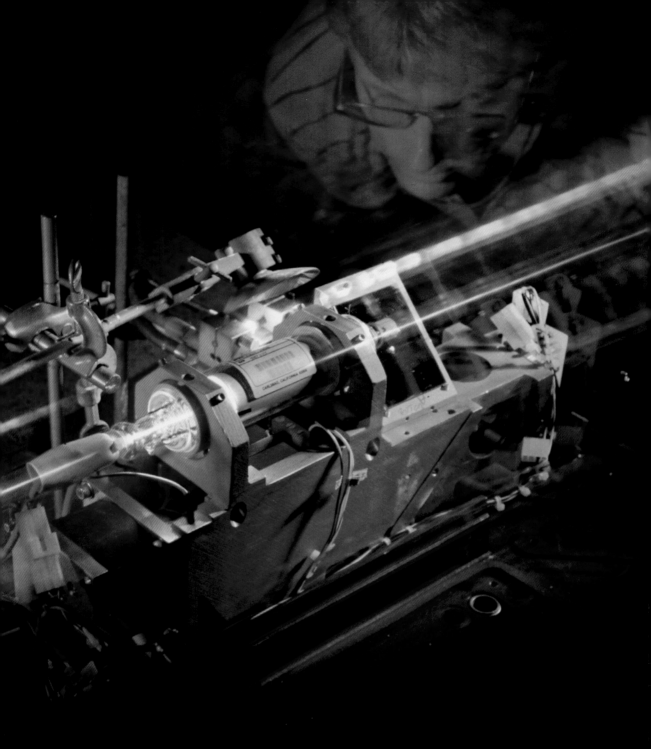

Bakelite®

Leo Hendrik Baekeland (1863–1944), Nathaniel Thurlow (1873–1948)

In the early twentieth century, organic chemistry began to furnish materials that had never been seen before. A few indications of this type of advance had appeared in earlier decades: polystyrene had been accidentally prepared, although its uses remained undiscovered for many years, and in the 1870s, a new substance called celluloid had been made during a search for an ivory substitute for billiard balls, but it used a natural polymer (cellulose) as a starting point.

Bakelite was something else again. It was discovered in 1907 by Belgian-born chemist Leo Hendrik Baekeland and his assistant, Nathaniel Thurlow, during a deliberate search for new materials. There had been references in the German chemical literature reporting on reactions between phenols and formaldehyde, but they mostly noted that the glassware used for those experiments was ruined. The new substance that formed, whatever it was, would dissolve in nothing known to science. For years, these properties were thought to be a sure sign of the product's uselessness, but by the early 1900s it was dawning on numerous researchers that there might be some real value in a material like this if only its formation could be controlled.

At the time, Baekeland was trying to soak wood in phenol/formaldehyde mixtures to see if he could strengthen it, but the tests did not go well. Some of the material that did not soak into the wood, though, formed a hard gum, which he made the subject of further experiments. Tweaking the conditions produced a resin that could be molded while it was a thick liquid, but then hardened quickly into a permanent form. This was Bakelite, the world's first thermosetting plastic. Although somewhat brittle, it was extremely versatile, and when it was found to be an excellent electrical insulator, its future was assured. The electrical appliance industry was expanding rapidly, and Bakelite was just what it needed for a wide variety of parts. Dubbed by Baekeland as "the material of a thousand uses," it was also used in many retail goods, including radios, telephones, costume jewelry, and game pieces. While other plastics have often taken their place, phenolic resins like Bakelite are still being made.

SEE ALSO Polymers and Polymerization (1839), Rubber (1839), Polyethylene (1933), Nylon (1935), Teflon (1938), Cyanoacrylates (1942), Ziegler-Natta Catalysis (1963), Kevlar (1964), Gore-Tex (1969)

This vintage 1930s radio is made out of Bakelite. Such items are now sought by collectors, since the Bakelite era overlapped with Art Deco design.

Spider Silk

Emil Hermann Fischer (1852–1919)

Spider silk has amazed humans for thousands of years. Far from being an exhausted topic, spiders have become more impressive as additional technology has been used to study them. The best varieties of silk have an unmatched combination of lightness, strength, and elasticity that could make them extremely valuable industrial materials. One could imagine bulletproof and wear-resistant lightweight clothing, incredibly strong ropes, nets, and parachutes, biodegradable bottles, bandages, and surgical thread—if only anyone could produce the material on a large scale. That's a feat that no one has quite managed, and no one can farm spiders the way that silkworms are farmed, either. Spiders produce smaller amounts of silk, and they keep their own schedules.

In 1907, German chemist Emil Hermann Fischer became the first chemist to publish an article on the composition of spider silks. He discovered that they were largely composed of proteins but had a very different selection of **amino acids** than silkworm silk. We now know a great deal about their composition—or compositions, because spiders produce a whole range of silks for different purposes. Their proteins have complex and unusual structures, with repeating regions of nearly crystalline beta sheets, linked together by looser, stretchy helical domains. They're not only unlike fabric silk, they're unlike most other proteins, with compositions biased toward the smallest amino acids while containing almost none of a few common ones. There are a number of other components as well—carbohydrates, lipids, and various other small molecules—whose purposes are not yet clear. One of the strangest things about the process is that spiders store "unspun" silk as a thick, semi-crystalline liquid ready for use, while pulling it from the silk glands apparently aligns the protein strands into finished thread.

Attempts have been made to engineer other organisms to make spider-silk proteins, from bacteria to goats (in their milk, in case you're wondering), and a number of companies are actively working in the area. But silk with the properties of the natural form remains elusive, and progress on mimicking the complex spinning process has been even slower. The spiders are keeping their secrets close.

SEE ALSO Amino Acids (1806), Alpha-Helix and Beta-Sheet (1951)

A "pumpkin" spider (Araneus trifolium) *poised on its dew-covered web. There is nothing else in the world quite like spider silk.*

pH and Indicators

Svante Arrhenius (1859–1927), Søren Peder Lauritz Sørensen (1868–1939), Hans Friedenthal (1870–1943), Pál Szily (1878–1945)

Swedish scientist Svante Arrhenius helped establish, in 1887, that acidic and basic solutions could be defined as having an excess or shortage of hydrogen ions, respectively. Water, he proposed, had a very small amount of H+ (hydrogen ion) and OH− (hydroxyl ion) in it, and disturbing this equilibrium was what made an acid or a base. In 1909, Danish chemist Søren Sørensen, acting on a suggestion by German scientist Hans Friedenthal, invented what we now know as the pH scale, which ranges from 0 to 14, from extremely acidic to extremely basic.

Mathematically, pH is the logarithm of 1 over the hydrogen ion activity in the solution. Plain water is pH 7 (neutral), and human blood is pH 7.4. (Anyone who tries to sell you on making your blood more acidic or alkaline to treat disease is peddling nonsense, by the way—the body works hard to keep your blood's pH right at 7.4.) Many chemicals will change color depending on whether they're in an acidic or basic solution. Their two ionized forms—meaning whether they've received or given up a hydrogen ion—are different enough to absorb different colors of light, and various compounds will cross from acidic to basic at various points along the pH scale.

Friedenthal and Hungarian chemist Pál Szily worked out a series of compounds that could determine pH, and Sørensen extended the list even further. One of the most well known is phenolphthalein. In acid solution, and out to a mildly basic pH of 8.2, it is colorless. But when it finally loses a proton, the new anionic (negatively charged) species is a brilliant purple-pink—a startling change that is always fun to watch. For many years, chemical analysis depended on such colorimetric techniques, using standardized solutions to carefully titrate a test sample until a visible change took place. Modern instruments, starting with the electrochemical pH meter, have moved most of these entertaining but now antiquated procedures solely into textbooks. The idea of measuring the strength of acids and bases, though, is still fundamental to chemistry, medicine, and biology.

SEE ALSO Soap (c. 2800 BCE), Hydrogen Cyanide (1752), Erlenmeyer Flask (1861), Acids and Bases (1923), Magnetic Stirring (1944).

This wide-range pH indicator paper has apparently just been used to test something that is very alkaline indeed.

Haber-Bosch Process

Fritz Haber (1868–1934), **Carl Bosch** (1874–1940), **Robert Le Rossignol** (1884–1976)

The production of ammonia is not a reaction that many people know about, but that only goes to show that the underpinnings of the world are not always visible. Beyond any doubt, the Haber-Bosch process has kept billions of people from starving to death. That's because nitrogen is an essential nutrient for all living things, including all food crops, but the nitrogen gas that we're surrounded by (78 percent of the air) is almost totally unreactive. Converting it into a biologically useful form (a process called nitrogen fixation) is done only by certain single-celled microorganisms. Almost all life on Earth depends on them to keep doing it.

But in 1909, human beings became the latest organisms to fix nitrogen, when chemists Fritz Haber, from Germany, and Robert Le Rossignol, from England, demonstrated a machine that took in air and produced ammonia, an essential precursor to nitrogen fertilizer. The process, which worked at higher pressures than any other industrial technique used at the time, is another example of **Le Châtelier's principle**. When four equivalents of gas (three of **hydrogen**, H_2, and one of nitrogen, N_2) are turned into only two equivalents of ammonia (NH_3), increasing the pressure shifts the equilibrium over toward the products. The nitrogen needed is, of course, everywhere, and the hydrogen is produced on site from methane (natural gas) in a process called steam reforming. These reactants flow over a bed of iron or ruthenium catalyst, all at high temperatures. The ammonia is condensed out in every pass in a vivid example of a continuous flow reactor.

German chemist Carl Bosch developed much better catalysts, and made the original process industrially feasible. In 1913, the first production plant opened, and today Haber-Bosch installations produce five hundred million tons of fertilizer and keep at least one third of the world's population alive. But during World War I, which started the next year, the ammonia was used to make nitric acid for explosives, and Haber himself turned to poison gas research (as did fellow Nobel Prize–winning chemist Victor Grignard). The pity of war, that such great minds turned their energies to destruction.

SEE ALSO Phosphate Fertilizer (1842), Nitroglycerine (1847), Le Châtelier's Principle (1885), Flow Chemistry (2006)

The Haber-Bosch process is a fundamental part of world agriculture. Until it appeared, only microorganisms fixed atmospheric nitrogen.

Salvarsan

Paul Ehrlich (1854–1915), **Sahachiro Hata** (1873–1938), **Alfred Bertheim** (1879–1914)

German physician Paul Ehrlich is considered the father of modern medicinal chemistry, and salvarsan (arsphenamine) is the prototype of all the drug development efforts that have come since. Ehrlich and his colleagues Sahachiro Hata (a Japanese bacteriologist) and Alfred Bertheim (a German inorganic chemist) were searching for compounds that would be more poisonous to disease organisms than to humans. They hoped to improve on a toxic, abandoned arsenic drug (atoxyl) created to kill trypanosomes (the parasites that caused sleeping sickness) that had, tragically, permanently blinded as many as 2 percent of the patients who had tried it. If a wide variety of organic arsenic compounds could be prepared and tested, Ehrlich thought, some useful relationship might arise to separate the activity from the toxicity.

This was the first organized search for such a structure-activity relationship (SAR), still the key step in all medicinal chemistry research programs. In 1909, the 606th compound they tried, arsphenamine, was not effective against the trypanosomes, but it proved very useful against syphilis bacteria. Until then, syphilis had been treated for the most part with mercury salts, which killed the bacteria but at grave risk to the patient. Compound 606, later branded as *Salvarsan*, was used on humans much sooner than any drug is in the modern era, and with far less testing. It still presented the risk of serious side effects, but nothing like the mercury treatment (or like untreated syphilis, for that matter). It was a frontline therapy for over thirty years.

It was also a difficult compound to administer. Like many organoarsenic compounds, it was unstable in air, and had to be quickly dissolved in sterile water and injected. To address this problem, Ehrlich and Hata developed another easier-to-handle compound, but it was not quite as effective. They were the first medicinal chemists to discover that many trade-offs of this sort were waiting for drug researchers. Getting all the properties you'd like into one molecule is not easy, but it was Ehrlich and his team who showed that it was possible at all, leading the way for the generations of drug researchers who followed.

SEE ALSO Mercury (210 BCE), Toxicology (1538), Cadet's Fuming Liquid (1758), Paris Green (1814), Sulfanilamide (1932), Streptomycin (1943), Penicillin (1945), AZT and Antiretrovirals (1984), Modern Drug Discovery (1988), Taxol (1989)

A Salvarsan treatment kit from 1912, containing the drug and all the equipment needed to prepare it for injection.

X-Ray Crystallography

William Henry Bragg (1862–1942), **Max von Laue** (1879–1960), **Paul Peter Ewald** (1888–1985), **William Lawrence Bragg** (1890–1971)

X-rays pass through some substances very easily but are absorbed by others. German physicist Wilhelm Roentgen's 1895 picture of his wife's hand, showing its bones, made that startlingly clear, as did the opaque appearance of her gold wedding ring. Scientists around the world sat down to discover what happened when X-rays encountered various kinds of materials.

In 1912, German physicist Max von Laue, building on the ideas of his colleague Paul Peter Ewald, was the first to realize that crystalline substances show startling X-ray behavior. Under some conditions, **crystals** produce strong diffraction patterns, bands and spots reflected out from the sample. The classic case is a sodium chloride crystal, made up of layer after layer of sodium and chloride ions. X-rays will hit these atoms and reflect, but from more than one layer at the same time. If the waves reflecting from the different layers are in synch ("interfere constructively"), then there will be a stronger signal from that direction. However, if the waves cancel each other out ("interfere destructively"), then there will be nothing to detect at all. This all depends on the wavelength of the X-rays, the angle they come in from, and the distance between the layers, as worked out by British physicists Sir William Henry Bragg and Sir William Lawrence Bragg (father and son, respectively), in what is now called Bragg's law.

And since the crystallographer sets the wavelength and the angles, the patterns obtained can be used to work back to the atom arrangements that caused them. A three-dimensional picture of the crystalline lattice is the result, with each atom located in space. The more complicated the crystal, the more complex the calculations, but modern hardware and software are so powerful (and modern X-ray sources are so bright) that crystallography has become an essential tool for confirming molecular structures. A compact X-ray diffractometer was even built into the Mars rover *Curiosity*.

SEE ALSO Crystals (c. 500,000 BCE), Carbonic Anhydrase (1932), Penicillin (1945), Nonclassical Ion Controversy (1949), Conformational Analysis (1950), Alpha-Helix and Beta-Sheet (1951), Ferrocene (1951), Green Fluorescent Protein (1962), Protein Crystallography (1965), B_{12} Synthesis (1973), Quasicrystals (1984), Coordination Frameworks (1997)

An X-ray diffraction pattern produced by a crystal of Rubisco enzyme (see **Photosynthesis**). *Simple substances give much less artistic patterns, but molecules as complex as enzymes need serious computational power to be solved.*

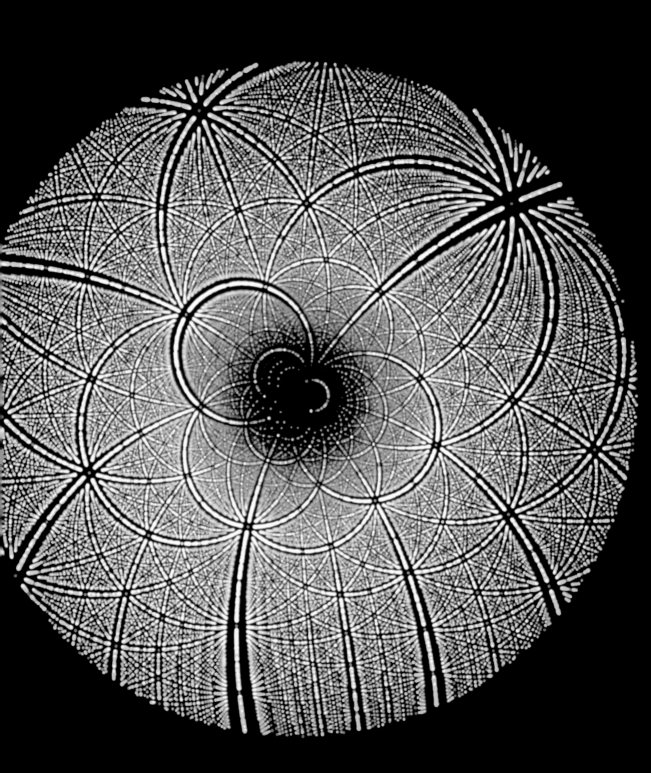

Maillard Reaction

Louis-Camille Maillard (1878–1936), John Edward Hodge (1914–1996)

You have performed the Maillard reaction, even if you've never taken a chemistry class. If you've toasted bread, grilled a hamburger, or even overheated a batch of popcorn, this is the reaction that gave you your desired (or undesired) products. French physician and chemist Louis-Camille Maillard was studying the synthesis of proteins and found that **amino acids**, their building blocks, would react with many types of sugar molecules under heating (he published the first paper on this process in 1912). Since there are amino acids and simple carbohydrates in virtually every kind of food that is cooked, this reaction takes place in every kitchen and over every fire.

But what does it produce? Early studies were inconclusive; there were just too many products to characterize. Maillard himself knew that he had produced a new brown substance that would not dissolve well in water, but he could not get much further with 1912 equipment. Since then, every time analytical instrumentation has improved, more compounds have been found in Maillard's mixtures. In any kind of real-world foodstuff, hundreds (if not thousands) of different end products are present, formed through a bewildering variety of condensation, dehydration, and cyclization reactions whose basic details were first worked out by American chemist John Edward Hodge in 1953. These are the sources of the flavors of cooked food, and organic chemists are invariably impressed when they see the range of compounds found in every toaster and frying pan.

In the 1980s, food writer Harold McGee repeated an interesting experiment first run in 1947 with the simplest Maillard reactions, heating some sugar syrup in a pan and sprinkling a number of pure, single amino acids into it. He reported that cysteine gave a noticeable smell of fried onions and broth, while lysine smelled like toasted bread. Phenylalanine began unappetizingly, like melted plastic, but then turned into "essence of almonds." If these simple reactions can produce such results, you can imagine the chemical complexities of a steak or a pan of roasted vegetables. Modern chefs employ the Maillard reaction at every turn, whether they realize it or not.

SEE ALSO Amino Acids (1806), Fougère Royale (1881), Fischer and Sugars (1884), Isoamyl Acetate and Esters (1962).

The browned crust on this steak is the result of Maillard reactions, which produce the flavors of any food cooked over high heat.

Stainless Steel

Pierre Berthier (1782–1861), **Elwood Haynes** (1857–1925), **Harry Brearley** (1871–1948)

Most simple iron alloys begin to rust quickly on exposure to water and **oxygen**. The process causing this is the oxidation of the metal to iron oxide, and it continues because the oxide is brittle and flaky, constantly exposing the metal underneath to further oxidation. It also reacts with the iron atoms next to it, converting them to iron oxide and spreading the reaction farther into the metal.

But if you add enough chromium to the mix (more than 10 percent), you can produce steel that remains shiny under most ordinary conditions. One reason this works, oddly, is that chromium is even easier to oxidize than iron is. You might guess that some sort of chromium rust would lead to another pile of flakes and powder, but the chromium oxide that forms is hard, dense, and unreactive, creating a protective layer, a molecular skin, on the surface of the metal, which keeps oxygen from getting to any of the metal atoms below.

French metallurgist Pierre Berthier first recognized this property in 1821, but the metalworkers of the time could only make brittle objects with the resulting steel. In the early 1900s, though, metallurgists in France, Germany, England, and the U.S. found several routes to stainless alloys, with the most useful and easily prepared ones discovered in 1912 by the metallurgists Elwood Haynes and Harry Brearley, who were American and English respectively. This started off a large, new industry and any number of patent fights, but by the 1920s stainless steel was becoming a familiar part of life in the industrialized world, used in a range of products, from cutlery and scalpels to aircraft engines and automobile trim.

The protective-layer principle (known as *passivation*) is also used in metal alloys for handling fluorine gas, which otherwise tends to react with everything. And it keeps some other reactive metals tamed, most notably aluminum. See the entry on **thermite** for what happens when aluminum finally does react with oxygen in the air, and be glad that—thanks to an invisible layer of aluminum oxide— this has no chance of happening to your roll of aluminum foil.

SEE ALSO Iron Smelting (c. 1300 BCE), Oxygen (1774), Isolation of Fluorine (1886), Thermite (1893)

The U.S. Steel building in Pittsburgh is seen in this bullet-riddled reflection. Stainless steel and other alloys that are essential to modern civilization have created huge industrial firms (and fortunes).

Boranes and the Vacuum-Line Technique

Alfred Stock (1876–1946), **Herbert Charles Brown** (1912–2004), **William Lipscomb** (1919–2011)

Again and again, compounds of boron that looked like useless lab curiosities have emerged as important reagents and proving grounds for chemical theory. There is no better example of this than the boranes, the class with a boron-hydrogen bond. Boranes form odd cages and clusters, in which two electrons seem to be shared across three atoms rather than the usual two-atom chemical bond. They undergo a huge number of reactions with other organic compounds. American chemists William Lipscomb and Herbert Charles Brown discovered these and other aspects of boron chemistry, and both were awarded Nobel Prizes for their efforts.

An early hindrance to borane and organoboron chemistry was the terrifically flammable nature of the pure compounds. They're extremely reactive to oxygen (forming boron-oxygen bonds is energetically quite favorable) and to water, making them tremendously dangerous if exposed to the open air. In the 1950s and 1960s, organoborons were studied as potential rocket fuels and were used to ignite the engines in the SR-71 spy planes.

In the early 1900s, German chemist Alfred Stock and his research group were the borane pioneers, developing much of what's now called vacuum-line technique to deal with the compounds' super-reactivity. In 1912, they set up an elaborate glass apparatus hooked to strong vacuum pumps, which allowed for compounds to be formed, distilled, and transferred without any exposure to air at all. The only problem with this system was that it depended on liquid mercury—and plenty of it—as a material for valves and vacuum pumps. Stock's later papers, chillingly, were on the effects of mercury poisoning in humans and were described from personal experience. He particularly warned against many organic compounds that incorporated the element, realizing that these could be even worse than exposure to the metal itself. His efforts did a great deal to make other chemists aware of better techniques and precautions.

SEE ALSO Mercury (210 BCE), Toxicology (1538), Glove Boxes (1945)

The SR-71 Blackbird *spy plane. Its engines (and their difficult-to-ignite fuel mixture) were started by triethylborane, which is reliably flammable and burns at a high temperature. For these reasons, boranes need to be handled with vacuum-line techniques in the lab.*

Dipole Moments

Peter Debye (1884–1966)

A fundamental characteristic of atoms in a molecule is that they are all surrounded by clouds of electrons. And these electrons aren't evenly distributed—quantum mechanics ensures that they're more likely to be found in some places than others. Only if you have a perfectly symmetrical bond (such as hydrogen-hydrogen), will both atoms have the same electron density, with the bonding electrons spread out evenly between them.

But there are a lot of uneven chemical bonds in the world, as American chemist Linus Pauling would show systematically (see ***The Nature of the Chemical Bond***). Before that, however, Dutch-American physicist and chemist Peter Debye gave science a way to think about them. An uneven distribution of charge is called a *dipole*. One end ends up with more negative charge, and the other with more positive. The "dipole moment," defined by Debye in a 1912 paper, is the strength of the charges times the distance between them, and it can be added up and measured for whole molecules as well. Solvents with a strong dipole moment have higher boiling points (attraction between charges on adjacent molecules makes them stick together more) and are much better at dissolving other polar species. Dimethyl sulfoxide (commonly referred to as *DMSO*) is an example of a high-dipole organic solvent—it will even dissolve many ionic compounds (salts), which makes it very useful for a range of chemical reactions. Dipolar interactions are very important in dissolving compounds and when compounds interact with each other (as when a drug molecule binds to its biological target in the body).

Debye himself emigrated from Germany to America as World War II was starting. In recent years, accusations surfaced that he might have been a spy for the Germans, which was surprising considering his efforts to help his Jewish colleagues. These charges do not appear to have been substantiated, and some new evidence suggests that he might have been a spy for the British instead. Debye made a number of contributions to physics and chemistry, but it now seems that he was carrying around even more in his head than anyone realized.

SEE ALSO Transition State Theory (1935), Reaction Mechanisms (1937), *The Nature of the Chemical Bond* (1939), Kevlar (1964)

Peter Debye, posing for a bronze bust commissioned by the Netherlands's Ministry of Education in 1937. A Nobel Prize increases the chances of this sort of thing happening.

Mass Spectrometry

Joseph John Thomson (1856–1940), Arthur Jeffrey Dempster (1866–1950), Francis William Aston (1877–1945), Ernest O. Lawrence (1901–1958)

Mass spectrometry is a technique originally used by physicists, then moved into chemistry, where it is equally at home. In every case, it depends on generating charged ions from atoms and whole molecules, and letting them fly through a small vacuum chamber in the presence of electric and magnetic fields. This causes the paths of the charged particles to bend, and that deflection depends on how heavy those particles are. What this gives you is a way to separate out the various compounds in a mixture by their molecular weights, and that, as it happens, is an endlessly useful thing to be able to do.

British physicist Joseph John Thomson was the first to notice this, when he deflected a stream of pure ionized neon atoms in 1913 and saw that it separated into two. He was seeing neon-20 and neon-22, the first physical evidence for the existence of stable **isotopes** (forms of the same element with different numbers of neutrons). His student Francis William Aston, a British physicist and chemist, went on to build a better instrument and found that many other elements were composed of isotopic mixtures. Canadian-American physicist Arthur Jeffrey Dempster's machine was better yet, and he discovered that an electrical spark could be used to ionize a sample inside the vacuum chamber. By the time of World War II, American physicist Ernest O. Lawrence was developing mass spectrometers that could be used to separate uranium isotopes not just to analyze them, but also to purify larger amounts for actual use—albeit with a huge amount of effort.

But by this time, chemists were finding ways to ionize entire molecules. Bombarding them with streams of electrons (the electron-impact method), or with streams of small ions (the chemical-ionization method), allowed a wide variety of compounds to "fly" under mass spectrometry conditions. Larger and larger molecules became open to analysis, at finer and finer resolution, providing unambiguous compound identification for a huge number of applications. Medical assays, drug research, geology, and forensic science all depend on it.

SEE ALSO Isotopes (1913), Radioactive Tracers (1923), Deuterium (1931), Gaseous Diffusion (1940), Gas Chromatography (1952), Electrospray LC/MS (1984), Fullerenes (1985), MALDI (1985), Isotopic Distribution (2006)

Mass spectrometers (calutrons) being used to produce enriched uranium for the Manhattan Project at the Y–12 plant in Oak Ridge, Tennessee. The operators, mostly young women with only high school educations, worked in shifts around the clock and out-produced the PhDs they replaced.

1913

Isotopes

Ernest Rutherford (1871–1937), **Francis William Aston** (1877–1945), **Frederick Soddy** (1877–1956), **James Chadwick** (1891–1974)

Early work on radioactivity produced significant advances in both physics and chemistry and opened whole new fields of research. In one breakthrough, British physicists Frederick Soddy and Ernest Rutherford proved that radioactive elements turned into entirely different elements as they decayed. Later Soddy showed that the radium that Marie and Pierre Curie had isolated from uranium ores was actually produced by radioactive decay of the uranium atoms. This led to the idea that there is more than one form of an element, even though their chemical properties seem identical.

In 1913, at the suggestion of his friend Dr. Margaret Todd, Soddy proposed the term *isotope* for these alternate forms (from the Greek *iso-* meaning "equal" and *topos* meaning "place," since these alternate forms occupy the same place on **the periodic table**). An element (which has a fixed atomic number) could somehow have more than one atomic mass. British physicist J.J. Thomson later showed that even nonradioactive elements could have isotopes, and he and British chemist Francis William Aston provided proof through early **mass spectrometry**. This led Rutherford to propose that some neutral particle with almost the same weight as a proton might be making up the difference. (The neutron was experimentally detected in 1932, by British physicist James Chadwick.)

Isotopes have been of great importance in chemistry ever since. **Kinetic isotope effects** have given information about reaction mechanisms. The different magnetic behavior of various isotopes in **NMR** (nuclear magnetic resonance) experiments led to a huge branch of analytical chemistry, and **radioactive tracers** are used to diagnose and treat disease. As mass spectrometry has grown more powerful, the nonradioactive "cold" isotopes have been used more and more as distinctive labels for such experiments, instead of the radioactive ones.

SEE ALSO Polonium and Radium (1902), Mass Spectrometry (1913), Radioactive Tracers (1923), Deuterium (1931), Technetium (1936), The Last Element in Nature (1939), Gaseous Diffusion (1940), Kinetic Isotope Effects (1947), Transuranic Elements (1951), DNA Replication (1958), Lead Contamination (1965), Enzyme Stereochemistry (1975), PET Imaging (1976), Iridium Impact Hypothesis (1980), Olefin Metathesis (2005), Isotopic Distribution (2006)

An arrangement of isotope possibilities. The number of protons increases off into the distance, and the number of neutrons increases to the right. The dark-blue blocks in the middle are stable isotopes, while the lighter blue are unstable ones that have been observed. Gray blocks are possible isotopes that have never yet been seen to exist.

Chemical Warfare

François-Auguste-Victor Grignard (1871–1935), **Fritz Haber** (1868–1934), **Winford Lee Lewis** (1878–1943)

World War I was a catastrophe from any perspective, and it triggered a chain of disasters that disfigured the rest of the twentieth century. Within a relatively short time, and despite international agreements banning the use of poison and poison weapons in warfare, its participants found themselves doing things that once would have seemed unimaginable—such as opening up the stopcocks on over 3,700 gas cylinders and sending a deadly stream of chlorine off into the spring breeze.

That was the scene near Ypres, Belgium, in April 1915. The French and Germans had both experimented with tear gas the year before, but to little effect. The German chemical industry, though, had plenty of chlorine, and German scientists began researching ways of delivering it in battle. On April 22, 1915, 168 tons of it were deployed in a drifting green cloud that opened a huge hole in the French line. The German troops were unable to exploit it, however, fearing the effects of the gas themselves.

This scene repeated itself many times, but in progressively less effective ways. The Ypres attack was the biggest success that gas warfare achieved, if "success" is the right word. Despite the use of chlorine, the still more poisonous (and harder to detect) phosgene, and the awful, persistent mustard gas (actually an oily liquid), advances in protective gear and gas masks kept things at a stalemate. Chemical weapons did not end the war, or even speed it up; they just made it even more horrible.

The chemistry behind these weapons is depressing. Everyone involved devoted time and effort to finding the most poisonously effective compounds and the most lethal ways of delivering them. At least twenty different agents were deployed in battle by all sides, and their uses were supervised by eminent chemists like the German Fritz Haber and Frenchman François-Auguste-Victor Grignard. American chemists, among them Winford Lee Lewis, developed still worse compounds, such as the arsenic-containing lewisite. Chemical warfare was not important in World War II, but it has made several horrible comebacks nonetheless in both warfare and terrorism, usually in the form of **nerve gas**.

SEE ALSO Greek Fire (c. 672), Chlor-Alkali Process (1892), Nerve Gas (1936), Bari Raid (1943)

World War I was already sufficiently inhuman before poison gas was introduced. Here, soldiers on the frontlines protect themselves with gas masks in 1918.

Surface Chemistry

Irving Langmuir (1881–1957), Katherine Blodgett (1898–1979)

American chemist Irving Langmuir spent the first half of the twentieth century at General Electric, working on the frontier between chemistry and physics. Among his many discoveries, Langmuir is remembered best for his work in surface chemistry (the study of processes that occur at interfaces between phases), particularly the behavior of thin films.

Everyone has seen these—an oil slick on the surface of a puddle is a perfect example. And the rainbow colors that develop are a tip-off, to a trained observer, that the thickness of such a film must be down in the range of the wavelengths of the light bouncing off of it. In 1917, Langmuir published a key paper on the behavior of oily films that were made from molecules with some sort of polar group at one end (such as a long-chain alcohol or a fatty acid, as found in **soap**). He provided evidence that these arranged themselves so that the polar groups buried themselves in the first layer of water molecules, while the rest of the long chain arranged itself more or less vertically on the surface. In the 1890s, British physicist John William Strutt along with the self-taught German experimenter Agnes Pockels (who carried out her work in a kitchen sink) had measured how far a given amount of oil would spread itself out, and thus how thick the molecular layers must have been. Langmuir extended this work to an understanding of how the individual molecular structures influenced the films they formed.

He and his colleague, American physicist Katherine Blodgett, later studied how these molecular monolayers could be deposited on solid objects. These Langmuir-Blodgett films turned out to be extremely useful, providing key insights into what we now know as nanotechnology (manipulation of materials on an atomic or molecular level). In general, surface chemistry is the key to understanding things as (seemingly) far apart as the walls of blood vessels, the behavior of oil spills, the fabrication of silicon computer chips, and the durability of fresh paint.

SEE ALSO Soap (c. 2800 BCE), Cholesterol (1815), Liquid Crystals (1888), Synthetic Diamond (1953), Single-Molecule Images (2013).

As with a soap-bubble film, the colors of an oil slick on water provide precise bounds on how thick the oil film is (i.e., not very thick at all, just a few wavelengths of light).

Radithor

The early years of radioactivity research were quite alarming. Many researchers (like the Curies) were injured by exposure to the new concentrated radioactive elements, but these substances were also tested on all sorts of skin diseases, cancers, and other conditions as potential medicines. Legitimate uses were found (such as treatment of some kinds of skin cancer), but there were plenty of other people with less scientific—and less altruistic—motivations. The number of radioactive "medicines" produced in the early twentieth century is probably beyond counting. The new phenomenon, it was thought, just had to have something to do with human health, and surely most people just weren't getting enough radioactivity. Time to sell them some.

How about a radioactive water dispenser? Radioactive toothpaste? Radioactive skin cream? All of these and more were advertised. A famous example is Radithor elixir, and its label was (unfortunately) accurate—there was guaranteed radium in every bottle. Beginning in 1918, Radithor was sold as a general tonic and health enhancer, and a prominent businessman of the time, steel company owner Eben Byers of Pittsburgh, was glad to serve as its spokesman. He drank up to three bottles a day, and look how successful he was, with all that energy!

Unfortunately, radium is in the same column of **the periodic table** as calcium, and it also concentrates in the bones. The main **isotope** found in Radithor (radium-226) is an alpha-particle emitter, which means that its radiation could be shielded by a barrier as insubstantial as a sheet of paper—if it's outside your body. Ingested, though, it has every chance to do its worst. Within two years, Byers came down with terrible bone cancer, but since he had consumed hundreds of bottles of Radithor, the wonder was that he lasted as long as he did. He'd been irradiated from within. His death in 1932 (and his burial in a lead-lined coffin) brought overdue scrutiny to the radioactive-cure industry. Soon, Radithor was among the products the FDA used to promote the passage of the 1938 Federal Food, Drug, and Cosmetic Act. There was a compelling case to be made.

SEE ALSO Toxicology (1538), Polonium and Radium (1902), Elixir Sulfanilamide (1937), Thalidomide (1960)

LEFT: *Sadly, the label on this bottle of Radithor is completely accurate in every detail, which should make you want to back away from it quickly rather than drink it.* RIGHT: *Lithograph of a song-sheet cover for "The Quack's Song," composed by F. C. Burnand and W. Meyer Lutz, c. 1900.*

THE QUACK'S SONG

Dean-Stark Trap

Ernest Woodward Dean (1888–1959), David Dewey Stark (1893–1979)

The Dean-Stark trap is a classic piece of chemical glassware, and an ingenious solution to a common problem in organic synthesis. It was invented in 1920 by American chemists Ernest Woodward Dean and David Dewey Stark, who worked in the Bureau of Mines's research station in Pittsburgh, studying the amounts of water in crude oil.

Many reactions are formally known as condensations, in which two reactants combine with the loss of a molecule of water, such as the formation of a peptide bond from two **amino acids**. All chemical reactions are reversible, though, in theory, and if the energies of the starting materials and the products are similar, they can shuttle back and forth endlessly once they reach their point of equilibrium. But if you could find some way to remove the water that's being formed from a condensation, the two reactants would then form more to replace it, and even a reaction that reverses very easily would eventually run to completion.

That's what a Dean-Stark trap does. If you run your reaction in solvents like benzene or toluene, you can take advantage of an azeotrope, a "constant-boiling mixture," where two solvents boil off together in an unchanging ratio. Toluene and water, for example, don't mix well. But they distill out as vapor that's 80 percent toluene by weight, and when this cools and condenses, the water separates back out as a bottom layer. The Dean-Stark trap allows this water to collect at the bottom of the tube—meanwhile, the excess toluene (or other azeotrope-forming solvent) gets a chance to run back into the original reaction flask, where it can boil off and take even more water with it. Eventually, all the water ends up in the trap, and the reaction is driven to the end (**Le Châtelier's principle** in action).

Every organic chemistry lab has a Dean-Stark trap for use in driving condensation reactions forward. It's the only reason that its inventors are remembered. Both of them had long careers in oil chemistry, but nothing brought them fame like the piece of glassware they invented together when they were still young researchers.

SEE ALSO Separatory Funnel (1854), Erlenmeyer Flask (1861), Soxhlet Extractor (1879), Le Châtelier's Principle (1885), Borosilicate Glass (1893), The Fume Hood (1934), Magnetic Stirring (1944), Glove Boxes (1945), Rotary Evaporator (1950), Isoamyl Acetate and Esters (1962).

A classic Dean-Stark trap after the flask has already been heated for several hours. Note the water layer in the trap section to the left.

1920

Hydrogen Bonding

Worth Huff Rodebush (1887–1959), **Wendell Mitchell Latimer** (1893–1955), **Maurice Loyal Huggins** (1897–1981)

Hydrogen bonds are the secret adhesive of the living world. They hold together the strands of DNA, help to determine the shapes of proteins, and occur in every kind of carbohydrate molecule. The active sites of receptors and enzymes, for example, almost invariably feature key hydrogen bonds in the protein's own structure and with the substrate molecules that bind there.

American chemist Maurice Loyal Huggins was the first to suggest the concept of hydrogen bonds, and his work inspired his colleagues Wendell Mitchell Latimer and Worth Huff Rodebush to publish a 1920 paper that used hydrogen bonds to explain properties of certain liquids. Almost one hundred years of work since then have still not revealed all their secrets, though.

So what's a hydrogen bond? That's not such an easy question to answer. Even the best minds (such as American chemist Linus Pauling) have found plenty to occupy them here. Hydrogen bonding is partly just the attraction between a positively charged hydrogen atom and a negatively charged atom, such as nitrogen or oxygen, on a nearby molecule. These don't have to be full charges. Oxygen and nitrogen atoms usually have extra electron density, making them clumps of partial negative charge. But this isn't just an ionic bond, because hydrogen bonds are directional—if they're not pointed in the right way, the attraction mostly disappears. It's like a ghostly form of a standard single bond, and it's strongest when the hydrogen itself is attached to an electron-rich atom like oxygen as well. Such oxygen-hydrogen and nitrogen-hydrogen compounds are found over a huge range of chemistry, and they're especially crucial in the behavior of many biomolecules.

Water is the best example, with two hydrogen atoms attached to a single oxygen. Water molecules are very good hydrogen bond donors and acceptors at the same time, which is what makes it such a weird substance. It has a much higher boiling point than such a tiny molecule should, and it freezes into a hydrogen-bonded crystal lattice that's actually less dense than the liquid. (Most liquids don't have ice that floats.)

SEE ALSO Hydrogen Sulfide (1700), Hydrogen (1766), *The Nature of the Chemical Bond* (1939), Alpha-Helix and Beta-Sheet (1951), DNA's Structure (1953), Polywater (1966), Computational Chemistry (1970), Polymerase Chain Reaction (1983), Recrystallization and Polymorphs (1998)

Hydrogen bonding is essential to the properties of water, which are essential to life on Earth.

Tetraethyl Lead

Thomas Midgley Jr. (1889–1944), Charles Franklin Kettering (1876–1958)

In the 1920s, automobile engines needed improvement. Their fuel had to burn more evenly so that their engines could run at higher compression, but attempts to realize this with the gasoline of the day led to premature ignition in the cylinders, the effect known as "knocking." A number of solutions were looked at—changes in oil refining, improvements in engine design, additives to the existing gasoline. The last was by far the easiest, and in 1921 American chemist Thomas Midgley Jr. at General Motors, working with Charles Kettering, discovered that the addition of tetraethyl lead (TEL) was a very effective anti-knock solution. At high temperatures, it released **free radicals** into the burning mixture, which significantly improved the combustion process.

The problem was that it also released lead, known for a long time as a danger to human health. Sales of Ethyl, the brand name given to the product, were suspended in 1925 after several deaths during its manufacture, but before its suspension, a 1924 press conference featured a particularly bizarre scene. Midgley maintained that the chemical was basically harmless, holding some under his nose and then washing his hands in it in front of reporters. In fact, he had already shown lead poisoning symptoms and had to recover from this incident as well.

The question became whether the small amounts in leaded gasoline were dangerous. Several reviews of TEL's effects concluded that there was no evidence that it was toxic in those concentrations—indeed, there are many substances that have a "no effect" level, where the body can clear them with no apparent harm. As slowly became clear, however, lead is not one of these substances. It accumulates in the body on repeated exposure, even at very low concentrations, and even those low levels have a real, detrimental effect on human health, particularly on the development of children. American geochemist Clair Cameron Patterson's work in the 1950s and 1960s on **lead contamination** was clinching evidence that TEL needed to be banned, a process that began in the industrialized world in the 1970s (improvements in **catalytic reforming** helped as well). As for Midgley, he went on to invent **chlorofluorocarbons**.

SEE ALSO Toxicology (1538), Free Radicals (1900), Catalytic Cracking (1938), Catalytic Reforming (1949), Thallium Poisoning (1952), Lead Contamination (1965).

Tetraethyl lead was an essential ingredient in high-performance aviation fuels during World War II and afterward, although the lead itself tended to foul the spark plugs.

Acids and Bases

Thomas Martin Lowry (1874–1936), Gilbert Newton Lewis (1875–1946), Johannes Nicolaus Brønsted (1879–1947)

The year 1923 saw the publication of two general theories about the nature of acids and bases. Danish chemist Johannes Nicolaus Brønsted and English chemist Thomas Martin Lowry, working independently, both defined acids as compounds that gave up hydrogen atoms and bases as compounds that accepted them. Now known as the Brønsted-Lowry acid-base theory, this definition was more general than that formulated by Swedish scientist Svante Arrhenius, who had thought of everything in terms of hydrogen ions and hydroxide ions in aqueous solutions. But American chemist Gilbert Newton Lewis flipped things around the same year in a paper that extended the concept beyond what anyone had imagined. Instead of positively charged hydrogen ions being the currency of exchange, he looked at acids as being acceptors of negatively charged electron pairs and bases as donors of them.

This idea came from his definition of a chemical bond, which was the sharing of two electrons between two atoms. A covalent bond, in Lewis terms, was what you had when each atom contributed an electron on equal terms, and a "coordinate" or "dative" bond was when one of the atoms donated both electrons to the new bond. The prototype Lewis acid is boron trifluoride—with three fluorines bound to it, the boron atom in the middle is parched for electrons and will vacuum up electron pairs wherever it can find them, tightly coordinating on atoms (like oxygen or nitrogen) that have them to spare.

This meant that "Lewis acids" could behave similarly to more traditional acids in many reactions, despite having no hydrogen atoms to donate. (Strong ones aren't much good in aqueous solution, though, since they go berserk sticking to all the water molecules and become mostly inactivated.) Lewis-acid based reagents are used in **catalytic reforming** in the oil industry, the **Friedel-Crafts reaction**, **Ziegler-Natta catalysis** for plastics, and more. Interestingly, while many electron-deficient metal compounds are good Lewis acids, they all seem to have slightly different personalities, with different affinities for oxygen and other atoms. A wide range of them can be tried to see which ones make a reaction go most quickly and cleanly.

SEE ALSO Aqua Regia (c. 1280), Sulfuric Acid (1746), Hydrogen Cyanide (1752), Friedel-Crafts Reaction (1877), pH and Indicators (1909), Catalytic Reforming (1949), Ziegler-Natta Catalysis (1963)

Lewis in his lab at Berkeley, where he was working when he died in 1946. He is widely considered one of the most influential chemists never to have won the Nobel Prize.

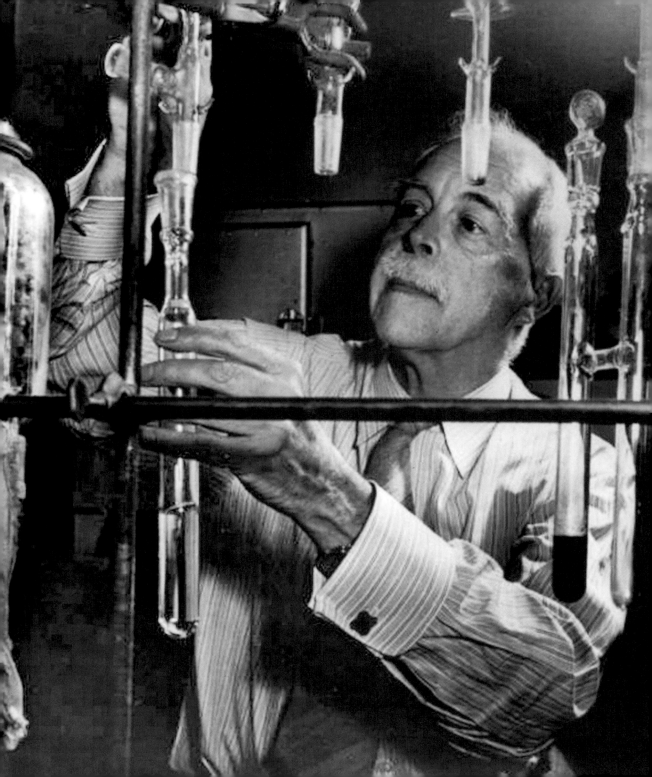

Radioactive Tracers

George Charles de Hevesy (1885–1966)

Radioactive elements have chemistry of their own, and one of their most well-known uses is as radioactive tracers, which allow chemists to track substances in living systems—an ability that revolutionized biology and biochemistry. The reactions, flow, and distribution of all sorts of biomolecules can be followed if radioactive forms of them are made by substituting their various atoms with radioactive ("hot") isotopes, whose distinctive signatures can be precisely measured. For example, tritium is often swapped for plain **hydrogen**, and carbon-14 for carbon-12. There are similar radioisotopes for sulfur, phosphorus, oxygen, and most other elements.

As the story goes, Hungarian chemist George Charles de Hevesy, who had begun to study this tracer technique with inorganic metal compounds, used it in 1912 to prove that his landlady in Manchester was recycling leftover food from his plate. He supposedly radiolabeled his leftover meat one night at dinner and brought an electroscope to the meal the next day to prove that the evening's hash was now mildly radioactive. In 1923 he fed a tracer compound to living plants with great success.

Radiolabeling is rarely used on leftovers these days (one hopes), but it is used to track the metabolism of drug molecules, unravel the synthetic pathways of natural products, and measure the uptake of soil components by plants, among countless other applications. In recent years, advances in **mass spectrometry** have made it easier to detect subtle differences in atomic mass, which has led to more use of "cold" isotopic labels, which show up only because of their unusual molecular weights. But de Hevesy was there first, too. He and his lab partner drank measured amounts of diluted "heavy water" (nonradioactive deuterium oxide) and then collected and distilled their own urine to see how much water the body contained and how quickly it was exchanged. Perhaps he did slip some isotopes into his leftover roast beef, after all!

SEE ALSO Aqua Regia (c. 1280), Polonium and Radium (1902), Mass Spectrometry (1913), Isotopes (1913), Deuterium (1931), Technetium (1936), Enzyme Stereochemistry (1975)

Radioactive tracers injected into the body can be used to generate gamma scans like this one, which help to detect bone tumors.

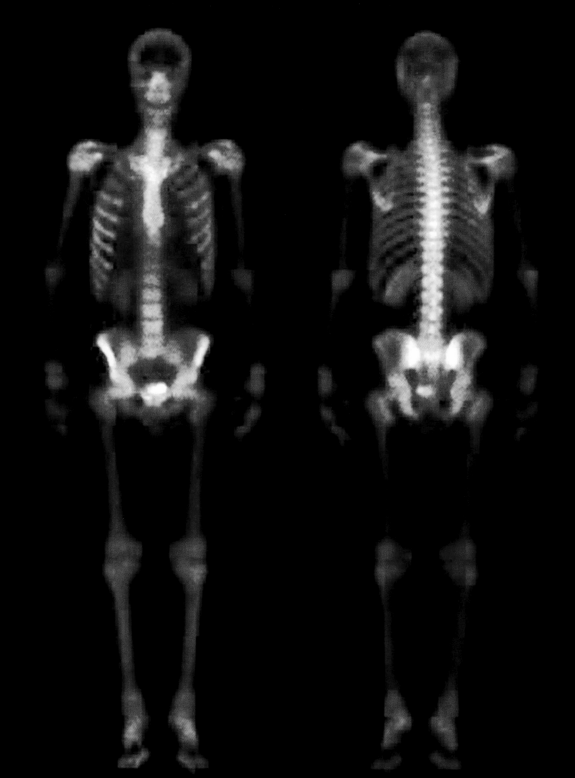

Fischer-Tropsch Process

Franz Fischer (1877–1947), **Hans Tropsch** (1889–1935)

In 1925, German chemists Franz Fischer and Hans Tropsch at the Kaiser Wilhelm Institute for Coal Research (now the Max Planck Institute for Coal Research), patented the process that bears their names. It takes carbon monoxide and **hydrogen** (two gases) and turns them into a stream of liquid hydrocarbons: oil and gasoline. Since the starting materials can be produced from natural gas or coal, the process has been used on an industrial scale in countries with few oil deposits but substantial reserves of those feedstocks, most notably Germany and South Africa.

During World War II, Germany set up a number of Fischer-Tropsch plants, although these still proved inadequate to their enormous needs. The synthetic fuels were used for tanks and trucks, and for freeing up oil refineries to produce the high-octane fuels needed for aircraft. In later years, South Africa faced economic boycotts from much of the world during its apartheid era and developed a great deal of expertise in making hydrocarbon fuels from its own coal. South African modifications to the Fischer-Tropsch process are still used.

The process relies on metal catalysts (usually iron or cobalt) at relatively high temperatures and has been improved many times since its discovery. Different conditions can produce different mixtures of hydrocarbons, allowing for some fine-tuning depending on which fractions are needed. Even under optimized conditions, the overall efficiency is only around 50 to 60 percent at best, which makes it very difficult to compete with production from oil reserves. It's also a relatively costly process to run, so it's a technology that's economically attractive only in special situations. Natural gas discoveries far from any pipelines or local customers have attracted Fischer-Tropsch plants in recent years, allowing condensed liquid fuels to be shipped out profitably. It will probably need an efficiency breakthrough of some sort to go into wider use.

SEE ALSO Thermal Cracking (1891), Catalytic Cracking (1938), Catalytic Reforming (1949)

Work on new Fischer-Tropsch catalysts continues to this day, as seen here in this fully automated, around-the-clock synthetic-fuels research lab in Perth, Australia.

Diels-Alder Reaction

Otto Paul Hermann Diels (1876–1954), Kurt Alder (1902–1958)

German chemist Otto Paul Hermann Diels and his student, Kurt Alder, reported a reaction in 1928 that was to make them famous, change the course of organic chemistry, and win them a Nobel Prize. It would also eventually feature in every second-year chemistry course in the world, so that people who've never had any further contact with the science may still have vague memories of the Diels-Alder reaction.

It was a strange transformation, and it immediately attracted the attention of both bench chemists and theoreticians. A four-carbon compound with two alternating double bonds, a diene, reacts with a two-carbon double-bond compound to form a new six-membered ring. Even in the early days, a chemist could draw **reaction mechanisms** to show where the two new bonds came from (and why the product ended up with a leftover double bond in the middle of the diene part), but why the reaction should happen at all was a mystery. Not only did it take place reliably (and across an impressive variety of starting materials), but it tended to do so with precision, favoring certain isomeric products in patterns that were surely meaningful but did not (for many years) make much sense. Factors that had large effects on some other reaction types, such as the physical crowding of the molecules involved, seemed to trouble the Diels-Alder process far less, and it was also strangely unresponsive to what sort of solvent it was run in.

The reaction turned out to work by overlapping of the electron clouds of each partner molecule, the highest-occupied and lowest-unoccupied molecular orbitals blending into a new bonding arrangement. Explaining this process led to a great deal of knowledge about bond making and bond breaking, and eventually to a general theory of "frontier molecular orbitals" that covers a wide class of reactions. The original Diels-Alder reaction is still an extraordinarily fast way to build complex ring systems and has been used in natural-product synthesis and in pharmaceuticals, where multiple ring systems occur often.

SEE ALSO Sigma and Pi Bonding (1931), Transition State Theory (1935), Reaction Mechanisms (1937), Dipolar Cycloadditions (1963), Woodward-Hoffman Rules (1965), B_{12} Synthesis (1973), Unnatural Products (1982)

A diagram of the Diels-Alder reaction appears on this redbrick walkway at the University of Kiel's Otto Diels Institute of Organic Chemistry in Germany.

Reppe Chemistry

Walter Reppe (1892–1969)

Born in Germany, Walter Reppe was a first-rate industrial chemist, but he was also a brave man. We know this because he developed a whole suite of large-scale reactions that used **acetylene** as a starting material, often running under high temperatures and pressure. This is not something to try at home—in fact, when Reppe began, his employer (the German chemical producer BASF) had standing orders forbidding anyone to compress acetylene, for the very good reason that it was ferociously explosive under these conditions. Reppe decided, though (as he put it later), "to break with all delivered opinions," because he thought that a lot of useful chemistry was waiting to be discovered.

He was right. But he had to build special equipment to handle the pressure and learn by experimentation how far acetylene could be pushed before it became deadly. Diluting the gas with extra (inert) nitrogen was a key safety measure, and eventually Reppe and his team were able to produce a variety of compounds where the acetylene triple bonds ended up as vinyl groups (unsubstituted carbon-carbon double bonds). Production of vinyl ethers, acrylic acids, and more was made industrially feasible during the 1930s, and this led to new polymers such as Plexiglas®. Four acetylenes could even be cyclized to give an eight-membered ring with four double bonds (cyclooctatetraene), which was predicted to be a non-aromatic compound by Hückel's pi-electron rules, and indeed was not.

Most of Reppe's chemistry has become outmoded, because acetylene itself has become outmoded as an industrial feedstock. It's much easier to come by from coal than from oil, and oil is now the starting material of choice for most applications. There's still no better way to make cyclooctatetraene, though, which is a component of several industrial catalysts. But Reppe's technique of adding a carbon monoxide molecule to acetylene (carbonylation) has been extended to many other starting materials and catalysts and is still in wide use today.

SEE ALSO Benzene and Aromaticity (1865), Acetylene (1892), Sigma and Pi Bonding (1931)

Acrylic polymers are used for transparent, underwater tunnels and many other demanding applications. Reppe's team pioneered many acrylic acid–containing materials.

1930

Chlorofluorocarbons

Charles Franklin Kettering (1876–1958), Thomas Midgley Jr. (1889–1944), Albert Leon Henne (1901–1967)

Few remember it now, but refrigerators used to be able to kill you. Virtually all commercial refrigeration units run on the cooling and heating cycle provided by the expanding and compressing of some working gas, but back in the 1920s, the choices for gases were a bit rough. Ammonia was widely used, as was sulfur dioxide, but both of those are quite toxic, not to mention corrosive to the refrigeration equipment itself. Propane was found in some models, but it is, of course, extremely flammable in the case of a leak. Something nonpoisonous and non-explosive was needed, and a General Motors research team under Charles Kettering was one of several looking at the possibilities.

American chemist Thomas Midgley Jr., the respected discoverer of **tetraethyl lead**, and his colleague, Albert Leon Henne, were put on the problem. They realized that halogenated carbon compounds had a lot to recommend them, since they were very hard to ignite. Even better, the smaller ones had convenient boiling points for refrigerant gases and were very compressible. The best properties were to be found in the compounds that mixed chloro- and fluoro-substituents, such as the nontoxic and noncorrosive dichlorodifluoromethane, which they first prepared in 1930. The company gave it the trade name *Freon* (later their base name for a whole series of similar compounds), and it was a hit. Soon it was being used as an inert propellant in spray cans and later in asthma medicine inhalers.

Midgley received many honors for his invention of chlorofluorocarbons (CFCs), including the distinguished awards named after English chemists Sir William Henry Perkin and Joseph Priestley, along with election to the National Academy of Sciences. No one realized yet the havoc CFCs were causing in the upper atmosphere, or the human costs tetraethyl lead was exacting closer to the ground. So although Midgley was an extremely successful and inventive research chemist, he may well have been the cause of more damage to the Earth's atmosphere than any single human being in history. The extent of that damage was apparent only decades after his own death.

SEE ALSO CFCs and the Ozone Layer (1974).

A young couple standing next to a refrigerator that must have looked quite modern at the time. With the round compressor on top, this refrigerator used either sulfur dioxide or methyl formate — both quite toxic — for cooling, but it would soon be replaced by a new breed using Freon.

Sigma and Pi Bonding

Erich Armand Arthur Hückel (1896–1980)

Carbon-carbon bonds come in more than just the tetrahedral single-bond arrangement. There are double bonds as well, and they have their own geometry. These lie in a flat plain, with the attached groups pointed out at about 120 degrees. And while carbon-carbon single bonds allow their groups to spin around like drive shafts, the double bonds, alkenes, don't let that happen at all. It's as if some sort of locking device prevents them from rotating.

German physicist and chemist Erich Armand Arthur Hückel developed a theory to explain why this is so. The two pairs of electrons involved in a double bond, he said in a landmark 1931 paper, were of two different kinds ("sigma" and "pi" electrons). Carbon-carbon single bonds were all sigma, and spent their time localized between the two atoms. But the pi electrons, Hückel calculated, formed two "clouds" above and below the sigma bond, and they weren't interchangeable. In a molecule like benzene, with alternating single and double bonds, the pi electrons blended together into doughnut-like rings above and below the plane of the molecule. His theory predicted that cyclic compounds with alternating double bonds would show aromaticity if the number of pi electrons was a multiple of four, plus two (benzene qualifies with six).

These ideas were quite powerful, and they explained a good deal of puzzling molecular behavior. The "pi clouds" around aromatic compounds affect how they interact with other molecules during drug binding and in other biological systems, for example. But Hückel's insights were not widely applied for some years (some think that his work was expressed in too mathematically rigorous a style for organic chemists to appreciate). Better and more precise modifications to Hückel's original work have appeared in the decades since his papers, especially as applied to **computational chemistry**, but the original ideas are as strong as ever.

SEE ALSO Benzene and Aromaticity (1865), Tetrahedral Carbon Atoms (1874), Diels-Alder Reaction (1928), Reppe Chemistry (1928), Computational Chemistry (1970), Single-Molecule Images (2013)

This computer-generated model of a benzene molecule shows pi orbitals as purple grids.

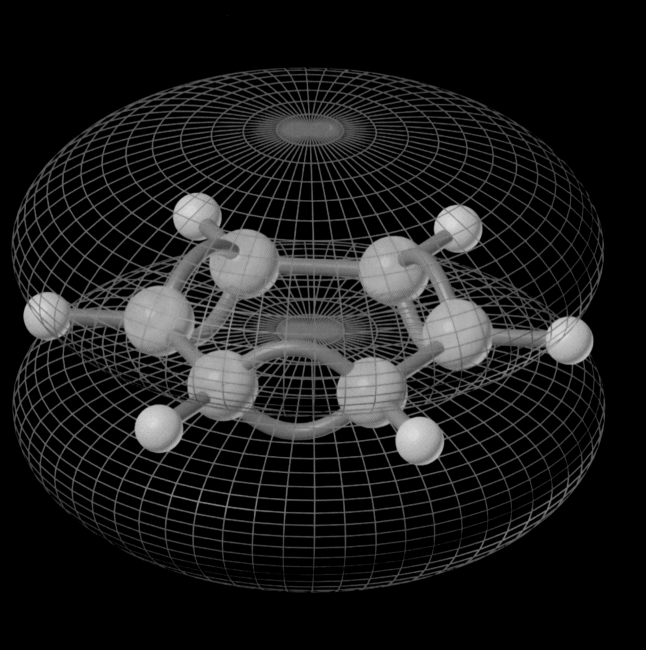

Deuterium

Gilbert Newton Lewis (1875–1946), **Harold C. Urey** (1893–1981), **Ferdinand Brickwedde** (1903–1989)

Deuterium ("heavy **hydrogen**") is all around us, but you have to look closely. With a single extra neutron, this **isotope** is perfectly stable (as opposed to hydrogen with two neutrons added, which is radioactive tritium). Almost all the deuterium in the universe was probably made during the big bang. The only nuclear process that produces it in any amount is hydrogen fusion inside stars, but that deuterium gets used up faster than it's made. Here on Earth, there are only about 150 deuterium atoms for every million ordinary hydrogen.

It was the improved mass spectrometer invented by British physicist and chemist Francis William Aston that uncovered the presence of deuterium in the 1920s. Its measured weight for hydrogen was a tiny (but real) bit lower than the weight found by chemical means, which suggested that normal hydrogen was contaminated with a bit of a heavier isotope. The name *deuterium* (from the Greek word for *second*) was suggested by American chemist Gilbert Newton Lewis, and Lewis himself was the first to prepare "heavy water," or D_2O. American physicist Ferdinand Brickwedde concentrated deuterium by a very tedious evaporation of liquid hydrogen in 1931, and his colleague, American chemist Harold C. Urey (formerly Lewis's student), detected it spectroscopically as deuterium's emission spectrum grew stronger. The heavier deuterium atoms tended to stay behind, so the liquid became steadily more enriched (the same technique was used to isolate the heavier isotopes of **neon** detected by British physicist Joseph John Thomson in the first **mass spectrometry** experiment in 1913).

Deuterium was discovered seven weeks before British physicist James Chadwick discovered the neutron, but its existence made the neutron seem almost mandatory. Urey was awarded a Nobel Prize for his work, but this caused a bitter rift with Lewis, who felt (with some justification) that he'd been passed over. Deuterium is now used in everything from nuclear weapons to **kinetic isotope effects**.

SEE ALSO Fractional Distillation (c. 1280), Hydrogen (1766), Flame Spectroscopy (1859), Mass Spectrometry (1913), Isotopes (1913), Radioactive Tracers (1923), Kinetic Isotope Effects (1947), Enzyme Stereochemistry (1975)

A Hubble Space Telescope image showing a portion of the Small Magellanic Cloud, a dwarf galaxy roughly 210,000 light-years from our own. Astronomers continue to use deuterium as a tracer for star and galaxy evolution.

Carbonic Anhydrase

William C. Stadie (1886–1959), **Francis John Worsley Roughton** (1899–1972), **Norman Urquhart Meldrum** (1907–1933)

When **carbon dioxide** dissolves in water, it can also react to form carbonic acid, which, in turn, can revert into water and carbon dioxide. That's not a particularly fast reaction in either direction—which is one reason that carbonated drinks don't just suddenly blow off all their CO_2 and go flat—but it becomes a speedy "diffusion controlled" reaction (with a rate that matches the time it takes for the individual molecules to get in and out of the enzyme's active site) in the presence of carbonic anhydrase, one of the fastest enzymes known. British chemists Norman Urquhart Meldrum and Francis John Worsley Roughton and American chemists William C. Stadie and Helen O'Brien discovered the enzyme almost simultaneously in 1932, solving a longstanding mystery about how carbon dioxide behaves in the blood. How could it be circulated and then released so quickly for exhalation by the lungs?

The answer to this mystery was in the extreme speed and versatility of carbonic anhydrase, which is present in large amounts in red blood cells. If there's a high concentration of carbon dioxide in the blood, the enzyme will produce bicarbonate, and if there's a lot of bicarbonate, the enzyme will turn it back into carbon dioxide gas. In addition to exchanging the gas into the lungs, both reactions help cells regulate pH (among many other functions), and its role in respiration has led to inhibitor drugs for it being prescribed for altitude sickness.

But carbonic anhydrase is well liked by chemical biologists and enzyme engineers for other reasons. It's easily (and cheaply) available in pure form and very stable on storage. Its structure has been thoroughly mapped out by **X-ray crystallography**, and its mechanism of action is very well understood. A zinc atom in the active site plays a key role, which made carbonic anhydrase the first confirmed "metalloenzyme," a very important class of proteins. A huge variety of inhibitor compounds are available, with many of their binding modes also determined by X-ray studies.

SEE ALSO Carbon Dioxide (1754), Amino Acids (1806), X-Ray Crystallography (1912), Sulfanilamide (1932), Protein Crystallography (1965), Enzyme Stereochemistry (1975), Engineered Enzymes (2010)

The alveoli of the lung: the small spaces where the exchange of oxygen and carbon dioxide takes place. The outflow of carbon dioxide is regulated by carbonic anhydrase in the bloodstream.

Vitamin C

James Lind (1716–1794), **Albert Szent-Györgyi** (1893–1986)

Scurvy was described as long ago as the fifth century BCE and was especially common in groups of soldiers and sailors on long journeys. Diet was long suspected as a factor, and even in the fifteenth century, the therapeutic effects of citrus fruit were already known. In 1747, Scottish physician James Lind ran possibly the first controlled clinical trial, testing his idea that scurvy was cured by acids. He fed scurvy-afflicted, scorbutic, sailors various acidic foods, but only the group fed citrus fruit responded (vinegar was not very effective!).

Diet was still not thought of as a cure by itself for decades, but even after the acceptance of citrus as a preventative, its active substance was unknown. In the 1920s, Hungarian physiologist Albert Szent-Györgyi was studying the enzymes that caused the browning of cut fruits and noticed that lemon juice stopped the process. Dilute acid didn't, though, and he made the intuitive leap that the same anti-scurvy compound might be responsible. Painstaking bench work, with this reaction as an indicator, led him to small amounts of a polar six-carbon acid compound that he called ascorbic, "no-scurvy, "acid, now known as vitamin C, a key player in numerous biochemical pathways. It is vital for an enzyme that synthesizes collagen protein (part of the structure of cartilage and blood vessels) and for the functioning of the immune system.

Szent-Györgyi turned to guinea pigs as an animal test subject, since (like humans) they cannot make their own vitamin C. Ascorbic acid did indeed heal diseased animals, but he wasn't able to isolate enough of it to conduct further testing—citrus fruits are full of similar compounds that wouldn't allow a clean separation. Szent-Györgyi later said he came home one night in 1932 in despair, and his wife served him some good Hungarian red peppers. He realized that peppers were one of the few sources he hadn't tested, and back in the lab, he discovered by midnight that they were full of easily purified ascorbic acid. Within three weeks, he had three pounds of it. And in 1937, he had the Nobel Prize.

SEE ALSO Wöhler's Urea Synthesis (1828)

Crystals of vitamin C in a polarizing microscope at 250× magnification.

Sulfanilamide

Ernest Fourneau (1872–1949), **Gerhard Domagk** (1885–1964), **Fritz Mietzsch** (1896–1958), **Josef Klarer** (1898–1953), **Daniel Bovet** (1907–1992)

People remember penicillin as the antibiotic "wonder drug," but the sulfa drugs, particularly sulfanilamide, were there first, and they created a sensation when they were discovered.

That discovery was not straightforward. German pathologist and bacteriologist Gerhard Domagk and his team, chemists Josef Klarer and Fritz Mietzsch, at German pharmaceutical giant Bayer had noticed how some dye molecules stained bacteria, and thought that there must be something about their structures that targeted the bacterial membranes. A search through hundreds of dyes (and related compounds) turned up a bright red one, in 1932, that showed weak activity in mice. This lead structure was modified into a series of analogs, one of which was truly potent. After human tests, it came on the market as Prontosil, the first broad-spectrum antibiotic ever found. But it worked only in people and animals; it did not kill bacterial cultures in the lab, although it did stain them red.

It would stain other things as well. During this testing period, Domagk's daughter, Hildegarde, became dangerously ill with a streptococcal infection from an accident with an embroidery needle. He gave her a large dose of Prontosil, which saved her arm and probably her life, but at the cost of staining her skin with a reddish hue that never disappeared. A French research team led by chemists Ernest Fourneau and Daniel Bovet shorly discovered that Prontosil was not the real drug. It was being cleaved inside the body to a much smaller active compound, sulfanilamide, which was cheap (and colorless, so the staining side effect was never a problem again). Sulfanilamide went into immediate use, followed swiftly by countless derivatives and analogs. "Sulfa" was the frontline antibiotic of World War II and was given to Winston Churchill during a bout of pneumonia, but bacteria were already starting to develop resistance to the whole class of drugs, due to the survival advantage of mutant enzymes that sulfanilamide did not inhibit.

Domagk was awarded the Nobel Prize in Physiology or Medicine in 1939, but he was arrested by the Gestapo for attempting to accept it. He finally got his medal in 1947.

SEE ALSO Perkin's Mauve (1856), Indigo Synthesis (1878), Salvarsan (1909), Carbonic Anhydrase (1932), Elixir Sulfanilamide (1937), Streptomycin (1943), Penicillin (1945), AZT and Antiretrovirals (1984), Modern Drug Discovery (1988), Taxol (1989)

A tube of Prontosil tablets, the first commercial "sulfa" antibacterial drug, which became available in the 1930s.

Polyethylene

Reginald Oswald Gibson (1902–1983), **Michael Wilcox Perrin** (1905–1988), **Eric Fawcett** (1927–2000)

The year 1933 marks the first industrial synthesis of polyethylene, but (unfortunately) not the first reliable one. It had originally been prepared in 1898 in an accident during German chemist Hans von Pechmann's work with pure **diazomethane**. No one was foolhardy enough to work with the explosive and toxic diazomethane on a larger scale, so this remained a chemical footnote until British chemist Reginald Oswald Gibson and British-Canadian physicist Eric Fawcett tried a high-pressure, high-temperature reaction between ethylene gas and benzaldehyde (the same compound that inspired the work of German chemists Friedrich Wöhler and Justus von Liebig on functional groups back in 1832). The white, waxy polymer they produced turned out to be all long chains of CH_2 (methylene) groups from the polymerized ethylene, and, with its resistance to chemicals and solvents and its malleability, it seemed as if it could be a very useful material.

Getting the reaction to work reproducibly, however, was very frustrating until British chemist Michael Wilcox Perrin figured out the right conditions in 1937. Traces of oxygen, as it happened, caused that first reaction's accidental success, and later routes used small amounts of more reliable **free radical** initiators under milder conditions. Polyethylene became a secret material of World War II when its use as an insulator in electronics (such as radar equipment) was discovered. By the end of the war, it was being made on a large scale, and its many forms (hard blocks, thin sheets, flexible panels) were beginning to be appreciated.

Today polyethylene is the most common plastic polymer in the world. Depending on how it's made (how long the chains are, whether any branching compounds are added to the mix, and so on), it can take on a huge variety of properties, from flexible (low-density polyethylene, or LDPE) to rigid (high-density polyethylene, or HDPE). Hundreds of millions of tons are made each year, and it's found in products as varied as squeeze bottles, trash bags, sporting goods, and toys. Research on it is still continuing—all in all, an impressive performance for something whose first preparations were all mistakes.

SEE ALSO Polymers and Polymerization (1839), Diazomethane (1894), Free Radicals (1900), Bakelite (1907), Nylon (1935), Teflon (1938), Cyanoacrylates (1942), Ziegler-Natta Catalysis (1963), Kevlar (1964), Gore-Tex (1969).

Versatile polyethylene is incorporated into countless products and materials, including puncture-resistant fencing gear.

Superoxide

Linus Carl Pauling (1901–1994), **Rebecca Gerschman** (1903–1986), **Edward W. Neuman** (1904–1955), **Irwin Fridovich** (b. 1929), **Joseph McCord** (b. 1945)

American chemist Linus Pauling's thoughts on the nature of chemical bonding led him, in 1931, to propose that a century-old formula for a class of **oxygen** compounds was wrong. It was known that if the alkali metals were burned in oxygen (a vigorous reaction), tetroxides were formed, such as K_2O_4. But Pauling realized that the product of this reaction could contain an O_2 anion (a negatively charged ion)—an oxygen molecule that had picked up an extra electron—and that the real formulas must be KO_2, NaO_2, and so on, for the rest of the alkali metals. He proposed the name *superoxide* for this species, and in a 1934 publication, his coworker, American chemist Edward W. Neuman proved him right by showing that potassium superoxide had the magnetic properties of **free radicals** (with that extra unpaired electron).

What no one realized was that superoxide wasn't just a curiosity of inorganic chemistry—it was an essential part of life for every oxygen-breathing creature. In 1954, American biochemist Rebecca Gershman proposed that superoxide might be active in living systems, and in the early 1960s, American biochemist Irwin Fridovich likewise proposed that superoxide was a free-floating species inside cells but was met with what he called "extreme skepticism." Then in 1968, American biochemist Joseph McCord discovered an enzyme whose only purpose was to scavenge it. The enzyme had actually been known for thirty years, but scientists did not understand what it did or why it was so abundant. McCord renamed it *superoxide dismutase*, and he and Fridovich went on to establish a whole new area of cell biology around it.

Superoxide is one of the most important reactive oxygen species (ROS) in cells, along with the hydroxy radical and hydrogen peroxide. These are known to be capable of causing much damage to biomolecules, which has been proposed as a mechanism for aging. But recent work has shown that ROS are essential for normal cell function and are responsible for the beneficial effects of exercise (i.e., the small amounts of ROS damage during such stress signals the muscle cells to grow and metabolism to increase).

SEE ALSO Oxygen (1774), Free Radicals (1900), Cellular Respiration (1937)

Superoxide salts are used in oxygen generators in closed environments like submarines.

The Fume Hood

1934

Over the centuries, chemistry laboratories—and in the early days, "laboratory" could translate to "kitchen" or "shed out back"—have not always featured the cleanest air. Modern ones, though, have ventilated workspaces called *fume hoods*, large enclosures with bench space and a sliding door or sash in front, with fans constantly pulling in fresh air from the room via an exhaust system. This clears the air of any vapors, and if most of the chemistry is done inside the hood to start with, the fumes never escape into the room. The sliding sash also provides protection in case a reaction goes vigorously wrong.

This is one reason why today's chemists find old lab pictures so alarming—working without forced-air ventilation seems primitive now. Several early chemists (those working on the **isolation of fluorine**, for example) would likely have been spared injury—even death—had they been able to perform their experiments in some type of fume hood. But these were not often used until the middle of the twentieth century, and the ones that existed were often built to order, sometimes by the chemists themselves. Thomas Edison is an early example of someone who recognized the unhealthy nature of some of his experimental by-products and attempted to ventilate the fumes through the fireplace chimney, sometimes even standing inside a room and working on a purpose-built shelf outside a window.

Fume hoods began to be standardized and manufactured in the 1930s, and as new buildings were put up, more and more exhaust fans began to appear on the roofs of the labs. (Looking for these is a good way to pick out a research building even now.) If the hoods fail in only one part of a building due to a power outage, the lab doors can become hard to open against the wind.

The other crucial use for a fume hood, of course, is to write chemistry structures on the glass of the front sash with markers. It's a safe bet that every working chemistry lab in the world has plenty of ideas (good and bad) scribbled on its hoods!

SEE ALSO Separatory Funnel (1854), Erlenmeyer Flask (1861), Soxhlet Extractor (1879), Isolation of Fluorine (1886), Borosilicate Glass (1893), Dean-Stark Trap (1920), Magnetic Stirring (1944), Glove Boxes (1945), Rotary Evaporator (1950)

A modern fume hood. Almost all chemical reactions (should!) take place inside these.

Transition State Theory

Michael Polanyi (1891–1976), **Henry Eyring** (1901–1981), **Meredith Gwynne Evans** (1904–1952)

What exactly is going on during a chemical reaction? We know the products that go in, and we can identify the products that come out (most of the time). And from those we can say that certain bonds must have been formed and broken along the way. But what really happens when bonds form and break?

This is a central question in physical chemistry, and one for the entire science. One very successful explanation has been transition state theory, developed simultaneously, and independently, in 1935 by American chemist Henry Eyring and the team of Hungarian-British polymath Michael Polanyi and British chemist Meredith Gwynne Evans. Imagine a chemical reaction as something that starts from one energy level (the starting materials), moves up over a sort of hill (the barrier representing the energy that has to be added to initiate a reaction), and then runs down to an even lower energy state for the products.

The very top of the hill, the highest energy state that exists along the course of the reaction, is the transition state. Getting up to and past it is what determines the rate of a reaction, and if you can find a way to somehow stabilize that transition state and bring its energy down, the reaction will run faster. American chemist Linus Pauling proposed that this was one of the main ways that enzymes (biological catalysts) exert their effects, and designing compounds to look like possible transition states has helped to design enzyme inhibitor compounds.

The transition state isn't something that can be isolated—it probably lasts only as long as a single molecular vibration. But it's critical. If a reaction's transition state is more polar, then running the reaction in a polar solvent will stabilize it and speed it up. If it has a smaller volume (the **Diels-Alder reaction** is an example of this), then running the reaction under high pressure will increase its rate, and so on. Classic transition state theory breaks down under more extreme conditions, but for most reactions, it gives a good picture of what might really be happening.

SEE ALSO Gibbs Free Energy (1876), Dipole Moments (1912), Diels-Alder Reaction (1928), Reaction Mechanisms (1937), Kinetic Isotope Effects (1947), Dipolar Cycloadditions (1963)

Forming a new crystalline product like this will only happen if the thermodynamics of the reaction are right and if the molecules can pass through the bottleneck chemists call the transition state.

Nylon

Elmer Keiser Bolton (1886–1968), **Wallace Hume Carothers** (1896–1937), **Julian Werner Hill** (1904–1996)

Modern polymer chemistry began with the discovery of **polyethylene** in the U.K. and nylon in the U.S. The natural polymer fibers (such as cotton and wool) clearly showed that there was a lot of potential in the area that no one knew how to realize. **Bakelite** had been a great success, but it could never be drawn out into threads. American chemist Wallace Hume Carothers, who headed a polymer lab at DuPont, had discovered a useful synthetic rubber (neoprene) and several polymers that could be spun into thread, but these had some severe disadvantages: they dissolved in dry-cleaning solution, for starters.

Carothers's team turned to different projects, but news of other chemists' polyethylene work motivated the director of DuPont's chemistry department, American chemist Elmer Keiser Bolton, to ask Carothers to try again. They used the same thread-drawing techniques that their team member American chemist Julian Werner Hill had tried before, this time using amide bonds between the polymer constituents, and a whole new class of materials emerged. Varying the chain lengths of the acid and amine components was a key part of the research, and when both the acid and amine chains were six carbons long, the material known as nylon was produced—and in only three years, the first industrial plant producing spun nylon thread was in operation. Like many other discoveries of the late 1930s, it spent its early years contributing to the World War II effort (as a substitute for silk in parachutes, among many other things). It was strong, hard to break or tear, and resistant to heat and solvents (but not strong acid). It's still the most common synthetic fiber in textiles, and also used in fasteners, machine parts, and cookware.

Carothers, sadly, did not live to see it. He had always been prone to depression, and eventually committed suicide—a great loss for the scientific community, even if he was unable (as he said) to value his own achievements. He had an eye for talent, too, hiring the Nobel Prize–winner Paul Flory, who became one of the great polymer chemists of twentieth century.

SEE ALSO Polymers and Polymerization (1839), Rubber (1839), Bakelite (1907), Polyethylene (1933), Teflon (1938), Cyanoacrylates (1942), Ziegler-Natta Catalysis (1963), Kevlar (1964), Gore-Tex (1969).

During World War II, salvaged nylon stockings were repurposed into parachutes for army fliers, tow ropes for glider planes, and other war material.

Nerve Gas

Gerhard Schrader (1903–1990)

German Chemist Gerhard Schrader was working for his country's chemical conglomerate IG Farben, experimenting with organofluorine compounds as insecticides. But many compounds that will kill insects will also kill humans, and two days before Christmas in 1936, he and his lab assistant unknowingly produced one of them. After the holidays, they were characterizing this new compound when they both noticed that they were getting short of breath and their vision was dimming. They left the lab quickly, a sound decision. They were extremely close to becoming the first victims of the compound we now call *tabun*, the world's first nerve gas.

Most of these compounds are actually volatile liquids rather than outright gases. But they all work the same way, by irreversibly binding to a crucial enzyme called *acetylcholinesterase*, which clears out a major neurotransmitter compound, acetylcholine, after it's released from the nerve cells. If that enzyme is taken out of commission, the acetylcholine builds up quickly and fatally. Schrader and his assistant were noticing these effects on the nerves controlling their lungs and their eyes, which are the first organs to be affected. Acetylcholinesterase inhibitors (ones that are reversible and less strong than the nerve gases) can have actual medical uses, and they are indeed very effective pesticides, but concerns about their effects on human health have seen their use restricted more and more over the years.

The Germans industrialized nerve gas production during World War II (a fearsomely toxic and dangerous process), but it did not become a factor in the war for a variety of reasons, one of which was the belief that the U.S. would rapidly reproduce the agent for use against Germany. Schrader and his group produced progressively more poisonous variants, though, as the war went on, and afterward both the U.S. and the Soviet Union made and stockpiled still more. The major nations of the world have since officially renounced these weapons, but—hideously and disgracefully—nerve gas has been used by others in small wars, and, tragically, in a terrorist attack by a deranged Japanese cult on the Tokyo subway.

SEE ALSO Greek Fire (c. 672), Toxicology (1538), Chemical Warfare (1915), Bari Raid (1943).

A chemical-warfare suit designed for use in World War II, but (fortunately) never used. Whether it would have been sufficient protection against a nerve agent, though, is not clear.

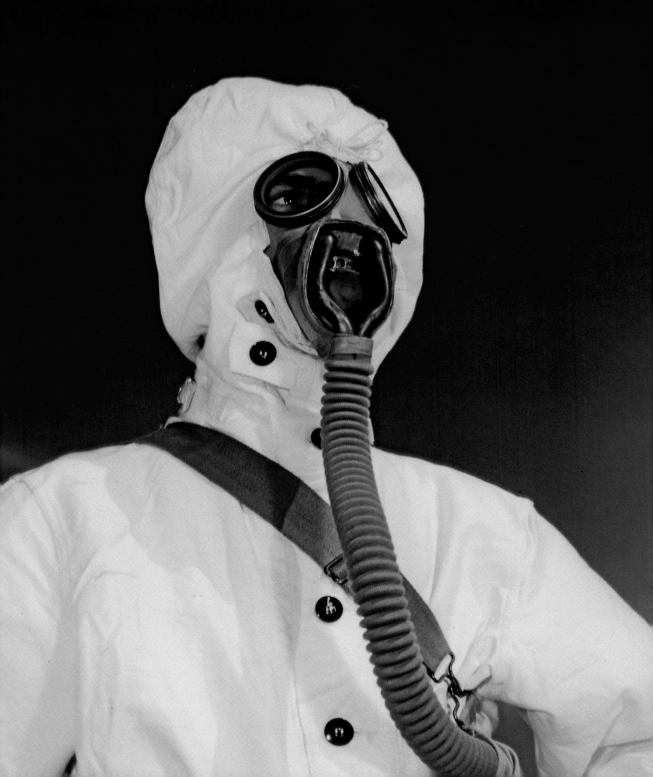

Technetium

Otto Berg (1873–1939), **Carlo Perrier** (1886–1948), **Walter Noddack** (1893–1960), **Ida Tacke Noddack** (1896–1978), **Emilio Gino Segrè** (1905–1989)

Technetium (from the Greek word for *artificial*) sits right in the middle of the periodic table, surrounded by other perfectly normal metals. But it has no stable **isotopes**—the longest-lived, technetium-98, has a half-life of 4.2 million years. That seems relatively stable on a human timescale, but in geologic time it means that it has had plenty of time to decay. Couple that with the fact that technetium isotopes are very uncommon from the decay of the heavier radioactive elements, and it's no wonder that the attempts to fill this hole in **the periodic table** came up empty for so long. The chemical literature of the nineteenth and early twentieth centuries is littered with technetium false starts and mistaken identities, and, in 1936, it ended up being the first element discovered through artificial means when the Italian team of Carlo Perrier, a mineralogist, and Emilio Gino Segrè, a physicist, bombarded molybdenum with **deuterium** nuclei.

Or so the textbooks said until recently. One of the rejected claims for technetium came from German chemist Ida Tacke Noddack and her coworkers Walter Noddack and Otto Berg, who discovered rhenium at the same time. They had produced X-ray spectra in 1925 for what they claimed was the elusive element, but it was only in the late 1990s that a team at the National Institute of Standards and Technology found that (given the Noddack team's methods of concentrating samples and their likely composition) they probably had really isolated what would later be called technetium, for which they'd proposed the name *masurium*. It's unlikely that they could have prepared enough of it in pure form to make a convincing claim, though.

Technetium-99 is a commonly used **radioactive tracer** in medicine, with several advantages. It has a very clean and convenient gamma-ray emission profile, with a half-life of about six hours (a useful timescale for human ingestion), and it decays to another very weakly radioactive form. It also gets cleared from the body rapidly. It's a valuable contributor for an element that seemed to resist being found for so long.

SEE ALSO The Periodic Table (1869), Polonium and Radium (1902), Isotopes (1913), Radioactive Tracers (1923), Deuterium (1931), The Last Element in Nature (1939)

A patient's hand gives off gamma rays after an injection of a radioactive technetium tracer. This one concentrates in the bone tissue and can be used to spot tumors that might otherwise be missed.

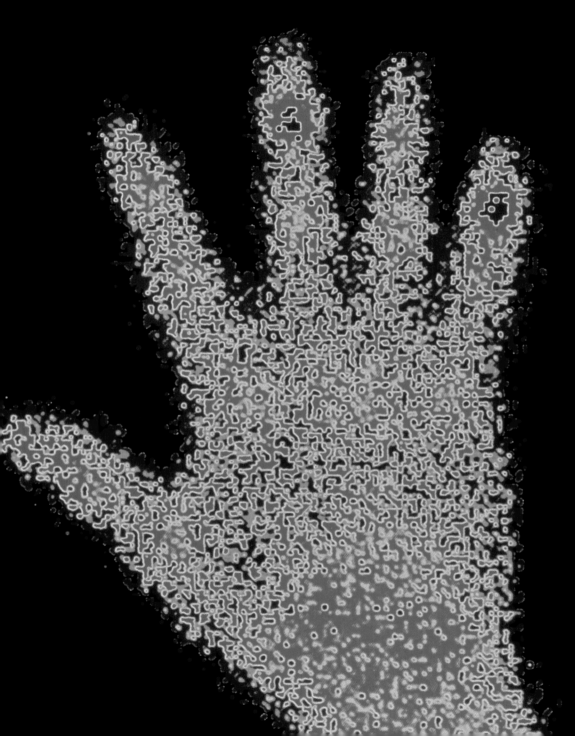

Cellular Respiration

Otto Fritz Meyerhof (1884–1951), **Albert Szent-Györgyi** (1893–1986), **Karl Lohmann** (1898–1978), **Fritz Albert Lipmann** (1899–1986), **Hans Adolf Krebs** (1900–1981), **Paul Delos Boyer** (b. 1918), **Peter Mitchell** (1920–1992), **John Ernest Walker** (b. 1941)

Every living creature needs energy, and they all use the same molecule to carry it: adenosine triphosphate (ATP), discovered in 1929 by German chemist Karl Lohmann, working with Otto Fritz Myerhof. ATP has a phosphate bond that requires a good deal of energy to form, energy that it gives back when that bond is cleaved. The molecule serves as a storable form of power that is ready on demand, an idea proposed by German-American biochemist Fritz Albert Lipmann in 1941. Throughout your body, untold billions of adenosine diphosphates and triphosphates are shuttling back and forth, providing chemical energy like so many battery packs to all sorts of proteins. The ATP-binding pockets built into them are a standardized motif that shows up over and over.

An enzyme called *ATP synthase*—discovered by British biochemist Peter Mitchell, and further explained by the American and British biochemists Paul Boyer and John Walker—allows cells to produce ATP at all times in specialized structures called *mitochondria*. These resemble bacteria, and it's no accident—mitochondria appear to have been bacteria at some point in the distant evolutionary past, moving into early cells and making a home. They're now specialized ATP factories. The first part of their chemistry was worked out in 1937 by German-born British biochemist Hans Adolf Krebs, building on the work of Hungarian physiologist Albert Szent-Györgyi (of **vitamin C** fame). It's a cycle of reactions starting with citric acid, which is generated again in the last step and sent around for more. Along the way, the two-carbon building block acetate (produced by breaking down carbohydrates and lipids) gets consumed, and **carbon dioxide** is released. The products of the Krebs cycle go straight into another series of enzyme reactions (called oxidative phosphorylation), which produce ATP while using up oxygen. This is where the food you eat and the oxygen you breathe end up, and where the carbon dioxide you exhale is made: in the nonstop furnaces of the mitochondria.

SEE ALSO Phosphorus (1669), Carbon Dioxide (1754), Oxygen (1774), Superoxide (1934), Photosynthesis (1947), Isoamyl Acetate and Esters (1962), Isotopic Distribution (2006)

The mitochondria are the green ovals in this computer-generated model of the main structures in a typical cell. In muscle cells, mitochondria are more numerous.

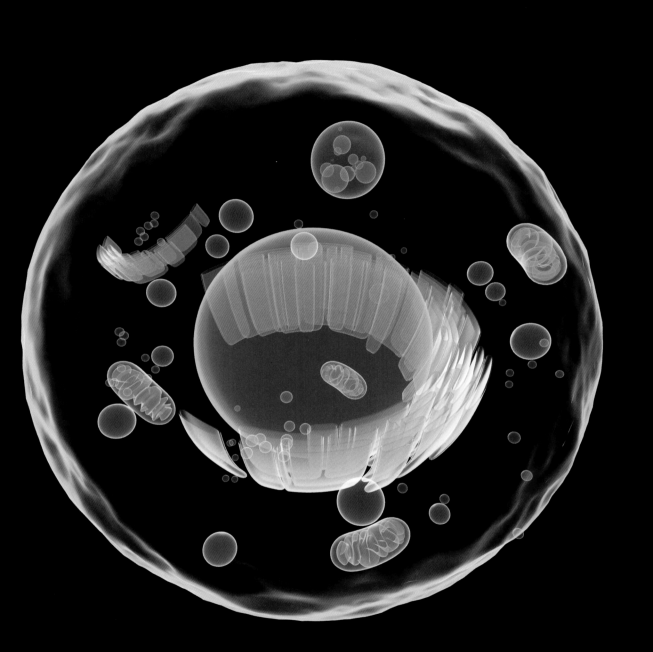

Elixir Sulfanilamide

Walter Campbell (1877–1963), **Harold Cole Watkins** (c. 1880–1939), **Frances Oldham Kelsey** (1914–2015), **James Stevenson** (d. 1955)

Here is a grim chapter in the story of a drug that was considered one of science's greatest benefits to humankind. **Sulfanilamide** was the wonder antibiotic of the 1930s, but it was difficult to dispense to children. Like many antibiotics, it had to be given in fairly large quantities, which meant painful injections or pills that children had trouble swallowing. Sulfanilamide is not soluble in water or alcohol, which ruled them out as vehicles for delivery as well. And like many drugs, it had a bitter taste that would have to be masked in a sweet syrup to be palatable.

Formulation work—turning an active compound into a useful mixture—is essential to drug development. In modern labs, all sorts of solutions, additives, coatings, and solid forms are investigated, but the process was primitive at best in the 1930s. Harold Cole Watkins, chief chemist of a small Tennessee firm, found that diethylene glycol (DEG) would dissolve sulfanilamide in useful amounts. In 1937, the company added color and flavor and began selling this as "elixir sulfanilamide," a name that implied that alcohol was the solvent. Within two weeks, reports began to come in of patients (many of them children) dying from kidney failure. James Stevenson, a physician in Tulsa, Oklahoma, was among the first to make the connection, alerting the American Medical Association, which tested the product and immediately warned the public. An FDA study led by American chemist Frances Oldham Kelsey showed that the DEG was the toxic agent, and the agency's Walter Campbell pressed the company to recall all supplies. Watkins attempted to show that diethylene glycol was not toxic by ingesting some himself—but he did not have the body of a child. Eventually, he committed suicide, apparently unable to live with what his product had done.

Six gallons of the syrup were given to patients across the country before the recall, causing over a hundred deaths. The FDA was greatly strengthened by a law passed the next year, mandating that all new drugs and formulations be tested for safety and reviewed by the agency before any products entered the market.

SEE ALSO Radithor (1918), Sulfanilamide (1932), Thalidomide (1960)

An actual bottle of the preparation that poisoned its customers, mostly children, in 1937.

Reaction Mechanisms

Robert Robinson (1886–1975), **Christopher Kelk Ingold** (1893–1970), **Edward David Hughes** (1906–1963)

British chemist Christopher Kelk Ingold's close studies of simple reactions led to the first real understanding of the order and direction of chemical bonds breaking and forming, taking what had been a disorderly heap of mysteries and giving it a clear system that is still used today. Consider a classic substitution reaction, described by Ingold and colleagues in a 1937 paper: If you take methyl bromide and react it with an iodide, you can produce methyl iodide. Somehow, a carbon-iodine bond has formed and the carbon-bromine bond has broken. Ingold's studies showed that the iodide comes in on the opposite side of the carbon from the bromide and sort of knocks it off from behind. The three hydrogen atoms of the methyl group flip over like an umbrella in a high wind. The transition state structure has the carbon-iodine bond about halfway formed, the carbon-bromine bond about halfway broken, and the hydrogen atoms pointing straight out from the middle, in midflip. One consequence of this mechanism is that it will flip the orientation of a chiral center if it takes place on a chiral (left- or right-handed) carbon.

In our sample reaction, the iodide is the "nucleophile," a negatively charged species that reacts with the "electrophile" partner. In Ingold's terms, the reaction mechanism would be called "Sn2," which stands for "substitution, nucleophilic, 2nd order" (which means that the rate is influenced by the concentrations of both reactants). The Friedel-Crafts reaction is a perfect example of an electrophilic substitution.

Ingold's "electron pushing" notation for showing reaction mechanisms is still the standard. Along with his nomenclature, it has been used by chemists ever since to help design reaction conditions and to predict their products. His work, much of it done over decades with his British colleague Edward David Hughes, made sense of a whole list of observations about what changed the rates and product distributions of various reactions. Sir Robert Robinson, a British organic chemist, was working on similar lines, but Ingold's system gained acceptance more quickly and became the standard (to Robinson's own dismay, at times).

SEE ALSO Tetrahedral Carbon Atoms (1874), Friedel-Crafts Reaction (1877), Dipole Moments (1912), Transition State Theory (1935), *The Nature of the Chemical Bond* (1939), Kinetic Isotope Effects (1947), Nonclassical Ion Controversy (1949), Dipolar Cycloadditions (1963)

The Royal Society of Chemistry designated Ingold's London laboratory a "National Chemical Landmark" in 2008.

RSC | Advancing the Chemical Sciences

National Chemical Landmark

Chemistry Department
University College London

During the period 1930-1970 **Professor Sir Christopher Ingold** pioneered our understanding of the electronic basis of structure, mechanism and reactivity in organic chemistry, which is fundamental to modern-day chemistry.

28 November 2008

Catalytic Cracking

Eugene Jules Houdry (1892–1962)

The thermal cracking of petroleum (breaking down larger hydrocarbons into smaller, more useful ones) was a significant advance in the field, but it had quickly become obvious that a better technique was needed, as the thermal process required a great deal of energy and still ended up producing too much tarry residue as a by-product. Chemists around the world searched for alternatives, but in France, pharmacist E. A. Prudhomme and mechanical engineer and chemist Eugene Jules Houdry came to the conclusion that heat was not enough—chemical catalysts were needed to break and reform carbon-carbon bonds more efficiently. Houdry experimented with a wide variety of different systems, first in France with Prudhomme and then in the United States, but the Great Depression, which reduced the demand for gasoline, also slowed down his research.

The mineral catalysts he discovered were refined several times, as were techniques for keeping them active (always a concern in large-scale catalytic reactions). By 1938, the process was developed enough that an older thermal cracking plant in Pennsylvania had been converted to the new process, which produced double the amount of gasoline from the same petroleum, startling the entire industry when it was revealed. This improved process proved vital only a few years later, during World War II. Furious at France's occupation by German forces, Houdry was in a perfect position to do something about it, as his technology provided huge quantities of high-performance aviation fuel from U.S. refineries.

After the war, he invented one of the first systems for catalytically reducing the pollutants in engine exhaust, the forerunner of today's catalytic converters. Both these processes illustrate how catalysis, a process in which a small amount of an active species can cycle through thousands (or millions) of turnovers in a chemical reaction, became the key to modern industrial chemistry. The search for new and better catalysts (for bond making and breaking, for pollution control, and to use as new and cheaper starting materials) remains a huge field of research, one whose possibilities are nowhere near their limits.

SEE ALSO Thermal Cracking (1891), Tetraethyl Lead (1921), Fischer-Tropsch Process (1925), Catalytic Reforming (1949)

A modern hydrocarbon cracking plant in Slovakia. All the gasoline and other fuels in the world are produced in such refineries.

Teflon®

Roy J. Plunkett (1910–1994)

There's no other element like fluorine, and there's no substitute for many of its derivatives either. The **chlorofluorocarbons** were the first industrially important ones, and they led indirectly to this next breakthrough. In 1938, American chemist Roy J. Plunkett was part of a group looking at new fluorinated gases for use as refrigerants and had produced tetrafluoroethylene as a starting material. He stored it in small cylinders in the cold, but when he hooked one up and opened the valve, nothing came out. The cylinder still weighed the same as before, though, so Plunkett and his lab assistant sawed it open, only to find, most unexpectedly, that it now contained a white powder that no one had ever seen before. The gas had spontaneously turned into a polymer, polytetrafluoroethylene (PTFE), soon to be known around the world under the trade name *Teflon*. It was later found that iron from the inside of the cylinder had served as a catalyst to start the polymerization.

PTFE is the completely fluorinated version of **polyethylene**, with every hydrogen replaced, and it's a very different substance indeed. It can stand both high and low temperatures and cannot be set on fire. It's extremely resistant to almost all reagents and solvents, although working organic chemists know a tiny weak point: a **Birch reduction** will turn a white PTFE-coated **magnetic stirring** bar black by breaking some of the carbon-fluorine bonds. It also has an extremely low coefficient of friction, which has led to its well-known use in cookware as well as in many industrial applications where it reduces the wear and energy consumption of machinery. Too much heat, though, can undo the polymerization and send volatile (and potentially toxic) fluorinated compounds back into the air.

Its first big role was far from the kitchen, though. Under total secrecy, the Manhattan Project used it as a chemically resistant material in the vicious reactions that produced uranium hexafluoride for **gaseous diffusion**. These days PTFE provides a breathable, waterproof material in **Gore-Tex**, insulates hundreds of miles of cables, and coats the inside of your clothes dryer, among almost uncountable other products.

SEE ALSO Polymers and Polymerization (1839), Bakelite (1907), Chlorofluorocarbons (1930), Polyethylene (1933), Nylon (1935), Gaseous Diffusion (1940), Cyanoacrylates (1942), Birch Reduction (1944), Magnetic Stirring (1944), Ziegler-Natta Catalysis (1963), Kevlar (1964), Gore-Tex (1969)

PTFE-coated pans are found in almost every kitchen, but they were an expensive novelty when they were first produced.

The Last Element in Nature

Marguerite Perey (1909–1975)

Science may be an endless frontier, but some parts do reach their limits. Francium was the last element to be found in nature. After that, all the new ones were synthesized through nuclear reactions. French physicist Marguerite Perey, who isolated francium from actinium while she was working in Marie Curie's lab, became the last human being who would isolate a new element from ore samples, as scientists had been doing for centuries. Thus, the classical age of element discovery came to an end.

It was an appropriate element to finish with, because it proved to be one of the most difficult. While Perey was purifying actinium from uranium ore in the lab, she noticed uncharacteristic radiation coming from the sample, and her further investigations led to the isolation of the new, highly radioactive element (Perey eventually died from cancer as a result of radiation exposure). In fact, it's no exaggeration to say that francium barely exists at all. Every one of its **isotopes** is radioactive—to be more accurate, they're all *wildly* radioactive. The longest-lived one has a half-life of twenty-two minutes, and the only reason it can be found at all is that it's continually being produced from the radioactive decay of actinium-227. A sample of uranium or thorium ore has a few atoms of francium scattered throughout it, but good luck detecting them before they disappear.

Chemists had suspected since the late 1800s that there was a metal beyond cesium, and **the periodic table** indicated that this must indeed be the case. (Before Perey's discovery, there had been false alarms for years.) The alkali metal column gets more violently reactive as you go down it, and there is every reason to think that the chemistry of francium is similarly dramatic. But no one knows. Producing and gathering enough francium atoms for a reaction takes serious equipment, and there's a limit to what any equipment can do. Francium's radioactivity is so fierce that it's probably impossible to produce even a speck large enough to see with the naked eye. It would blow itself into vapor from its own heat of radioactive decay.

SEE ALSO The Periodic Table (1869), Polonium and Radium (1902), Isotopes (1913), Technetium (1936), Transuranic Elements (1951)

Francium's electronic configuration. The single electron in the outermost shell would make it very chemically reactive—if its nucleus would let it last long enough for anyone to notice.

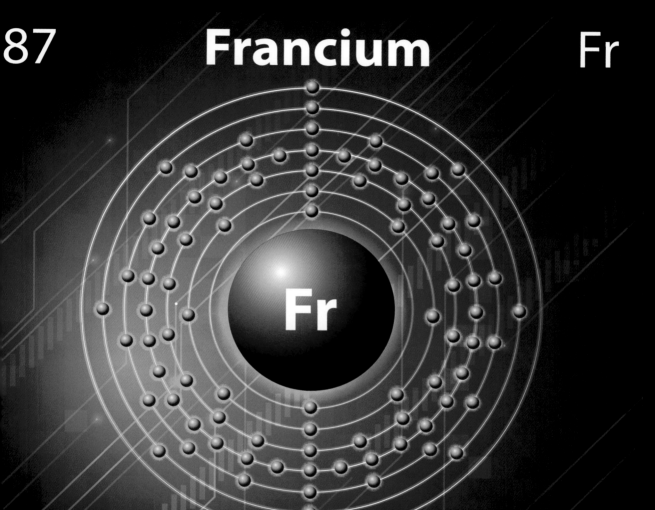

The Nature of the Chemical Bond

Linus Carl Pauling (1901–1994)

Considered one of the greatest chemists ever, Linus Pauling made fundamental discoveries in several different fields, including quantum chemistry and molecular biology. The year 1939 marks the publication of his book *The Nature of the Chemical Bond*, a perennially influential landmark text.

In the mid-1920s, Pauling studied in Europe with the founders of quantum mechanics, and he resolved to apply this new physics to chemical bonding. This was ambitious—a strict mathematical treatment rapidly gets unworkable as the systems get more complex, and a full understanding of chemical bonding beyond the dihydrogen molecule (H_2) was beyond the abilities of physics. But Pauling rapidly made a name for himself with a series of groundbreaking papers on molecular orbitals, X-ray structures (especially **crystals** of ionic solids), and chemical bonding in general.

Among other things, he showed that there was a continuum between pure ionic interactions (two charged ions attracting each other, as in a salt) and a pure covalent bond (where electrons are shared equally between the uncharged atoms, as in dihydrogen or in a plain carbon-carbon bond). Pauling introduced the idea of electronegativity—the tendency of an atom to gather electron density (negative charge) to itself—and proposed that all chemical bonds varied between the two extremes. Since Pauling, every chemist has thought in these terms, which have helped greatly in designing new reactions.

Physicists had already shown that electrons around a nucleus were arranged in discrete energy levels, or orbitals. Pauling found that in chemical bonding, these could mix (or hybridize, as chemists call it), and that these hybrid orbitals accounted for the shape of **tetrahedral carbon atoms**, the geometry of carbon-carbon double bonds, and all facets of organic chemistry. It also extends to inorganic and organometallic compounds and helps explain their shapes and reactivities as well.

Pauling won a well-deserved solo Nobel Prize in Chemistry in 1954 for this work, and his work for arms control and disarmament won him another (for peace) in 1962. In later years, he became well known for his advocacy of **vitamin C**, which many thought crossed the line into the realm of crankery.

SEE ALSO Tetrahedral Carbon Atoms (1874), Dipole Moments (1912), Hydrogen Bonding (1920), Vitamin C (1932), Reaction Mechanisms (1937), Noble Gas Compounds (1962), Computational Chemistry (1970), Unnatural Products (1982), Single-Molecule Images (2013)

Dr. Linus Pauling—without a doubt one of the greatest chemists of the twentieth century—holding arrangements of water molecules in a classroom at the California Institute of Technology.

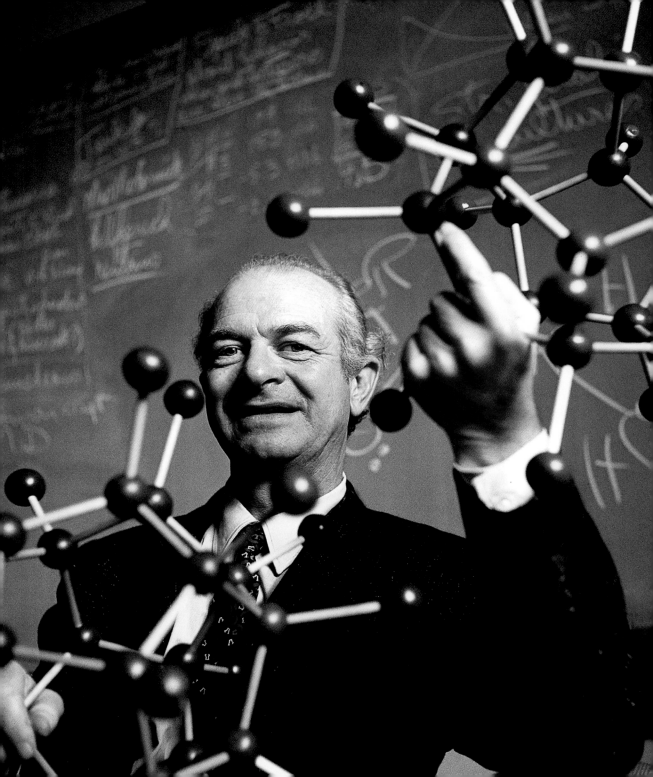

DDT

Othmar Zeidler (1859–1911), Paul Hermann Müller (1899–1965)

DDT was discovered in 1939 by Swiss chemist Paul Hermann Müller after a long search for an effective compound against disease-carrying and crop-destroying pests. It's still probably the most (in)famous insecticide in the world, even though it's used much less than it was during its peak in the late 1950s. Müller had noted that insects absorb compounds differently than higher animals, and he hoped to find a compound that was selectively toxic, easy to synthesize, and long-lasting. DDT (abbreviated, understandably, from dichlorodiphenyltrichloroethane) looked like the answer. It had been synthesized in 1874, by Austrian chemist Othmar Zeidler, but its biological effects had been completely missed, probably because it seemed to have little activity on humans or other mammals. However, the compound was toxic to a huge range of insect pests and other arthropods including potato beetles, lice, biting flies, and mosquitoes.

Extremely stable, long-lasting, and capable of being produced by the ton, DDT was widely used late in World War II to control malaria-carrying mosquitoes, and over the next twenty-five years it eradicated malaria outright from the United States and many other countries. Müller received a Nobel Prize in 1948, and little wonder. His discovery ended up saving at least half a billion people from dying of malaria alone, and it increased food crop production as well. But the problems with DDT became apparent through the 1950s. The compound was actually *too* stable and persistent. It could last for many years in the environment, and this allowed it to build up in the tissues of animals further up the food chain from its insect targets. Rachel Carson's 1962 book, *Silent Spring*, made the case against the use of the compound in detail. While not all of its claims were accurate, its main point certainly was. DDT buildup was especially harmful to birds, causing defects in their eggshells that caused population declines in bald eagles and many other species.

DDT production was cut back and eventually banned in most of the world, although in recent years it's made a small comeback in less polluting applications, such as indoor-only mosquito control in poor tropical regions.

SEE ALSO Toxicology (1538), Paris Green (1814)

An unexpected side effect of DDT is a disruption in some birds' abilities to build up calcium in their eggs. Birds of prey and waterfowl, as well as many songbirds, were especially affected.

Gaseous Diffusion

Thomas Graham (1805–1869), **Francis Simon** (1893–1956), **Nicholas Kurti** (1908–1998)

In 1848, Scottish physical chemist Thomas Graham proposed what is now known as *Graham's law*: the rate at which gases escape through porous barriers varies with the square root of their mass. If you compare two pure gases, one with a molecular weight that's four times as heavy as the other, the lighter gas will diffuse through a barrier (or escape through a very small pinhole) twice as fast as the heavier one. This was later found to be a result of the general kinetic theory of ideal gases, and it remained just another physical proof of that theory for almost one hundred years.

In the early 1940s, the scientists of the Manhattan Project were producing the first atomic bomb, and gaseous diffusion became an essential step. They had to enrich uranium to produce uranium-235, the critical isotope in an atomic fission reaction (the splitting of an atomic nucleus to release enormous quantities of energy). However, since isotopes have almost exactly the same chemical reactivity, effects like Graham's law that depend on molecular weight had to be enlisted. Uranium is very far from being a gas, but strenuous chemistry converted it to gaseous uranium hexafluoride. The difference in weight between uranium-235 and uranium-238 isn't much, but using multiple cascading centrifuges with semipermeable membranes allowed the gaseous isotopes to be separated.

The gaseous diffusion process was first developed in 1940 by German-born physical chemist Francis Simon and Hungarian-born physicist Nicholas Kurti, both of whom fled from Germany to England upon Adolf Hitler's rise to power. It was perfected in Oak Ridge, Tennessee—a town that ended up using a significant fraction of all the electricity being generated in the U.S. by the end of the war. Gaseous diffusion alone still didn't enrich the uranium enough, so the material from the centrifuges was passed through an early type of **mass spectrometry** apparatus to get material suitable for a bomb. But gaseous diffusion is a suitable technique to make low-enrichment uranium for nuclear power plants, and a new plant, the first in many years, is being built in the state of Ohio for just that purpose.

SEE ALSO Ideal Gas Law (1834), Maxwell-Boltzmann Distribution (1877), Mass Spectrometry (1913), Isotopes (1913)

General Leslie Groves, commander of the Manhattan Project, speaks to a crowd at Oak Ridge, Tennessee, in 1945. The city did not even exist in early 1942, but two years later it was consuming almost 15 percent of the electric power output of the entire country, much of which ran giant centrifuges.

Steroid Chemistry

Russell Marker (1902–1995)

Russell Marker is a legendary figure in the history of medicinal chemistry. He was a professor at Penn State, researching steroid chemistry, when he realized that the structure of a plant steroid, sarsasapogenin, had been assigned incorrectly. The main part had the common four-ring steroid system, but its side chain was reactive enough to be removed, meaning that with just a few more chemical reactions, the steroid ring could be converted to progesterone. The process eventually came to be known as the *Marker degradation*.

Progesterone and all the steroids were known to be extremely powerful and important substances in human biochemistry, but they were terribly expensive and difficult to prepare. Marker knew that sarsasapogenin was too scarce to be used as a starting material, but he found another useful plant steroid, diosgenin, that might be suitable. Plant-collecting trips and deep dives into the botanical literature turned up no likely candidates for a reliable source of diosgenin—until Marker came across a huge Mexican yam, in the same genus as some of his other finds, whose root was said to weigh up to two hundred pounds.

In 1942, Marker traveled to Veracruz, Mexico (where this plant was said to grow), and with the help of local country-store owner Alberto Moreno he managed to get a fifty-pound root onto the return bus. Back at Penn State, he found that the root was indeed a good source of diosgenin, but established drug companies weren't interested in a Mexican yam-harvesting operation. So Marker set up his own, enlisting Moreno to help round up ten tons of roots. These were eventually processed into three kilos of pure progesterone using his chemistry, resulting in what was easily the largest batch in the world at the time. Marker's various commercial ventures ended up starting a huge Mexican steroid industry, and his semi-synthetic progesterone went on to be used in oral contraceptives and as a precursor to the anti-inflammatory drug **cortisone**.

SEE ALSO Natural Products (c. 60 CE), Cholesterol (1815), Conformational Analysis (1950), Cortisone (1950), The Pill (1951), Isotopic Distribution (2006).

The "elephant foot" yam (Dioscorea mexicana), which Russell Marker hauled to Penn State in order to start the synthetic-steroid industry.

Cyanoacrylates

Harry Wesley Coover Jr. (1917–2011), Fred Joyner (1922–2011)

Two common themes in mid-twentieth-century chemistry—the technological greenhouse of World War II and the accidental discovery of useful polymers—converged yet again when American chemist Harry Wesley Coover Jr. was experimenting with the transparent acrylate plastics that were already a major part of aircraft design. In 1942, he discovered that incorporating a cyano (nitrile) group where a methyl group would be in the usual polymer didn't lead to a better clear plastic for gunsights or airplane cockpits—it led to the stickiest mess Coover had ever seen. That idea went back into the desk drawer, not to be revived until 1951, when Coover and his lab mate, researcher Fred Joyner, were working at chemical company Tennessee Eastman.

High-performance clear plastics were still being sought, and the cyanoacrylates still looked too gooey to be of any possible use. But when Joyner inadvertently ruined an expensive piece of equipment, permanently bonding two lenses together with the cyanoacrylate he was testing for use as a temperature-resistant coating for jet cockpits, Coover realized that they were looking at a potential new kind of glue. Cyanoacrylates became the popular instant glues that can be used to bond all sorts of materials. Small amounts of water were enough to start the polymerization, and the water molecules that were already on most surfaces under ordinary conditions were plenty.

Coover himself made a memorable appearance on a popular game show of the time *I've Got a Secret*, being raised into the air while holding onto a metal bar he had glued together only a minute before. He thought up many uses for cyanoacrylates over the years—one based on the observation that the glue tended to stick a person's fingers together at every opportunity. That suggested that it could serve as a medical adhesive to seal wounds, and medics actually used it successfully in the field during the Vietnam War. Cyanoacrylates are now used for the same purpose in both the veterinary and medical fields, and the glue is sold in the form of liquid bandages for home use as well.

SEE ALSO Polymers and Polymerization (1839), Bakelite (1907), Polyethylene (1933), Nylon (1935), Teflon (1938), Ziegler-Natta Catalysis (1963), Kevlar (1964), Gore-Tex (1969)

U.S. soldiers gathered around an acoustic guitar in Vietnam, January 1968. Cyanoacrylate glue—useful for the repair of such instruments—was first utilized in battlefield surgery during this war, saving many lives in an application that no one would have imagined at first.

LSD

Albert Hofmann (1906–2008)

The discovery of LSD (lysergic acid diethylamide) is one of the most famed stories in all of medicinal chemistry. Chemist Albert Hofmann was working for the Swiss pharmaceutical company Sandoz Laboratories, preparing derivatives of lysergic acid, which is a natural product from ergot, a fungus that grows on grains such as rye. These contaminated grains had long been known to induce strange neurological effects and even convulsions (ergotism) in people who consumed food prepared from them. Such compounds would surely have some biological activity.

Hofmann had already synthesized what the world now knows as LSD five years before, but he began reinvestigating it in 1943. He had no way of knowing that the compound was incredibly potent and that being exposed to mere microgram levels of it (which happened by accident, in the normal course of his work) would lead to profound effects. He said later that he'd felt unusual during the afternoon, and when he got home from work he hallucinated on his couch for two hours. Thinking that the compound he'd been working on might have brought on these effects, and that such low doses were unlikely to be toxic, he intentionally took 250 micrograms a few days later. Famously, he started to feel the effects while he was riding home on his bicycle, and while Basel is a perfectly presentable city, it had surely never been seen the way Hofmann saw it that afternoon.

His colleagues were skeptical at first that any compound could be active at such low doses, but LSD spoke for itself. Its effects were later found to come from tight binding to a number of brain receptors (particularly the serotonin subtype called 5-HT2a). The drug gradually became well known after Sandoz introduced it for psychiatric use in 1947, and it became particularly identified with the 1960s drug culture. Though LSD is still illegal for sale in the U.S., the altered perceptions it brings on have been the subject of many studies, among them trials to treat post-traumatic stress disorder, anxiety in patients with terminal cancer, and alcoholism.

SEE ALSO Natural Products (c. 60 CE), Morphine (1804), Caffeine (1819)

Sold in a variety forms, including tablets of various shapes and elaborately designed blotters, LSD's extraordinary potency (millionths of a gram for an adult human dose) makes transporting and ingesting it much easier than most other drugs.

When people think of natural products in chemistry, they often picture complex molecules derived from exotic jungle plants or tropical reef creatures. But some of the strangest (and most useful) compounds come from soil microorganisms that could be living in a crack in a city sidewalk or under someone's rosebush. Ukrainian-born American biochemist Selman Abraham Waksman was one of the great microbiologists in this field, and he knew that soil bacteria and fungi were engaged in a constant struggle for resources. He and his coworkers at Rutgers University started a large antibiotic screening program in 1939, looking for compounds that these species might be using to kill off their rivals. The idea was to culture as many microbes as possible, testing extracts from each one against human disease organisms, and then isolating the active compounds responsible.

The U.S. drug company Merck later entered into an agreement with the group to test (and commercialize) promising compounds. The best, discovered in 1943 by American microbiology graduate student Albert Schatz, was later named streptomycin, and it became an extremely valuable broad-spectrum antibiotic. In later years, bacteria that were already becoming resistant to penicillin were mowed down by streptomycin, and it was the first successful treatment for tuberculosis, which until then was a slow death sentence. The clinical trial that proved this also became well known. It was the first curative trial to be run against a *placebo* (Latin for "I shall please," indicating a medicine that is more to please the patient than to benefit their health; they were pills that did nothing, and "do nothing" was unfortunately the standard of tuberculosis care at the time). The trial was also double-blinded, meaning that neither the patients nor the investigators knew which group was getting which treatment. This is still the standard for clinical trials in drug research today.

Waksman's team discovered at least nine antibiotics, with streptomycin and neomycin remaining in use today. Similar screening efforts are still used to search for active natural products, although the success rate has understandably decreased since this highly productive first pass through the soil organisms.

SEE ALSO Natural Products (c. 60 CE), Salvarsan (1909), Sulfanilamide (1932), Penicillin (1945), AZT and

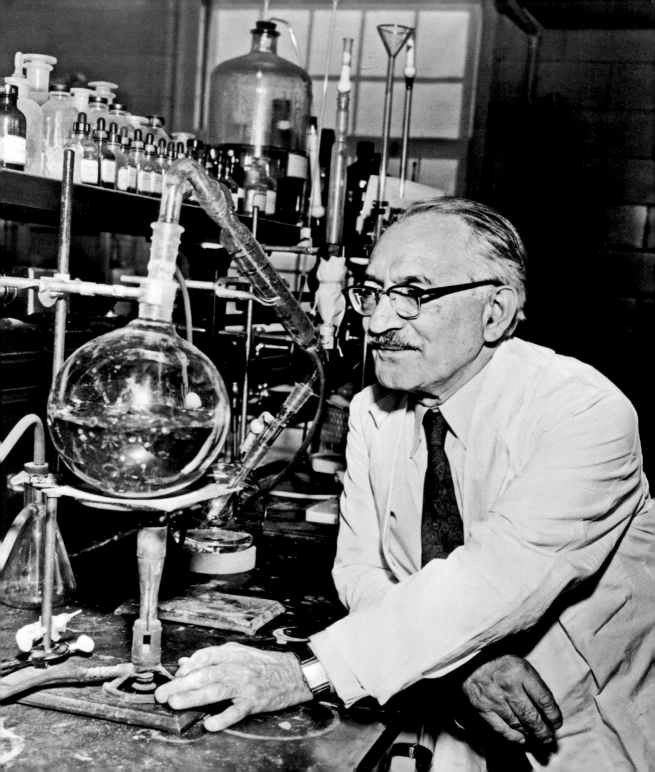

Bari Raid

Louis Sanford Goodman (1906–2000), **Alfred Gilman Sr.** (1908–1984), **Stewart Francis Alexander** (1914–1991)

Chemotherapy, in its classic form, is an easy concept to understand. Cancer cells divide more quickly than normal ones, so if you can find something that kills cells when they divide, you'll have an anticancer agent. That's a crude summary of the process, but the idea can be an effective one.

Its origins are even cruder. In 1943, a German air raid on the southern Italian port of Bari had the inadvertent and hideous side effect of releasing large quantities of mustard gas. This **chemical warfare** agent was (in theory) banned by international treaty, but the Allies wanted to have it on hand as a countermeasure in case Germany deployed a chemical weapon of its own. The U.S. cargo ship S.S. *John Harvey*, destroyed in the raid, was secretly carrying mustard gas bombs, and treatment of the victims of the raid was severely impaired because no one was told that the chemical could have been involved. In the aftermath of the bombing, the Army sent Lieutenant Colonel Stewart Francis Alexander, a doctor specializing in chemical warfare, to investigate its effects on the victims. He noted that they showed signs of having had populations of fast-dividing cells in their bodies selectively destroyed and proposed that this might be used as a type of cancer therapy.

Pharmacologists Louis Sanford Goodman and Alfred Gilman Sr. were already working for the military, investigating possible medical applications of such chemical warfare agents, and they quickly took up Alexander's suggestion. Analogs of mustard gas with a nitrogen, instead of a sulfur, atom in the center turned out to shrink lymphomas in both mice and humans, which was the first evidence that there could be such a thing as a drug against cancer. Oncology has always been an any-weapon-to-hand endeavor, and "nitrogen mustards" are still used today in some cases, prolonging lives in a strange echo of two world wars.

SEE ALSO Greek Fire (c. 672), Toxicology (1538), Chemical Warfare (1915), Nerve Gas (1936), Antifolates (1947), Thalidomide (1960), Cisplatin (1965), Rapamycin (1972), Modern Drug Discovery (1988), Taxol (1989)

An ammunition ship explodes in the harbor of Bari, Italy, in this photo from a German news agency, 1943.

Birch Reduction

Humphry Davy (1778–1829), **Charles August Kraus** (1875–1967), **Arthur John Birch** (1915–1995)

English chemist Sir Humphry Davy, whose **electrochemical reduction** technique was the first to provide metallic sodium and potassium, was the first person to see what happened when these metals contacted liquid ammonia. To his surprise, they showed unexplained deep-blue colors that went on to turn bronze and gold colored as they became more concentrated. The chemical species behind these colors remained a subject for speculation until American chemist Charles August Kraus suggested, in 1907 while working on his PhD, that they might be due to the electrons liberated from the metals, rather than the metallic elements themselves. A range of reactive metals all gave the same blue color, even in some other liquid amines, and loss of electrons was what they all had in common.

Kraus was correct. The color is an "electride" salt, containing a free electron surrounded by a solvent shell of ammonia molecules, and it's that strong solvation that keeps the two species apart and allows the salt to exist. In 1944, Australian organic chemist Arthur John Birch reported the reaction that bears his name, which involves the use of these electron solutions to reduce a wide range of organic molecules. Even aromatic compounds could have their structures (and their aromaticity) broken up under these conditions, producing six-membered ring compounds with patterns of double bonds that are very difficult to make by any other route. All sorts of unusual reductions can be performed with such a strong reducing agent, and although the reaction is not often done on an industrial scale, it's a valuable weapon in more specialized organic chemistry.

You can even use the blue color as an indicator—add lithium in portions to an ammonia solution of your starting material until the blue color stops disappearing, and you know that you've reduced everything there is to reduce. Letting the reaction warm to room temperature boils away all the liquid ammonia, leaving the small-molecule products of the reaction and some lithium salts, which is a perfect situation in which to use a **separatory funnel** to extract the mixture.

SEE ALSO Electrochemical Reduction (1807), Separatory Funnel (1854), Benzene and Aromaticity (1865), B_{12} Synthesis (1973).

Sodium dissolves in clear, liquid ammonia, giving the distinctive solvated-electron blue of a Birch reduction.

Magnetic Stirring

Arthur Rosinger (1887–1969)

There's hardly a chemical reaction in the world that doesn't need to be stirred (although one would have to call **thermite** an honorable exception). Many solid reagents have to dissolve to take part in a reaction, and many products are likely to "crash" (precipitate) out of solution. Even though some solid reagents are not supposed to dissolve, they still need to be stirred vigorously in the mixture in order to be effective (otherwise, they sit in a mound on the bottom of the flask!).

Mechanical mixing in chemistry takes many forms, but on the research scale, one of the most common is the magnetic stirrer. It's an ingenious device that is hard to improve on, and 1944 marks the appearance of the first patent for one, registered by American chemist Arthur Rosinger. (A Scottish chemist, Edward McLaughlin, invented it independently a few years later.) Both of these inventions describe essentially the same system as is used today: A small bar magnet is coated in an inert material (**Teflon** is commonly used, and sometimes glass). This is placed in the reaction flask, which is positioned over a rotating magnet, causing the stirring bar to rotate with its motion. The speed can be adjusted to fit the reaction conditions, and stirring bars of all different sizes and shapes can be introduced for different sorts of vessels. These vessels can then be sealed off to keep oxygen or water vapor out of the reaction as the stirring continues to be driven from outside. The rotating magnet is usually protected under a porcelain (or other nonmagnetic material) "stir" plate, which also allows for a heating element to be built into the same system.

There are only a few potential problems. Very thick reactions can overwhelm the force of the stir bar, especially on small scales. Also, the magnetic stirrer idea doesn't translate well to the larger scales of process chemistry, either, since it becomes more difficult to build and handle appropriately sized stirring plates. (In both these cases, a mechanical paddle stirrer, introduced from above on a rotating shaft, is preferable.) But for research-sized reactions, the convenience of magnetic stirring is obvious.

SEE ALSO Separatory Funnel (1854), Erlenmeyer Flask (1861), Soxhlet Extractor (1879), Thermite (1893), Borosilicate Glass (1893), Dean-Stark Trap (1920), The Fume Hood (1934), Teflon (1938), Glove Boxes (1945), Rotary Evaporator (1950).

The spinning magnet bar at the bottom of this beaker is a familiar sight in chemistry and biology laboratories around the world.

Penicillin

Alexander Fleming (1881–1955), **Howard Walter Florey** (1898–1968), **Ernst Boris Chain** (1906–1979), **Dorothy Crowfoot Hodgkin** (1910–1994), **Norman George Heatley** (1911–2004), **Edward Penley Abraham** (1913–1999), **John Clark Sheehan** (1915–1992)

The discovery of penicillin by Scottish biologist/pharmacologist Alexander Fleming in 1928 wasn't a complete accident—he often left culture plates around just to see what might grow in them. This time, he noticed that some mold near one of the bacterial cultures was surrounded by a zone of dead staphylococci. It was years, though, before real development began. At first, just isolating enough penicillin was nearly impossible, and whether it would work in humans was unknown, but World War II again provided the impetus. Australian pharmacologist Howard Walter Florey, German-born British biochemist Ernst Boris Chain, English biochemists Norman George Heatly, Sir Edward Penley Abraham, and others at Oxford found ways to scale up production and began animal studies. Human trials followed, and penicillin saved its first patients in 1942. Several U.S. pharmaceutical companies took on the job of production, which involved advances in fermentation and a search for a productive strain of the mold. (The best was found on a rotting cantaloupe from Peoria.) In 1942, half the world's supply was used to treat one patient (whose urine was collected to recycle the precious drug), but, by the middle of 1945, millions of doses were being produced.

Chemically, penicillin remained a mystery during most of this development. Its structure was very difficult to determine chemically, but British biochemist Dorothy Crowfoot Hodgkin's **X-ray crystallography** studies showed that the compound contained a four-membered "beta-lactam" ring—as she, Chain, and American organic chemist R. B. Woodward had suspected. That ring was the key to its activity. Very few such structures were known, but a huge amount of effort was directed onto their chemistry, with American organic chemist John Clark Sheehan making important progress throughout the 1950s. He developed the first practical method for synthesizing penicillin and created synthetic and semisynthetic forms of it that could combat bacteria that had become resistant to the original drug.

SEE ALSO Natural Products (c. 60 CE), Salvarsan (1909), X-Ray Crystallography (1912), Sulfanilamide (1932), Streptomycin (1943), AZT and Antiretrovirals (1984), Modern Drug Discovery (1988), Taxol (1989).

As advertised in this World War II poster, penicillin did indeed save uncounted lives. In today's world, we can hardly imagine the awe in which it was held back then, but resistance to the drug (and other antibiotics) puts us at risk of finding out.

Glove Boxes

It goes without saying that some of the reagents used in a chemistry lab must be handled carefully, especially reactive compounds that instantly burst into flame when exposed to air. Vacuum-line techniques are one way to handle these substances, especially the liquids, but sometimes what you really need is a lab bench located in a place with no oxygen.

Enter the idea of the glove box: a large, enclosed box, typically with a glass or plastic front for visibility, whose air has been replaced with an inert gas such as nitrogen or argon. Chemists perform their work inside the box by placing their arms into rubber sleeve gloves built into the box's front wall, but they have to plan their operations carefully. Taking things in and out of a working glove box is not done casually. There's usually an air-lock arrangement involving a small transfer box that allows the operator to bring items inside after purging them of outside air, but that takes a while.

The glove box seems to have first been used during the Manhattan Project. Those models were built from scratch out of plywood, glass, and rubber when the scientists realized just how difficult and dangerous their work was turning out to be. The practice seems to have been carried from there into laboratories around the world. Some industrial applications employ large, multiuser glove boxes with portholes for viewing. Glove boxes are installed on the International Space Station for microgravity experiments with hazardous reagents as well.

Many organometallic compounds are sensitive to oxygen or moisture, so inorganic chemists are most likely to have well-maintained glove boxes. Keeping a glove box well maintained and operational requires dedication and a steady supply of inert gas. The classic test for this is to switch on an incandescent light bulb inside the box and then carefully break the bulb open. If it continues to burn—that is, if the atmosphere inside the glove box is as inert as that inside a commercial light bulb—then you're in business.

SEE ALSO Separatory Funnel (1854), Erlenmeyer Flask (1861), Soxhlet Extractor (1879), Borosilicate Glass (1893), Boranes and the Vacuum-Line Technique (1912), Dean-Stark Trap (1920), The Fume Hood (1934), Magnetic Stirring (1944)

A large-scale glove box. Smaller versions (often with separate entries for each arm) are in use wherever research on air-sensitive compounds is the main focus.

Antifolates

Sidney Farber (1903–1973), Yellapragada Subba Rao (1895–1948)

Pediatric pathologist Sidney Farber had seen how folate supplements (a form of vitamin B) could restore normal blood production in some types of anemia, and in 1947 he decided to see if that was true for leukemia, too. Unfortunately, giving folic acid to leukemia sufferers actually made the disease accelerate, so Farber wisely jettisoned that hypothesis in favor of the opposite one: if extra folic acid made leukemia worse, perhaps blocking the folate pathway could help slow it down. Indeed, a diet specifically chosen to be deficient in folic acid did seem to help.

Indian biochemist Yellapragada Subba Rao was also working on the medicinal chemistry of folate analogs at Lederle Laboratories in New York, trying to develop a better folic acid supplement for anemia patients. Some of the compounds he produced were close enough to the natural substance to be enzyme inhibitors in the biochemical pathway, and he provided one of these, aminopterin, to Farber to test for his "antifolate" idea. This compound showed dramatic effects in children with leukemia. Ten out of the sixteen patients given the treatment showed temporary remissions, an unprecedented response. In fact, it was so unheard-of that Farber's results were initially simply not believed. But other folate analogs showed efficacy, too, and other clinics were able to reproduce the study.

The most widely active and well-tolerated antifolate, methotrexate, also a product of Farber and Subba Rao's collaboration, went on to be tried against numerous forms of cancer (successfully against some), and along with several other antifolates, it is still in the therapeutic arsenal. It's a strong inhibitor of the enzyme dihydrofolate reductase, which is part of a pathway whose end products are the purine and pyrimidines that form the "rungs" in the ladder-like structure of DNA. Rapidly dividing cells—in this case, cancer cells—have to produce more DNA, since each new daughter cell gets a full copy, so starving a cancer cell of the materials it needs to produce all that DNA can be fatal to it. Chemotherapy would attack such targets for the rest of the century and beyond.

SEE ALSO Toxicology (1538), Bari Raid (1943), DNA's Structure (1953), Thalidomide (1960), Cisplatin (1965), Rapamycin (1972), Modern Drug Discovery (1988), Taxol (1989)

Fluorescent-tagged animal cells in a disease model of leukemia. The cell on the right is dividing, an all-too-common sight in cancer-cell cultures.

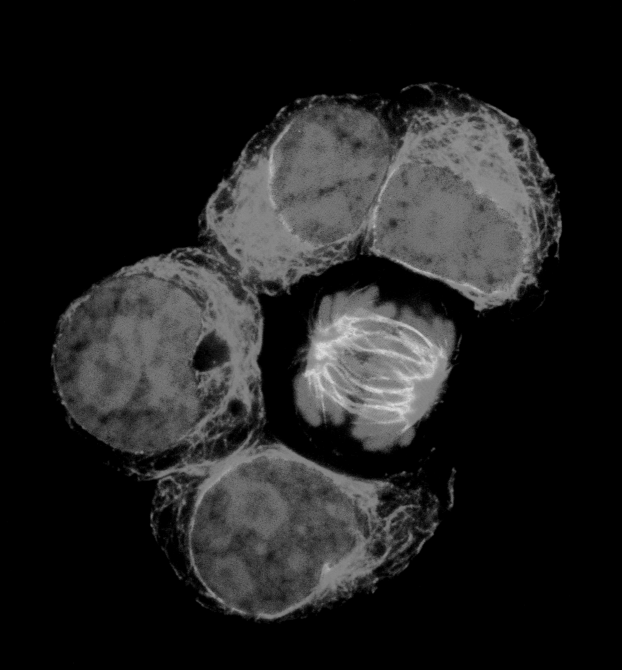

Kinetic Isotope Effects

Jacob Bigeleisen (1919–2010), Maria Goeppert Mayer (1906–1972)

American chemist Jacob Bigeleisen and German-born American physicist Maria Goeppert Mayer both worked on the Manhattan Project, using **photochemistry** to try to isolate uranium-235. The technique was not successful, but it did lead to the development of isotope chemistry. The topic can seem fearsomely complex, but you can create a model of one of the most important effects with a couple of tennis balls, a twangy spring, and some duct tape to hold them together. Experimenting with this model, you'll find that the balls vibrate together at a particular frequency. If you make one of the balls heavier, though, the vibrations slow down.

That's exactly what happens when you compare a carbon-hydrogen bond and a carbon-**deuterium** bond. Deuterium, an isotope of **hydrogen**, weighs about twice as much as plain hydrogen, and the carbon-deuterium bond vibrates more slowly (and has a lower energy state) than the carbon-hydrogen bond. This means that the bond to deuterium is harder to break, and if breaking it is a key step in a chemical reaction, then that reaction will run more slowly than it does with the "normal" carbon-hydrogen bond. This is the primary kinetic isotope effect. Bigeleisen and Goeppert Mayer published the first evidence of it in 1947, and chemists have used it in many ingenious ways to understand the details of **reaction mechanisms**.

There's also a secondary effect, observable even if no carbon-hydrogen (or carbon-deuterium) bond is being made or broken. The transition state of the reaction is sensitive to a deuterium replacement nearby on the molecule, but the effect is small. The primary kinetic isotope effect can cause a reaction to run five to ten times more slowly, but the secondary effect might change the reaction rate by only a few percent and requires careful experimental design to notice.

If you measure the isotope effects with atoms that are closer in weight—say, carbon-13 and carbon-12, whose weights differ by only about 8 percent—the resulting isotope effects are subtle, indeed. But such experiments are among the best ways to learn the intimate details of a reaction, and the key to designing better ones.

SEE ALSO Hydrogen (1766), Photochemistry (1834), Isotopes (1913), Deuterium (1931), Transition State Theory (1935), Reaction Mechanisms (1937), Dipolar Cycloadditions (1963), Enzyme Stereochemistry (1975), Isotopic Distribution (2006)

Maria Goeppert Mayer, who helped launch the entire field of isotope effects. She and Jacob Bigeleisen had worked out many of the details during their work for the Manhattan Project.

Photosynthesis

Melvin Calvin (1911–1997), **Samuel Goodnow Wildman** (1912–2004), **Andrew Alm Benson** (1917–2015), **James Alan Bassham** (1922–2012)

Photosynthesis is the quiet, unnoticed chemistry that is keeping everyone and everything on Earth alive. Our planet's atmosphere didn't even have much oxygen until photosynthetic microbes began cranking it out as a waste product (gradually killing off the planet's original microbial inhabitants or driving them into hiding). Photosynthesis not only produces the oxygen we breathe, but it also helps regulate the amount of **carbon dioxide** in the air. And as if making our atmosphere breathable were not enough, photosynthesis underpins the global food chain for almost every living organism, including humankind.

The strange thing is, the whole process depends on what appears to be one of the clunkiest enzymes ever seen. In 1947, Samuel Goodnow Wildman reported his discovery of a large, extremely abundant enzyme in spinach leaves that turned out to be a key player. Referred to by the laboratory nickname *Rubisco*, which is short for *ribulose biscarboxylase oxygenase* (and no wonder), it is an essential part of the Calvin cycle—the plant world's equivalent of the Krebs cycle of **cellular respiration**—discovered by American biochemist Melvin Calvin, working with compatriots James Alan Bassham, a chemist, and Andrew Alm Benson, a biologist. (Instead of mitochondria, plant cells have other ancient interlopers called chloroplasts to perform this work.)

Rubisco, probably the most abundant protein on Earth, can account for up to half the protein weight of a plant. The reason there's so much of it is that it's an incredibly slow enzyme. Instead of zipping through thousands of molecular changes per second, it processes *three*. It may be that this bizarrely low rate is a trade-off against its ability to tell carbon dioxide from oxygen; this is still an open question. After several billion years of evolutionary pressure, odds are that there must be good reasons for such an important enzyme to be so strange, but many research groups are putting that to the test by seeing what happens if they try to improve it for use in **artificial photosynthesis**.

SEE ALSO Carbon Dioxide (1754), Oxygen (1774), Cellular Respiration (1937), Isotopic Distribution (2006), Engineered Enzymes (2010), Artificial Photosynthesis (2030).

The green chloroplasts—where the Rubisco enzyme does its slow, strange work—are clearly visible inside these plant cells.

Donora Death Fog

1948

The history of industrial chemistry has not been an unbroken string of successes. A little-remembered incident from 1948 showed what could happen when not enough attention was paid to the environment surrounding chemical plants. In late October of that year, a temperature inversion trapped air around the valley town of Donora, Pennsylvania, for four days. Large steel and zinc works in the town regularly emitted a variety of toxic fumes, and though the zinc plant was also responsible for killing all the vegetation for a half mile around it, people didn't complain very much. But during this weather pattern, which essentially formed a lid over the valley, the poisonous exhaust fumes were trapped in the town at ground level. A toxic brew of sulfur dioxide, hydrogen fluoride, and fluorine, among other pollutants, combined with the smoke from the factories to make a nearly impenetrable smog. The residents of the town began coughing and having trouble breathing, and (remarkably, by today's thinking) many thought some sort of asthma epidemic had been triggered.

But the yellow "death fog" was doing more harm than that to the citizens of Donora. The Donora fire department depleted its supplies of oxygen, going from house to house helping distressed residents, and the Red Cross set up an emergency center to coordinate the work of the town's doctors. By the time rainstorms cleared out the clouds on the fifth day, twenty people and hundreds of animals had died, seven thousand more residents were ill, and surely many more would have succumbed had the weather not changed. The survivors had lingering health problems for years, and the entire incident, which got national attention, helped to make the country aware of the perils of air pollution.

Over the next decades, legislation and public pressure forced a huge change of attitude regarding factory emissions and air pollution. By now, the idea of a valley full of metalworking plants spewing out untreated waste for the inhabitants to breathe seems almost barbaric, but unfortunately it still occurs in many places around the world.

SEE ALSO Lead Contamination (1965), Bhopal Disaster (1984)

A nurse wears a surgical mask while making her rounds during the deadly Donora smog incident.

Catalytic Reforming

Vladimir Haensel (1914–2002)

The **catalytic cracking** technology for gasoline, developed in the 1930s by Eugene Houdry, was a vast improvement over Vladimir Shukhov and William Burton's **thermal cracking** process of forty years earlier, but it still needed the toxic **tetraethyl lead** for best performance. The platforming process developed by German-born American chemical engineer Vladimir Haensel solved that and several other problems at the same time. Haensel's technique gives higher octane numbers without lead and also produces valuable **hydrogen** gas, which is used to strip the sulfur out of the crude petroleum, turning it into **hydrogen sulfide** for use in the **Claus process**. It can also produce large amounts of aromatic by-products like benzene, which was a big factor in petroleum's supplanting coal and coal tar as a feedstock for industrial chemistry.

Getting this process to work, though, was an uphill struggle. Haensel's idea was to use platinum as a catalyst, which made sense chemically but was considered borderline crazy from an economic standpoint. Platinum metal was (and is) significantly more expensive than gold, and a reforming column packed with solid platinum beads, even if such a thing could be built, would have to be surrounded by armed guards. But Haensel realized that all the activity was in the surface layer of the metal, which meant that comparatively tiny amounts of platinum, appropriately spread out, could do the job. The solid support for the platinum was alumina, which was both inexpensive and also provided a Lewis acid surface that performed part of the catalysis.

Platforming (*plat* for platinum and *form* for reforming, or producing gasoline by cracking hydrocarbons) and the improvements made on it are the foundation for the modern oil industry (and for several others that use the starting materials it generates, in particular the plastics industry). Haensel himself turned to another application of platinum, leading the research programs that developed catalytic converter technology for cleaner emissions. In a way, Haensel's work cancels out the effects of Thomas Midgley's tetraethyl lead and **chlorofluorocarbons** on the atmosphere, since Haensel reduced acid rain (by making sulfur extraction much easier), helped eliminate lead pollution from gasoline, and drastically reduced smog formation by cleaning up engine exhaust.

SEE ALSO Hydrogen Sulfide (1700), Hydrogen (1766), Claus Process (1883), Thermal Cracking (1891), Tetraethyl Lead (1921), Acids and Bases (1923), Fischer-Tropsch Process (1925), Catalytic Cracking (1938).

A catalytic converter installed on a turbocharged engine. Vast amounts of exhaust gases are cleaned by this technology, resulting in carbon dioxide and water.

Molecular Disease

Linus Carl Pauling (1901–1994), James Van Gundia Neel (1915–2000), Harvey Akio Itano (1920–2010), Vernon Ingram (1924–2006), Seymour Jonathan Singer (b. 1924)

Links between some genetic mutations and disease are now taken for granted, but American chemist Linus Pauling and his coworkers—including American cell biologist Seymour Jonathan Singer and American chemist Harvey Akio Itano—were the first to establish the concept. Pauling's 1949 paper was titled "Sickle Cell Anemia, a Molecular Disease," and in that same year, American geneticist James Van Gundia Neel showed the disease's inheritance pattern in human beings.

Geneticists had shown that entire enzymes could appear and disappear via genetic defects, but Pauling and colleagues took the idea a step further. People with sickle cell anemia have an altered form of hemoglobin, the **oxygen**-carrying protein in red blood cells. Mutant hemoglobin has reduced oxygen capacity and twists the cells into the characteristic sickle shape. (It also confers some resistance to malaria, as later research showed, which helps explain why the condition has persisted in tropical populations.)

Pauling's student Itano eventually determined that the altered protein had a tiny difference in charge at different pH values, which allowed it to be separated from regular hemoglobin by careful **electrophoresis** (a technique in which proteins are made to move through a gel matrix under the influence of a strong electrical voltage difference). Once protein sequencing and other techniques became available, German-American biologist Vernon Ingram found, in 1956, that a sickle cell hemoglobin has a single change: a valine **amino acid** is replaced by a glutamic acid, which alters the structure of an entire region of the protein.

Every human being ever born carries some of these single-point mutations, and the vast majority of them are silent and harmless. But at certain points, in certain proteins, tiny variations can kill. Pauling's pioneering studies showed how this could work, demonstrating the application of chemistry techniques to biomolecules and helping to found the science of molecular biology.

SEE ALSO Hydrogen Cyanide (1752), Oxygen (1774), Amino Acids (1806), Sanger Sequencing (1951), Electrophoresis (1955)

"Sickled" red blood cells. The mutated hemoglobin aggregates into rodlike structures inside the cells, changing their shape.

Nonclassical Ion Controversy

Saul Winstein (1912–1969), **Herbert C. Brown** (1912–2004), **George Olah** (b. 1927)

The year 1949 inaugurated a controversy that was to last for decades, eventually involving several Nobel laureates and nearly all of the world's most eminent physical organic chemists. And it all started with one tiny compound: the norbornyl cation.

It had been clear for some time that positively charged carbon atoms (carbocations) were intermediates in many reactions. But the norbornyl system, a small bicyclic ring, behaved oddly in one of the standard carbocation reactions. Two different isomers of the starting material gave the same product, which made sense if they were going through the same intermediate, but one mysteriously reacted much faster than the other.

Two explanations were advanced. American chemist Saul Winstein proposed a strange-looking "nonclassical" carbocation, with the single positive charge distributed across three carbon atoms. This species would form faster from one isomer than from another. But London-born American chemist Herbert C. Brown (later a Nobel laureate for his work in organoboron chemistry) rejected this idea, believing that no exotic species had to be invented. He held that two standard carbocations were interconverting so quickly that experiments couldn't distinguish them, and that the rate differences were due to crowding around the reaction center ("steric hindrance," in the language of chemistry).

Sorting out these ideas strained the chemical and analytical techniques of the time and forced a number of new ones to be invented. A crucial method was found by Hungarian-American chemist George Olah (later a Nobel winner himself), who discovered powerful acid solutions that would allow carbocations to persist long enough to be studied by **NMR** (nuclear magnetic resonance). His experiments suggested that Winstein was right, and most of the field ended up believing the nonclassical structure, but Brown never threw in the towel.

In 2013, an **X-ray crystallography** structure of a norbornyl cation was obtained. It showed Winstein's nonclassical structure, which was, by that time, what most chemists had expected to see. The controversy was long over by then, but attempts to resolve it had resulted in advances in analytical techniques and a much greater understanding of chemical reactions and structures.

SEE ALSO Friedel-Crafts Reaction (1877), X-Ray Crystallography (1912), Reaction Mechanisms (1937), NMR (1961).

George Olah, whose NMR techniques provided strong evidence that the nonclassical ion was, in fact, real. The bitter controversy over this issue advanced physical organic chemistry in many ways.

Conformational Analysis

Hermann Sachse (1862–1893), **Ernst Mohr** (1873–1926), **Odd Hassel** (1897–1981), **Derek Harold Richard Barton** (1918–1998)

One of the early triumphs of **X-ray crystallography** was the determination of the structure of diamond in 1918. As shown by German chemist Ernst Mohr, it turned out to be a three-dimensional network of **tetrahedral carbon atoms** (just the way van 't Hoff had pictured them). The way that every atom in the lattice was bonded to its neighbors definitely explained diamond's hard and inert nature—it's as tied-down a structure as anyone could imagine.

If you trace out six-membered rings of carbons in the lattice, you get a characteristic one-end-up, one-end-down shape known to chemists as a "chair" conformation. (The six-membered ring with both ends pointing up is the "boat" conformer.) It was not widely appreciated that six-membered rings by themselves would behave the same way, although German chemist Hermann Sachse had tried to point this out in 1890. Norwegian physical chemist Odd Hassel showed, in 1943, that it was basically impossible for a single-bonded carbon ring like this to be flat: it had to have three-dimensional geometry.

Hassel went on to work out many of the details of six-membered ring conformations, which are extremely common structures across organic chemistry. Not all organic chemists were willing to believe that these structures were important under real reaction conditions, though, until 1950, when English organic chemist Sir Derek Harold Richard Barton explained many patterns of reactivity by showing how they influenced the products that could be formed. He drew many examples from **steroid chemistry**, which was being explored in detail and yielding results that could be explained only by paying close attention to the various ring geometries. Reagents have easier (or harder) paths in space to get to their reacting partners depending on what side of a flexible six-membered ring they approach from, for example.

Barton's theories eventually won him great acclaim and, in 1969, a Nobel Prize. But he always referred to the 1950 publication explaining the link between molecular conformation and reactivity as his "lucky paper," insisting that others could have figured out the same principles by closely studying the literature (and their own reactions!).

SEE ALSO Tetrahedral Carbon Atoms (1874), X-Ray Crystallography (1912), Steroid Chemistry (1942), Cortisone (1950), Synthetic Diamond (1953).

This three-dimensional model shows the conformation of a six-membered ring of carbon atoms.

Cortisone

Edward Calvin Kendall (1886–1972), **Philip Showalter Hench** (1896–1965), **Tadeus Reichstein** (1897–1996), **Percy Lavon Julian** (1899–1975), **Kenneth Callow** (1901–1983), **Lewis Sarett** (1917–1999), **Max Tishler** (1906–1989), **John Warcup Cornforth** (1917–2013)

The complex steroid ring system was an excellent proving ground for synthetic chemistry, and the products were in great demand, so **steroid chemistry** and steroid biology were exciting fields for discovery during the 1950s. When American chemist Edward Calvin Kendall, American physician Philip Showalter Hench, and Swiss chemist Tadeus Reichstein discovered the structures of cortisone and the adrenal corticosteroids and realized their far-reaching effects on the body, their value as drugs became very clear. Indeed, cortisone was successfully tried early on to treat rheumatoid arthritis.

But synthesizing the steroids strained the abilities of industrial chemists. In the 1940s, American chemist Percy Lavon Julian found routes from soybean compounds, following up on Russell Marker's yam-derived syntheses, and he later famously improved the cortisone route by avoiding a step using the highly toxic (and highly expensive) osmium tetroxide. In England, Australian-born chemist Sir John Warcup Cornforth and biochemist Kenneth Callow found a progesterone route for the drug firm Glaxo from compounds in the sisal plant.

Pharmaceutical company Merck's early cortisone process (discovered by American chemist Lewis Sarett and requiring over thirty steps) probably set the contemporary record for the longest industrial synthetic route—one that would be tried only for a very high-value product. One intermediate was a bright red dinitrophenylhydrazone, similar to the compounds produced by Emil Fischer from carbohydrates and in **Sanger sequencing** from proteins. A widely told story has American chemist Max Tishler, Merck's director of process chemistry, coming into a lab to find a red liquid spill and shouting, "I hope that's blood!" In 1951, though, a team at pharmaceutical company Upjohn found that microbial cultures oxidized progesterone to a perfect starting material for cortisone. They stunned Mexican pharmaceutical firm Syntex with an order for ten *tons* of progesterone and competed with Merck's cortisone chemistry, producing a drug that is still used to treat a wide array of conditions.

SEE ALSO Natural Products (c. 60 CE), Cholesterol (1815), Steroid Chemistry (1942), Conformational Analysis (1950), Sanger Sequencing (1951), The Pill (1951), Modern Drug Discovery (1988), Taxol (1989)

*An old-fashioned assay for cortisone in blood samples, 1952. (Note the forest of **separatory funnels**.)*

Rotary Evaporator

Lyman C. Craig (1906–1974)

Organic chemists use a lot of solvents in their work, but eventually those solvents have to be removed to isolate the reaction products to dry solids. Distillation is the way to do that, and it used to be a tedious, time-consuming process. But modern chemists have sped it up with devices called *rotary evaporators*, or *rota-vaps*; most don't even think of what they're doing as distillation anymore.

The apparatus, first described in 1950 by American biochemist Lyman C. Craig and colleagues, takes advantage of several effects at the same time. A flask of liquid is spun around on its axis to keep it mixing and to keep a thin film of liquid exposed at all times. The whole flask can then be lowered into a warm-water bath while it turns, and the system is put under a vacuum to lower the boiling point of the solvents inside it. Together, these make short work of most common organic solvents—they distill off rapidly and are collected (by a cold condenser) into a removable trap for disposal. If a strong vacuum pump is attached, even high-boiling solvents can be stripped off.

There are still a few things to look out for. If the flask is filled too high, if it's not rotating, or if it's heated up too quickly, the liquid in it can "bump"—suddenly blasting up and flooding the condenser—which generally means that the whole evaporating process has to be done again. Forgetting to turn on the vacuum pump will lead to a flask full of valuable material slipping off into the water bath, an irritating event that every chemist has experienced at some point. And naturally, if the product of a reaction is low-boiling itself, it will be stripped right out of the flask along with the solvent. But when a rota-vap is working well (which is most of the time), it is one of the most useful tools in an organic chemistry lab.

SEE ALSO Purification (c. 1200 BCE), Separatory Funnel (1854), Erlenmeyer Flask (1861), Soxhlet Extractor (1879), Borosilicate Glass (1893), Dean-Stark Trap (1920), The Fume Hood (1934), Magnetic Stirring (1944), Glove Boxes (1945), Acetonitrile (2009).

The rota-vap, a constant companion to organic chemists everywhere. Whatever that red compound is will soon be concentrated and dried along the inner walls of the flask.

Sanger Sequencing

Archer John Porter Martin (1910–2002), **Richard Laurence Millington Synge** (1914–1994), **Frederick Sanger** (1918–2013), **Hans Tuppy** (b. 1924)

Present-day chemists and molecular biologists are accustomed to manipulating proteins with confidence, figuring out their **amino acid** sequences and producing variations at will. But before 1951 the structures of proteins was a mystery. With twenty amino acid options per position, the number of possible proteins gets out of control very, very quickly, and there were many important open questions. Did a given protein always have a particular sequence and a particular structure? Perhaps just the active sites of a protein were fixed, and the rest of it could vary? No one would know until amino acid sequences could be determined with certainty.

In 1951, Sanger and coworkers, in particular Austrian biochemist Hans Tuppy, showed the world how it could be done, sequencing the "B chain" of insulin. They were helped by the revolutionary 1943 discovery by English chemist Archer John Porter Martin and English biochemist Richard Laurence Millington Synge that amino acids, small peptides, and other molecules could be separated by **chromatography**, letting the solvent soak upward from a mixture spotted near the bottom of a sheet of filter paper.

Sanger invented a reaction that would make a brightly colored dinitrophenyl (DNP) derivative out of the amino end of a peptide chain (the end of the chain with a free amine, or NH_2, group) without disturbing the rest of it, but then stay attached even after the amino acids were broken apart. Sequencing of short peptides was now possible by working backward from the amino end, albeit with painstaking work and with allowance for amino acids that reacted outside the usual patterns. For insulin, Sanger and his team broke up the protein into an assortment of shorter pieces. These were separated and run through the sequencing scheme, and then the whole protein was pieced together like a puzzle.

The result proved that proteins do have specific sequences that give them their shapes and properties—a critical advance that won Sanger a Nobel and suggested that cells must somehow contain exact codes for all their proteins. Sanger later invented techniques to sequence DNA and RNA, and won a second Nobel for that!

SEE ALSO Amino Acids (1806), Chromatography (1901), Molecular Disease (1949), Sanger Sequencing (1951), DNA's Structure (1953).

A molecular model of the insulin protein. Sanger's methods not only provided the composition of a protein, but they also helped chemists to think of them as actual organic compounds that could be studied and manipulated using chemical techniques.

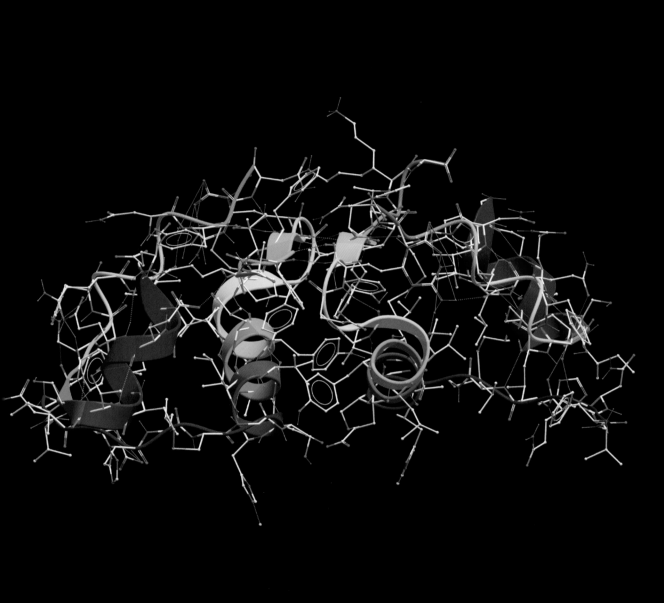

The Pill

Katherine McCormick (1875–1967), Gregory Goodwin Pincus (1903–1967), Min Chueh Chang (1908–1991), George Rosenkranz (b. 1916), Carl Djerassi (1923–2015), Frank Benjamin Colton (1923–2003), Luis Ernesto Miramontes Cárdenas (1925–2004)

One of the most important targets for the early steroid researchers was finding a compound that worked like progesterone but could be taken as a pill (oral progesterone itself was broken down too quickly in the body). The hormone was known to prevent ovulation, and an orally active mimic might work as a human birth-control pill. Mexican pharmaceutical company Syntex's expertise in steroid chemistry made it a major force in this endeavor, but its founder, Russell Marker, had left the company, taking his Mexican yam methods with him. That left Hungarian-born chemist George Rosenkranz to try to reverse-engineer the chemistry (which he did).

Austrian-born chemist Carl Djerassi and his group at Syntex, including Mexican chemist Luis Ernesto Miramontes Cárdenas, made norethindrone (only a few months after the Syntex **cortisone** synthesis had been revealed), and American chemist Frank Benjamin Colton, at pharmaceutical firm G. D. Searle, synthesized the similar norethynodrel. Clinical trials of norethynodrel were conducted with funding from Katherine McCormick, who was both a biologist and a wealthy heiress, and the drug proved highly effective. Both compounds were used as components of competing first-generation birth-control pills.

The subject of birth control was (and is) controversial. Some drug companies stayed out of the field altogether, or refused to explicitly work on birth control even if they did do steroid research. But the demand was certainly there, and during the early 1960s the first oral contraceptive pills were approved by the FDA.

The world has never been the same. "The pill," whose first iteration was co-invented by American biologist Gregory Goodwin Pincus and Chinese-born American biologist Min Chueh Chang, was not widely available as a contraceptive at first, and its distribution was regulated by the courts. But for the first time in human history, a medicine had made pregnancy optional, allowing women to have control over their fertility and thus the rest of their lives.

SEE ALSO Cholesterol (1815), Steroid Chemistry (1942), Cortisone (1950), Modern Drug Discovery (1988)

It's a wonder that such small pills (steroid derivatives can be very potent) can create such widespread changes in our world.

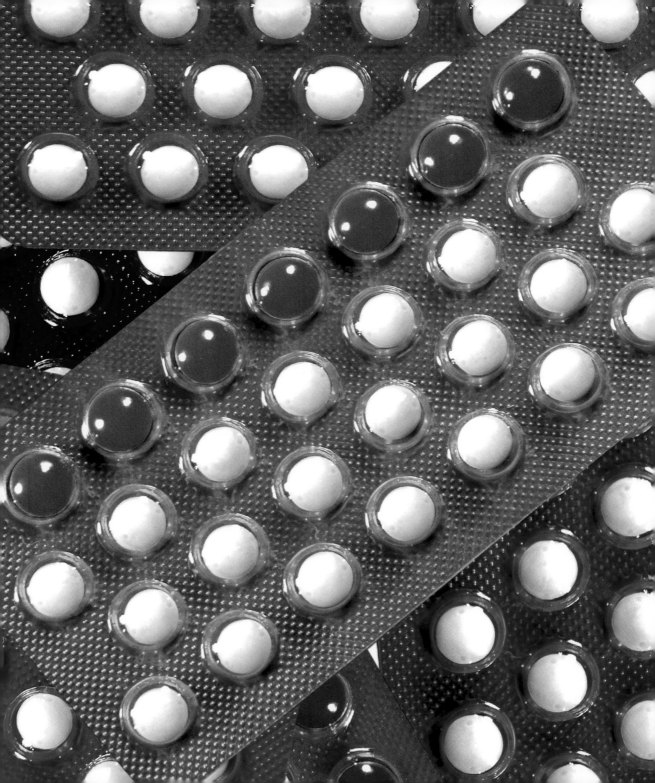

Alpha-Helix and Beta-Sheet

Robert Corey (1897–1971), **William Astbury** (1898–1961), **Linus Pauling** (1901–1994), **Herman Branson** (1914–1995)

Techniques like **X-ray crystallography** and **NMR** (nuclear magnetic resonance) have shown that although every protein is different, there are structural motifs and themes that appear over and over. The two most important are the alpha-helix and beta-sheet.

The alpha-helix looks a bit like what you would get if you held out a spool of ribbon and let some of it drop down. It's a spiral staircase of **amino acids**, gradually corkscrewing around and forming favorable hydrogen bonds. Some amino acids are much more helix-prone than others, and the parts of a protein that curl up like this can be predicted pretty well by just their amino acid sequence. American chemist Linus Pauling had the idea for it in 1948 while in bed with a cold—he drew a chain of amino acids out on a sheet of paper and tried curling it around to figure out how the hydrogen bonds that held the molecule together could be formed. He then gave American physicist and chemist Herman Branson the task of determining which helical structures were possible.

The beta-sheet, proposed by English molecular biologist William Astbury and refined by Pauling and American biochemist Robert Corey in 1951, is a different animal: a tightly hydrogen-bonded network of zigzagging amino acids that are, roughly, lying on top of each other. In fact, sometimes beta-sheet structures are so well matched that the protein becomes an insoluble aggregate, as happens with the amyloid protein in Alzheimer's disease. For this reason, there has surely been evolutionary pressure away from too much beta-sheet formation, but it's still a very important motif, giving local order to many classes of protein.

These structures are fundamental to the shapes of all proteins, and thus to all life. A typical structure might have a few alpha-helices bundling together at angles to each other (like a bunch of baguettes standing on end in a tall basket) and connected by loops with the rest of the chain forming beta-sheet regions off to one side. Sometimes protein structures are drawn in a ribbon-type diagram just to emphasize the twists and turns of the backbone.

SEE ALSO Amino Acids (1806), Spider Silk (1907), X-Ray Crystallography (1912), Hydrogen Bonding (1920), NMR (1961).

Rhodopsin, the light-sensitive pigment in the retina, sits in the cell membrane as a bundle of alpha-helix units. Such alpha-helix interactions are a key part of protein structure.

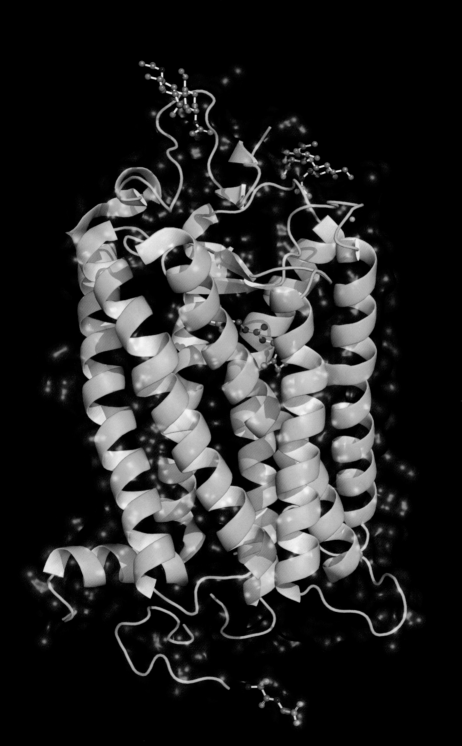

Ferrocene

Samuel A. Miller (1912–1970), **Robert Burns Woodward** (1917–1979), **Geoffrey Wilkinson** (1921–1996), **Ernst Otto Fischer** (1918–2007), **Peter Ludwig Pauson** (1925–2013), **Thomas J. Kealy** (1927–2012)

Two different research groups (chemists Peter Ludwig Pauson and Thomas J. Kealy in the U.S. and chemists Samuel A. Miller, John Tebboth, and John Tremaine in the U.K.) were investigating the reactions of a five-membered ring compound called cyclopentadiene in 1951 when, in the presence of iron salts, a new compound formed that was clearly not what anyone had planned. Both groups expected colorless liquid products, but this new material was a bright-orange, crystalline solid, which turned out to be a window into a new sort of chemical structure.

The molecule, named *ferrocene*, was found to contain one atom of iron for every two cyclopentadienes, but how these were arranged was a subject of great disagreement. Two eminent organic chemists, American Robert Burns Woodward and Englishman Geoffrey Wilkinson, proposed that the iron atom must be "sandwiched" between the cyclopentadienes. The bonding between the iron atom and the two rings was spread out evenly across both rings—a type of chemical bonding no one had yet thought of. This makes ferrocene the most perfectly balanced and stable example of the "metallocene" compounds—each cyclopentadiene gets enough extra electrons to be aromatic, and the iron atom completely fills its outer electron shell.

In Germany, chemist Ernst Otto Fischer had worked out the same structure, later completely confirmed by **X-ray crystallography**. He began to prepare similar compounds with other metals, as the ferrocene story set off a wave of organometallic chemistry research. Fischer and Wilkinson both received the 1973 Nobel Prize for their work.

Currently, metallocenes are used as industrial catalysts and as reagents for organic chemistry. It seems strange that such a widely applicable type of compound had been missed for so long, but it was revealed that cyclopentadiene distilled through an iron apparatus was known to clog the pipe with some worthless yellow-orange stuff. Thus, a Nobel Prize was scrubbed out with a brush and thrown away.

SEE ALSO Benzene and Aromaticity (1865), Coordination Compounds (1893), X-Ray Crystallography (1912), Unnatural Products (1982).

A digital model of a ferrocene molecule, showing the (relatively large) iron atom sandwiched between the flat cyclopentadienes.

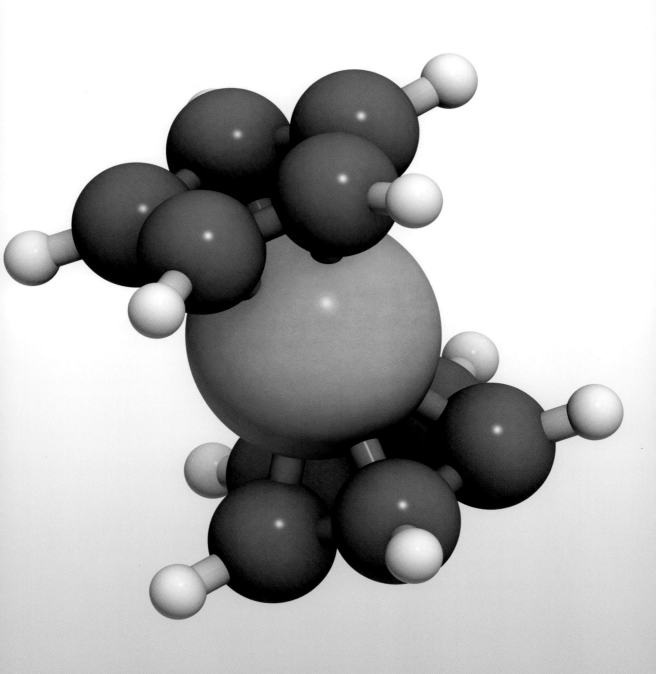

Transuranic Elements

Edwin Mattison McMillan (1907–1991), **Glenn Theodore Seaborg** (1912–1999), **Philip Abelson** (1913–2004)

The eighteenth and nineteenth centuries were the heyday for the discovery of new elements. Substances known since antiquity (gold, silver, copper) were recognized as fundamental building blocks, and new elements were teased out of ore samples and the air itself. By the time **polonium and radium** were discovered in the early twentieth century, the lighter elements of **the periodic table** had almost all been filled in, and it became clear that uranium was probably the heaviest element with isotopes that could be considered stable. The "transuranics"—elements with atomic numbers greater than 92—were all "hot" (universal scientific shorthand for *radioactive*), some of them spectacularly so, with extremely short half-lives. They had to be made through exotic heavy-atom collisions and were characterized by the radiation and decay products they produced as they fell apart.

American chemist Glenn Theodore Seaborg (a student of Gilbert Lewis of acid-base fame) made this his life's work and was involved in the discovery of nine of these exotic and dangerous substances. American physicists Edwin Mattison McMillan and Philip Abelson had discovered neptunium in 1940, so Seaborg and his team started with plutonium and became part of the Manhattan Project's effort to produce enough of it for the nuclear bomb dropped on Nagasaki in 1945. After the war, Seaborg continued the string of new-element discoveries (leading to a Nobel Prize in 1951), and also served as chairman of the Atomic Energy Commission, among many other posts. Most unusually, he lived long enough to see an element named after him: seaborgium, a fleeting transuranic with atomic number 106. Seaborg predicted that an "island of stability" might show up around atomic number 118, but stability looks to be a relative term. This work is now so specialized that only three labs in the world account for almost all the discoveries in recent decades: Seaborg's own scientific descendants at Berkeley, a German team in Darmstadt, and a Russian one in Dubna, north of Moscow. If the island of stability exists, one of these groups will find it.

SEE ALSO The Periodic Table (1869), Polonium and Radium (1902), Isotopes (1913), The Last Element in Nature (1939)

Glenn Seaborg and a representation of the element that bears his name. Only a relative handful of actual seaborgium atoms have ever existed on Earth at the same time.

Gas Chromatography

Archer John Porter Martin (1910–2002), **Anthony Trafford James** (1922–2006)

Chromatography began in the early twentieth century, with a liquid phase running over a solid one and the compounds separating out as they interacted with the solid medium. During the 1930s, researchers experimented with heated columns filled with gas-phase samples, and for several years gas chromatography (GC) technology evolved faster than liquid chromatography did. In a modern instrument, the sample is vaporized into a long, narrow, heated tube and swept along by a carrier gas (usually nitrogen and/or helium, chosen for their nonreactivity).

The pioneers of GC had used solid packing materials to separate their mixtures, but British chemists Anthony Trafford James and Archer John Porter Martin published a paper in 1952 demonstrating gas-liquid partition chromatography (GLPC). Allowing the gaseous samples to contact a thin layer of liquid while they moved through the column gave dramatically better separations—the components of the mixture dissolved in and out of the liquid phase, often a very high-boiling polymer, and this partitioning effect spread the mixtures out quickly and thoroughly. A very thin hollow column whose inside surface is coated with the liquid phase is often used (capillary GC), and while some GC columns are still packed with a solid support, it's usually there just to be coated with the liquid phase.

GC columns have been hooked up to a number of different detectors to read out the peaks as they come through, including sending the gas stream through a much higher-tech version of a flame test (identifying compounds by their emission spectra). **Mass spectrometry** (MS) was a natural fit, and, by the 1960s, the GC/MS became one of the most powerful analytical techniques available. Applications for this versatile technique include detection of performance-enhancing drugs in human samples; detection of illegal narcotics, explosives, and chemical warfare agents; and atmospheric analysis (even on Saturn's moon Titan where a GC/MS instrument was carried to its surface by the *Huygens* lander in 2005).

SEE ALSO Flame Spectroscopy (1859), Chromatography (1901), Mass Spectrometry (1913), Electrospray LC/MS (1984)

The inside of a modern GC, showing the coiled columns, which stay inside a heated cabinet and can be swapped out as needed.

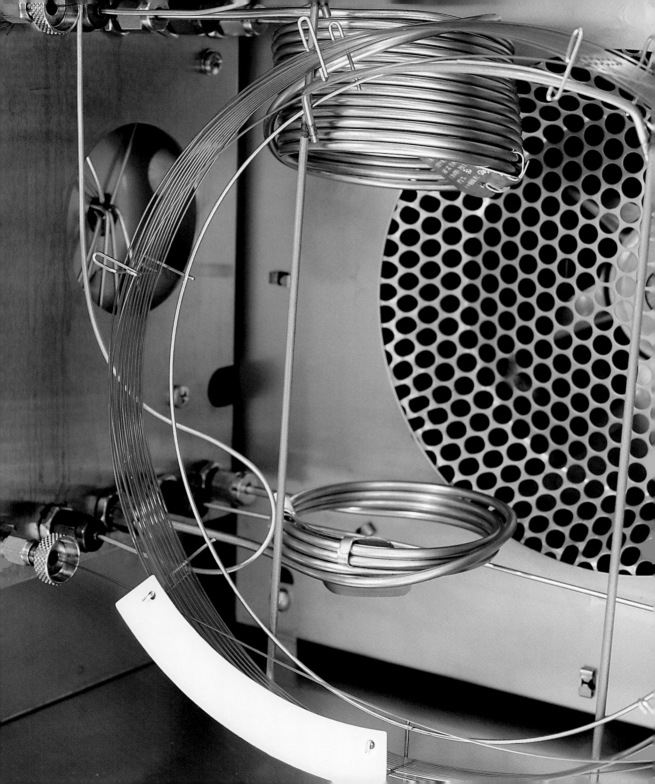

Miller-Urey Experiment

Harold C. Urey (1893–1981), Stanley Miller (1930–2007)

For thousands of years, humankind has been trying to understand the origin of life. Biochemistry had to start somewhere, and presumably was much simpler at the beginning. But what did that beginning look like, and how did it get going? Could it happen again on other planets? How similar, then, would it be to what we know? None of these questions have a good answer yet.

A major step forward was taken in 1952 by American chemists Stanley Miller and Harold C. Urey. The idea was to take a believable "prebiotic" atmosphere and subject it to heat and the equivalent of lightning to see what compounds might form. They sealed an apparatus with water, methane, ammonia, and **hydrogen**; heated the water; fired electrical sparks across the vapor; and then cooled the system again to send the condensate back into the water layer. The process was set to cycle repeatedly, and it began to make colored compounds during the first day. After two weeks, over 10 percent of the methane had been turned into more complex compounds, and analysis of the mixture was startling. At least eleven of the twenty key **amino acids** were present, along with many simple carbohydrates and a variety of other molecules. Modern re-analysis of the sample has shown that all of the major amino acids were produced, some of them originally below the limits of detection.

Many similar experiments have been run since, using all sorts of different possible early atmospheres and conditions. Almost all of them produce rich brews of simple organic compounds, including many of what we would now call the building blocks of life. These arise through the formation of reactive molecules like **hydrogen cyanide** and formaldehyde from the original gases, which can go on to produce complex structures. The mixture of compounds in samples like the **Murchison meteorite** can be quite similar to those produced by these experiments, and spectroscopic studies have found many of these compounds around other stars as well as in comets and interstellar nebulae. The universe appears to be swimming in small biomolecules.

SEE ALSO Hydrogen Cyanide (1752), Hydrogen (1766), Amino Acids (1806), Murchison Meteorite (1969), Tholin (1979).

A re-creation at NASA of the Miller-Urey experiment. Note the dark mixture already forming in the chamber. Such brews of organic molecules seem to have many opportunities to form in the universe.

Zone Refining

William Gardner Pfann (1917–1982)

Impurities lower a substance's melting point. That's why salts of various kinds are used to melt ice—salty water has a freezing point well below that of pure water. It's also the reason that carefully determined melting points are considered an excellent test of purity. In the old days of organic synthesis, before modern instrumentation, the final test of whether a natural product compound had truly been synthesized was a "mixed melting point." That involved mixing together the newly synthesized sample with a small amount of an authentic sample and checking the melting point before and after. The melting point of this mixture would stay the same only if its two components were identical.

Lowering the melting point is also the basis for an ingenious technique that's been used to purify metals and other substances. If a long rodlike sample is pulled slowly through a narrow heating zone, any impurities in the sample are swept along in the molten band (whose melting point goes down more and more as the impurities accumulate), and much cleaner (purer) material is left behind. At the end of the process, the impurities have concentrated at one end of the sample, which can be sawn off. In 1952, American chemist and materials scientist William Gardner Pfann developed the technique, known as *zone refining*, at Bell Labs to purify semiconductor materials such as germanium and (later on) higher-melting silicon, producing material that was one thousand times cleaner than had ever been tested.

Zone refining is still an excellent method for producing extremely pure samples of high-value materials, such as metals and semiconductors, although some materials are much better candidates for it than others. The best are substances whose solid and liquid phases have quite different properties. Even in these cases, the impurities have to be able to move through the melt zone at decent rates, and those rates need to be similar, but zone refining is still one of the best purification techniques available when those conditions are met.

SEE ALSO Purification (c. 1200 BCE), Beryllium (1828), Polywater (1966)

Ultra-pure silicon and material with precisely known amounts of the right impurities are essential for computer chip production.

Thallium Poisoning

The year 1952 saw a run of murders in Australia with the usual sorts of motives—family disputes, unfaithful spouses, and so on—but with one bizarre twist: they all were committed with a chemical murder weapon, the otherwise obscure element thallium. Thallium salts were commonly available as rat poison at the time, but they are just as toxic to people, for the same reasons. The ions of thallium and potassium (an essential element for life) are about the same size, and the protein channels that move potassium in and out of the cell will move thallium in as well. But its reactivity is very different from potassium's. Like other toxic heavy metals such as lead and **mercury**, thallium forms a stable bond with sulfur. This leads it to react irreversibly with many sulfur-containing molecules in the body (such as one of the **amino acids,** cysteine, important in the functioning of many enzymes).

Constant dosing of thallium leads to various symptoms, including hair loss and nerve damage, which can be hard to recognize as thallium poisoning unless foul play is suspected. It's not for nothing that thallium sulfate was once known as "inheritance powder." The antidote, effective if given in time, is large oral doses of the compound **Prussian blue**, which binds to thallium and removes it from the body via the digestive system.

Thallium shows up in several old mystery stories and is still occasionally used with criminal intent (though it is no longer available as an over-the-counter rat poison). It was a favorite murder weapon of the infamous Saddam Hussein regime, and in 1971 seventy people were poisoned by a man in England. A recent thallium case involved a New Jersey chemist who was convicted in 2013 of murdering her husband. Unfortunately for her, thallium poisoning is no longer difficult to detect. Atomic emission spectroscopy—the modern version of the flame test—can easily detect toxic levels of the element, and because it is not part of the biochemistry of any normal living creature, thallium's presence arouses immediate suspicion: the chemical equivalent of a large knife sticking out of the back.

SEE ALSO Mercury (210 BCE), Toxicology (1538), Prussian Blue (c. 1706), Amino Acids (1806), Paris Green (1814), Flame Spectroscopy (1859), Tetraethyl Lead (1921), Lead Contamination (1965).

Veronica Monty (right), accused of using thallium to poison her son-in-law, arrives at court in Sydney. She was acquitted but later poisoned herself . . . with thallium.

DNA's Structure

Francis Harry Compton Crick (1916–2004), **Maurice Hugh Frederick Wilkins** (1916–2004), **Rosalind Franklin** (1920–1958), **James Dewey Watson** (b. 1928)

Even people who know very little about biochemistry have probably heard of molecular biologists James Dewey Watson and Francis Harry Compton Crick, for the very good reason that almost everyone has heard of DNA. The story of the 1953 discovery of DNA's double-helix structure has been told many times, and not always without controversy (the contributions of English physicist Maurice Hugh Frederick Wilkins and English chemist and X-ray crystallographer Rosalind Franklin are sometimes considered to have been given short shrift). But from any angle, it is surely one of the key scientific breakthroughs of the twentieth century, lending insight to so many different questions, including the mechanism of heredity, the foundation of every person's individual biochemical identity, the differences that set one species apart from another (and the differences that make us human), the deeper similarities among all forms of life on Earth, and the timeline of evolution itself.

But the discovery of DNA's double helix represents something else as well: the founding of the science of molecular biology, whose early practitioners had started to make an important transition from being biologists to being something very much like chemists. DNA (deoxyribonucleic acid) is a real molecule that does real chemistry. It is held together by hydrogen bonds, for example, and it undergoes **photochemistry** in the presence of strong sunlight (which is why sun exposure can lead to skin cancer). The new hybrid field of chemical biology has used DNA's unique properties (and the suite of unique enzymes that have evolved to process it with such speed and accuracy) to invent completely new reactions and techniques, the first of which was the **polymerase chain reaction**, which can be used in genetic testing, tissue typing for organ transplant, and cancer treatment.

In the end, though, the DNA molecule is just a beautiful thing. The pairing of the bases across the double helix, the unzipping mechanism of its replication, the way it's stored in the cell as a coiled coil of coils—all of these and more make it one of the most extraordinary and elegant chemical substances that human beings have ever encountered.

SEE ALSO Photochemistry (1834), Hydrogen Bonding (1920), Antifolates (1947), Electrophoresis (1955), DNA Replication (1958), Polymerase Chain Reaction (1983).

A molecular model of DNA's twisted structure. A vast array of extremely efficient proteins bind to DNA, help to copy it, check it for errors, and constantly repair it.

1953

Synthetic Diamond

Charles Algernon Parsons (1854–1931), Howard Tracy Hall (1919–2008)

So what's the big deal about diamonds, anyway? They are just carbon, one of the more common elements on the planet, and although they are the most decorative of the carbon allotropes, other minerals (such as moissanite) sparkle as much or more. But diamonds offer other unique features—such as hardness, ability to transfer heat, and resistance to pressure—that make them valuable enough to be worth the effort to make them in the lab for use in electronics and nanotechnology applications.

The preferred term is "laboratory grown" rather than "synthetic," because "a synthetic diamond" sounds like some sort of cheap substitute for the real thing. Artificial diamonds, though, have the same carbon latticework as the ones mined from ancient volcanic vents. One of the techniques used replicates those same high pressures and high temperatures (HPHT), but that wasn't easy to realize. A number of supposed routes to diamonds were reported starting in the late 1800s under all sorts of violent and hazardous conditions, but none of these could ever be reproduced, and, in 1928, British engineer Sir Charles Algernon Parsons showed, after decades of research, that they almost certainly did not work. It wasn't until 1953 that HPHT diamonds were made without a doubt by ASEA, a large electrical manufacturing company in Sweden, and by a team led by American physical chemist Howard Tracy Hall at General Electric in the U.S., using tremendous pressures at temperatures of thousands of degrees. The diamonds produced were fine, gritty, and impure, but they were at least industrial-abrasive grade. Variations of this method eventually produced gem-quality diamonds in the early 1970s.

Another route to diamond formation is chemical vapor deposition (CVD), a process first achieved by American scientist William G. Eversole in 1952 that grows diamonds by slow accumulation from extremely hot carbon vapor and can produce large gem-quality **crystals**. But if CVD could be used to deposit thin diamond films in a controlled manner on the surface of objects as an industrial or decorative coating, that would be a much more profitable application, and a huge amount of research on this idea continues today.

SEE ALSO Crystals (c. 500,000 BCE), Surface Chemistry (1917), Conformational Analysis (1950), Fullerenes (1985), Carbon Nanotubes (1991), Graphene (2004)

A synthetic diamond, which is probably even more interesting to materials scientists than it is to jewelry shop owners.

Electrophoresis

Arne Wilhelm Kaurin Tiselius (1902–1971), Oliver Smithies (b. 1925)

The electrochemistry work of the early nineteenth century led to one of the most widely used techniques in biochemistry and molecular biology by the middle of the twentieth. That's because biomolecules carry charged groups as part of their structures (the acidic and basic **amino acids** and the phosphate groups on DNA and RNA). Such groups carry their molecules through a solution toward an oppositely-charged electrode. Swedish biochemist Arne Wilhelm Kaurin Tiselius pioneered the use of this effect for analytical chemistry in 1937, showing that proteins migrated through a buffer solution according to their size and charge. It wasn't an easy technique, though, and it couldn't separate closely related compounds well. Chemists needed a way to standardize the mixtures and slow the process down, leading to experiments with thicker solutions and even jelly-like materials.

British-born American geneticist Oliver Smithies reported in 1955 that starch gels were an excellent system for this purpose, a finding that immediately popularized the technique (although starch has largely been replaced by other materials, such as the polymer acrylamide). These days, premade gel strips and plates are loaded into an electrophoresis machine, the sample is added, and a voltage is applied, causing the various parts of the sample mixture to migrate and resolve into bands along the gel, very much like **chromatography**. Once the run is finished, the bands are stained with dyes or other reagents to make them visible. The standard blue dye for protein gels, Coomassie stain, would be the color of protein chemistry's flag if it had one.

Since DNA and RNA molecules are all negatively charged, they migrate through the gel according to their size—a phenomenon that has been extremely useful in identifying them. These gels are often run with a standard mixture of DNA or RNA pieces of known size loaded along one side so that the unknown bands can be compared to the bands on the standard ("ladder") mixture. Modern molecular biology is all about manipulation of proteins and nucleic acids, and so many gels are run in a typical research lab that the field would probably grind to a halt without them!

SEE ALSO Amino Acids (1806), Chromatography (1901), Molecular Disease (1949), DNA's Structure (1953)

Ms. Edna Ardales, a researcher from the International Rice Research Institute, uses UV light to review an electrophoresis gel from a DNA sample in 2007. Such "sequencing gels" have been replaced by faster techniques now, but the technique behind them is still in constant use.

The Hottest Flame

Aristid V. Grosse (1905–1985)

All flames are not created equal. Their temperatures depend on several factors, for example, the bonds broken and formed in the combustion reactions, the amount of oxidant present (usually **oxygen** itself), and how well it mixes with the fuel. The familiar propane/oxygen flame can reach about 3,600 degrees Fahrenheit (2,000 degrees Celsius), and an **acetylene** flame can reach almost 6,000 degrees Fahrenheit (3300 degrees Celsius), but both these hydrocarbons have limits. Their **hydrogen** is oxidized to water vapor, which absorbs a large amount of heat. Above about 3,600 degrees Fahrenheit (2,000 degrees Celsius), water itself breaks down, absorbing still more heat as it does. To reach the highest temperatures, a flame would need to consist of oxygen and a substance without hydrogen. Alternatively, a strong oxidizer other than oxygen might do the trick.

Two of the hottest flames known fit these conditions perfectly. If you are so foolhardy as to mix hydrogen and fluorine, you will set off a vigorous reaction indeed (and produce clouds of hot, corrosive, toxic hydrogen fluoride as well, so be warned). This flame can reach just above 7,200 degrees Fahrenheit (4,000 degrees Celsius), although measuring it will be a challenge, since it will destroy most temperature probes. But the hottest flame of all is produced by the exotic compound dicyanoacetylene (also known as *carbon subnitride*). Its firepower was discovered by A. D. Kirshenbaum and Latvian-American chemist Aristid V. Grosse, who were studying the effects of high-temperature flames for aerospace applications, and described it in a 1956 paper. Dicyanoacetylene is not very stable—an explosion hazard under the best of conditions. But it has no hydrogen in its structure, and its combustion products are carbon monoxide and nitrogen gas, both of them very stable. Its flame with pure oxygen reaches 9,008 degrees Fahrenheit (4,987 degrees Celsius), temperatures within striking distance of the surface of the sun (9,900 degrees Fahrenheit, 5,500 degrees Celsius). But there could be something even fiercer: If anyone has reacted dicyanoacetylene with fluorine, it has not been reported. Stand back!

SEE ALSO Hydrogen (1766), Oxygen (1774), Gibbs Free Energy (1876), Isolation of Fluorine (1886), Acetylene (1892), Thermite (1893).

This acetylene cutting-torch flame is hottest at the tip of the bright white "inner cone" near its opening. Very few people have ever seen hydrogen/fluorine or oxygen/dicyanoacetylene flames, which is probably a good thing.

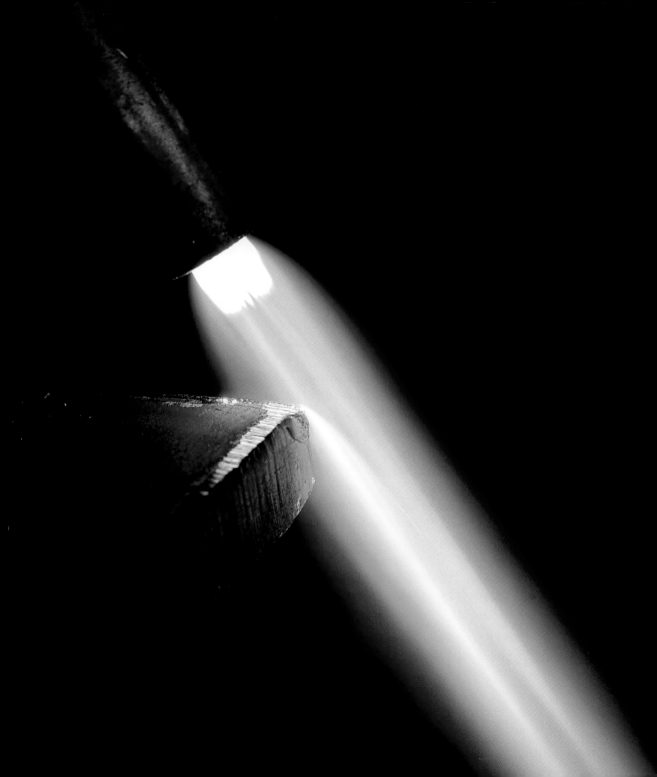

Luciferin

William David McElroy (1917–1999), **Bernard Louis Strehler** (1925–2001), **Emil H. White** (1926–1999)

If you're a chemist (or any kind of scientist), you should always be ready to ask "How does that work?" How, for example, do fireflies give off light, over and over, in a controlled fashion? Why do their species have different color lights? What chemicals are emitting the light, and can they be used outside of fireflies?

American biochemists William David McElroy and Bernard Louis Strehler (a graduate student) worked out the answers to these questions in the 1940s and 1950s, and it was no small feat. The technology for natural-products chemistry at the time required milligram amounts of the material (at least), and individual fireflies contain far less than that. So McElroy took out newspaper advertisements that Johns Hopkins, where he was a professor, would pay one penny per firefly. It was money well spent—the children of Baltimore, far more efficient firefly catchers than grad students and postdocs would have been, and with more time to devote to the task, brought in fifteen thousand specimens.

Extracting dried firefly abdomens was tricky, but about nine milligrams of a compound called luciferin were eventually isolated, and its properties were described in a 1957 paper. It was a small aromatic ring compound containing nitrogen and sulfurs, as worked out by American chemist Emil H. White in 1961, and it showed **fluorescence** under ultraviolet light (although not the yellow-green firefly color, since fireflies were unlikely to be shining ultraviolet light on luciferin from inside their abdomens. But if luciferin was oxidized, it produced the characteristic glow. This is an example of chemiluminescence, formation of a high-energy chemical species that sheds this energy by emitting light. Fireflies have an enzyme (luciferase) that does this oxidation reaction.

Further work showed that all firefly species used the same luciferin, but that different forms of the enzyme seemed to cause the different colors (some fireflies glow orange). Luciferin can be combined with an oxidizing reagent to emit its glow on demand—thus, its use in glow sticks sold for parties and festivals. Additionally, luciferin is used as a glowing protein "tag" for biochemical assays.

SEE ALSO Natural Products (c. 60 CE), Fluorescence (1852), Green Fluorescent Protein (1962)

The distinctive yellow-green of firefly luminescence in a forest near Nuremberg, Germany, is shown in this long-exposure photograph.

DNA Replication

Matthew Meselson (b. 1930), **Franklin Stahl** (b. 1929)

The discovery of **DNA's structure** raised important questions about how the molecule transmits genetic information. Somehow, when a cell divides, a full double-stranded copy of the original DNA ends up in each new cell—but how? The whole course of future genetic research hinged on knowing the exact process. James Watson and Francis Crick proposed that the double helix was somehow unwound and separated, with each strand serving as the template for a new helical partner. This was the "semiconservative" hypothesis, where each new DNA had one of the older DNA strands in it. The "conservative" hypothesis proposed that the new double-stranded DNA was made completely fresh, with enzymes able to read the sequence off the original DNA and make a new set from scratch. Finally there was the "dispersive" hypothesis, which suggested that the double-stranded DNA was broken up every so often to build on another chunk of fresh DNA, resulting in DNA that was a mosaic of old and new up and down each strand.

To distinguish between these ideas, American geneticists and molecular biologists Matthew Meselson and Franklin Stahl designed one of the most perfect experiments ever, a powerful example of isotopic labeling. They grew bacteria on an exclusive diet of heavy nitrogen-15 compounds for several generations, which caused the bacterial DNA to become totally labeled with the heavy isotope. This extra weight could be checked by the way it centrifuged down through a gel compared to normal bacterial DNA. Then they switched the labeled bacteria to normal nitrogen-14 food. After one cell division, their DNA was one band in the centrifuge tubes, between the heavy and light weights. That ruled out conservative replications, which wouldn't have made any "half-heavy" DNA (only all-heavy and all-light). They then let the cells divide once more, and they found two DNA bands—one "half heavy" as before, and one in the normal "light" position; exactly what you'd expect from the one-new-strand semiconservative idea, but not from the dispersive one (which would have given yet another new intermediate weight). One experiment cleanly separated all three ideas at once. A thing of beauty.

SEE ALSO Isotopes (1913), Sanger Sequencing (1951), DNA's Structure (1953), Polymerase Chain Reaction (1983).

A schematic of DNA replication as we now understand it, thanks to Meselson and Stahl's experiment.

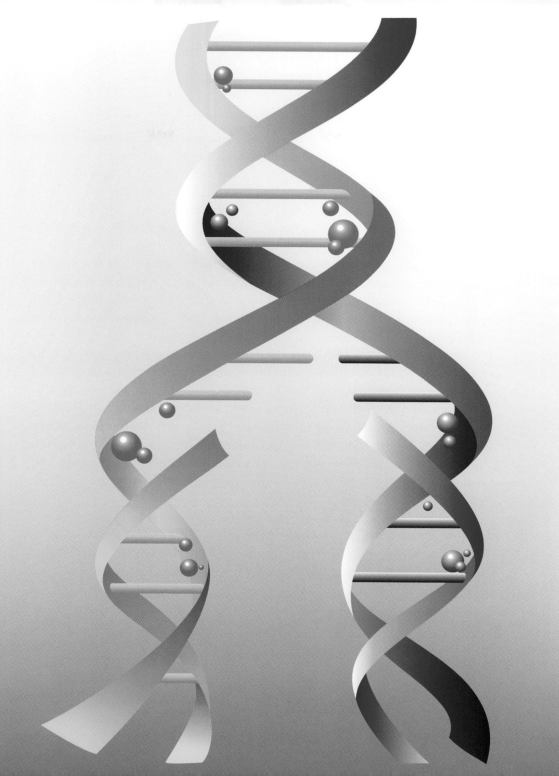

Thalidomide

Frances Oldham Kelsey (1914–2015)

Thalidomide was discovered in the 1950s in Germany and was widely prescribed there and in countries around the world to relieve nausea in pregnant women. In 1960, the American drug company Richardson-Merrell applied to the FDA for approval to sell the drug in the U.S. for the same purpose. The FDA asked for more safety and toxicity data, and during this period reports emerged about thalidomide causing pain and muscle weakness. Frances Oldham Kelsey, the pharmacologist at the FDA who had worked on the **elixir sulfanilamide** incident, was concerned about the drug's safety and had the agency deny the application, saying more study was needed. Later that year, the first reports of birth defects linked to use of the drug appeared. When the magnitude of the disaster became clear (over ten thousand babies around the world were affected, only about half of whom survived), it led to stricter drug approval standards worldwide (and a presidential award for Kelsey in the U.S.).

In retrospect, lack of knowledge about how drugs crossed the placenta was a major factor in the tragedy. Some other issues that are widely cited, though, turn out to be myths. Animal testing could not have provided a warning because the potential of a compound to cause birth defects varies widely from species to species. Thalidomide, for example, does not cause defects in either rats or mice (and causes completely different problems in rabbits). And while thalidomide has **chirality** (left- and right-handedness) and can be separated into enantiomers (molecules that are mirror images of each other), nothing can be said about which one is more dangerous, because the two rapidly get interconverted in the body.

Further research has shown that thalidomide inhibits blood vessel formation in developing tissue and has several other effects as well. It's an effective drug for some complications of leprosy, and the FDA approved it for this use in 1998, under extremely tight controls. In 2006, it was approved as part of a therapy regimen for multiple myeloma.

SEE ALSO Toxicology (1538), Chirality (1849), Radithor (1918), Elixir Sulfanilamide (1937), Bari Raid (1943), Cisplatin (1965), Rapamycin (1972), Taxol (1989).

FDA medical officer Frances Kelsey's refusal to approve the application to market thalidomide in the early 1960s attracted national attention.

Drug Detective

● Her skepticism and insistence on having "all the facts" before certifying the safety of a sleep-inducing drug averted an appalling American tragedy — the birth of many malformed infants.

O O O O O O

She resisted persistent petitions of commercial interests who presented data supporting claims the inexpensive drug was harmless. The facts finally vindicated Dr. Kelsey, as evidence piled up to show the drug — thalidomide — when taken by pregnant women, could cause deformed births.

Her action won her the President's Award for Distinguished Federal Civilian Service.

FRANCES O. KELSEY, M.D.
Food and Drug Administration

The Federal Civil Service

Four Score Years of Service to America

Resolution and Chiral Chromatography

LeRoy H. Klemm (1919–2003), **William Pirkle** (b. 1934), **Ernst Klesper** (b. 1927), **Vadim Davankov** (b. 1937), **Yoshio Okamoto** (b. 1941)

If you're going to make a pure chiral compound, you can begin with chiral atoms in your starting material (a carbohydrate or amino acid, for example) and find a way to carry these groups through to your final product. Starting with a chiral center also offers a chance to use **asymmetric induction** to set another chiral center on a nearby carbon. If your starting material has no chirality but can have a chiral atom once it undergoes a new reaction, then you can use a chiral reagent for that step, ranging from fairly simple molecules all the way up to **engineered enzymes**. The last option is resolution—finding a way to physically separate the mixture of left-hand and right-hand isomers from each other. Unfortunately, that means discarding up to half of your original mixture, but sometimes there's just no other way.

The classic way to resolve a compound is to make a salt of it (if possible) with a chiral acid or base and use recrystallization to purify it. But **chromatography** was a natural fit for resolution—if you could make a chiral solid phase for your column, then a mixture of chiral isomers might well separate out as it flowed through. Early attempts used ground sugar or starch, but, in 1960, American chemist LeRoy H. Klemm added chiral compounds to the surface of silica—a more common (and useful) packing material. The idea still didn't become practical until the modified solid phases for **reverse-phase chromatography** were tried. American chemist William Pirkle is considered a founder of chiral HPLC, attaching amino acids to the column packing in 1979, while Yoshio Okamoto in Japan later used modified carbohydrates. Amino acid/metal complexes were introduced by Vadim Davankov in Russia. **Carbon dioxide**, compressed to a **supercritical fluid**—first discovered by German chemist Ernst Klesper—turns out to be a particularly useful solvent system for chiral columns..

Since the majority of biomolecules are chiral, as are many drugs and natural products, having ways to analyze and purify chiral isomers is crucial. Separating them by simply pumping them through a column is an excellent solution to the problem.

SEE ALSO Carbon Dioxide (1754), Supercritical Fluids (1822), Chirality (1848), Tetrahedral Carbon Atoms (1874), Asymmetric Induction (1894), Chromatography (1901), HPLC (1967), Reverse-Phase Chromatography (1971), Shikimic Acid Shortage (2005), Engineered Enzymes (2010)

Cellulose fibers under a microscope. This plant-derived material—found in wood, paper, and cotton—is made up of long chains of carbohydrates, and modified versions of it coat the insides of today's chiral chromatography columns.

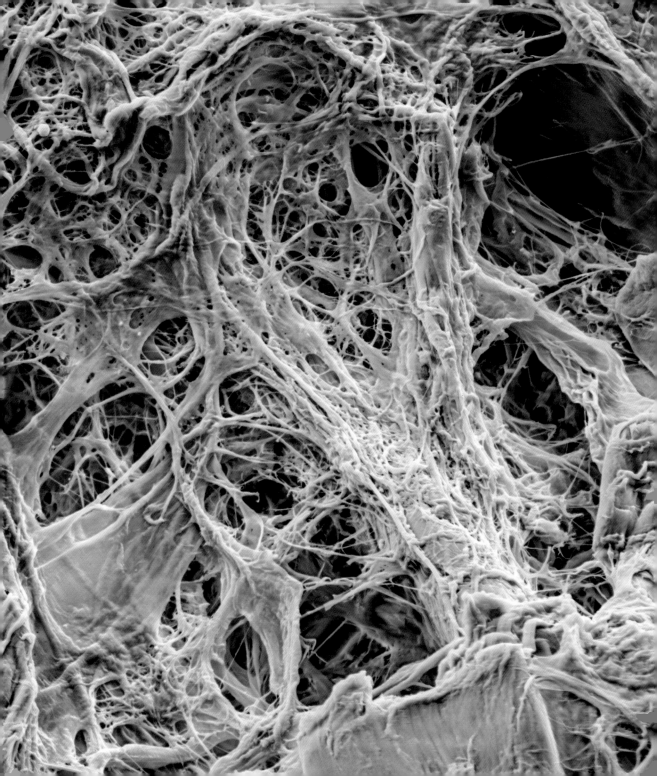

NMR

John Dombrowski Roberts (b. 1918), **Martin Everett Packard** (b. 1921), **Rex Edward Richards** (b. 1922), **James T. Arnold** (b. 1923), **James N. Shoolery** (b. 1925)

NMR (*nuclear magnetic resonance*) is one of the most powerful analytical techniques in chemistry. You can learn more about an unknown compound in ten minutes on a modern NMR machine than a chemist used to be able to learn in a year. The physics behind it are complex, but essentially, an NMR spectrum shows every different sort of hydrogen atom in a compound (depending on what sort of group they're attached to). This is extremely useful for determining compound structures, since all organic molecules have hydrogen atoms attached to them in different positions. Each of these hydrogen peaks is split into multiple lines depending on how many other hydrogens are attached nearby, and the distance between those lines is related to the angles between the atoms. But NMR isn't just for hydrogen. Fluorine atoms work very well, as does carbon-13 (the minor stable isotope of carbon), and many other nuclei. More complicated NMR experiments display all the proton-carbon connections simultaneously, which can quickly solve the structures of complex organic molecules.

NMR came from the cutting-edge physics of the 1930s, and its applications to chemistry weren't obvious. Many physicists didn't think there were any, but American biochemist Linus Pauling's advice at the time (to British chemist Sir Rex Edwards Richards, a 1950s NMR pioneer) was to ignore them. Still, building reliable NMR instruments, with their large, finicky electromagnets, took real perseverance. American physicist Martin Everett Packard first recorded the NMR spectrum of an organic molecule, and James T. Arnold built what was, in 1956, the world's best magnet to study more of them. The year 1961 marked American chemist James N. Shoolery's Varian A60, the first commercial NMR that seemed built to get results. American chemist John Dombrowski Roberts showed his fellow chemists how useful it could be for analysis of compound structures, and the field has never looked back.

SEE ALSO Natural Products (c. 60 CE), Helium (1868), Liquid Nitrogen (1883), Nonclassical Ion Controversy (1949), Alpha-Helix and Beta-Sheet (1951), Protein Crystallography (1965), Fullerenes (1985)

A chemist loads a sample into a modern NMR machine. The samples are contained in thin glass tubes, mounted in plastic "spinners." These are taken down into the center of the magnet by compressed air, spinning rapidly in order to give a sharper signal.

Green Fluorescent Protein

Osamu Shimomura (b. 1928), **Martin Chalfie** (b. 1947), **Douglas Prasher** (b. 1951), **Roger Yonchien Tsien** (b. 1952)

In 1962, Japanese chemist and marine biologist Osamu Shimomura reported isolating the protein that gives a species of jellyfish its glowing, unearthly blue-green color. He went on to describe other similar proteins and to show how they generate their own **fluorescence**. This is the sort of obscure discovery that might look to a non-scientist as somewhat less useful than painting rocks, but in fact Shimomura's work led directly to a Nobel Prize, has made possible a series of fundamental discoveries in medicine and biology, and is still in constant use in research labs around the world.

Green fluorescent protein (GFP) can be spliced into all kinds of DNA sequences to serve as a marker in living cells. If you want to know if a new gene is being expressed in a cell culture, tagging it with a GFP will tell you immediately—your cells start glowing. No one realized this was possible until American molecular biologist Douglas Prasher's work in 1992, but at the time his project did not look successful. As funding was running out, he sent samples of the GFP gene to other labs, including that of American biochemist Martin Chalfie, where it was discovered that a close relative of the wild-type protein could be folded into its functional shape and fluoresced in bacteria without needing any other jellyfish-derived additions. Its structure was solved soon afterward by **X-ray crystallography** (it's a barrel-shaped collection of beta-sheets), which led to more ideas about how to engineer the protein to make it brighter and longer lasting. Researchers began trying these in a variety of species, and the lab of American biochemist Roger Tsien (among others) extended the fluorescent spectrum of the proteins so that several could be used at the same time, each one emitting a distinct color.

Shimomura, Tsien, and Chalfie received the Nobel Prize in 2008, after which it was discovered that lack of funding had caused Prasher to leave research and drive a shuttle bus in order to scrape up a living. Shortly thereafter, Tsien gave him a job in his lab at the University of California in San Diego!

SEE ALSO Amino Acids (1806), Fluorescence (1852), X-Ray Crystallography (1912), Alpha-Helix and Beta-Sheet (1951), Luciferin (1957).

Aequorea victoria, one of several jellyfish species that contain related fluorescent proteins. From such simple creatures have come Nobel Prizes and huge research efforts across all of biology.

Noble Gas Compounds

Linus Carl Pauling (1901–1994), Neil Bartlett (1932–2008)

The outer electron shells of the noble gases were full, and they had no need to react with anything. No one had ever seen them form a chemical compound, and it was thought that no one ever would. But there were some doubters, the most famous being American biochemist Linus Pauling. He pointed out in 1933 that the noble gas xenon was a heavy enough atom that its outer electrons weren't quite so tightly held, and that given the right opportunity, it might be able to react. He suggested fluorine as the most likely partner, since its bonding properties are at the other end of the scale—it will bond with almost everything.

His suggestion still seemed pretty implausible to most chemists, so no one explored it further until the early 1960s. British chemist Neil Bartlett at the University of British Columbia had made a strange red salt while working with a new and interestingly reactive compound, platinum hexafluoride, and to his surprise, the new compound contained a positively charged oxygen molecule. PF_6, a red-colored gas, was apparently a stronger oxidizing agent than oxygen itself (not a common thing to see), and Bartlett realized that it might be strong enough to oxidize xenon and produce the world's first noble gas compound. He set up an apparatus where he allowed the two gases to mix, and they immediately reacted to deposit an orange solid. Bartlett recalled later that he was overcome with excitement and ran around the building trying to find someone to share the news with, only to find that it was so late that everyone had either gone home or off to dinner!

Even when his colleagues heard about his experiment, not all of them believed it. But Bartlett published his results quickly, and other labs around the world were able to replicate the reaction. There were, it turned out, a whole series of xenon (and krypton) fluorides and oxyfluorides—compounds that probably could have been made years before if anyone had had the nerve to believe that they could even exist.

SEE ALSO Isolation of Fluorine (1886), Neon (1898), *The Nature of the Chemical Bond* (1939)

Crystals of pure xenon tetrafluoride. The compound is stable, but because it reacts with water or moist air, it needs to be kept in an inert or well-dried atmosphere.

Isoamyl Acetate and Esters

Isoamyl acetate is a simple member of a huge class of compounds known as *esters*. They're formed from the condensation of a carboxylic acid and an alcohol—in this case, the five-carbon compound isoamyl alcohol reacts with acetic acid (the ingredient in vinegar), giving off water as a by-product. (If you're forming esters, you'll want to have a **Dean-Stark trap** handy, as taking away the water helps to force the reaction to completion.)

Acetate esters are extremely important in biology (a particular one, acetyl-CoA, is a vital intermediate in the synthesis of ATP—adenosine triphosphate, an essential component of metabolism—and in thousands of other processes). They show up in polymers and plastics (which is where the term *polyester* comes from), in pharmaceuticals, as industrial solvents, and in interstellar space. They are truly everywhere. And everywhere humans encounter the smaller, more volatile members of the ester family, we notice their smell. For once, it's a pleasant encounter. Even chemists admit that the science has its share of otherworldly stinks, but esters are fruity and floral (since they are, in fact, produced by plants as attractants). They're often used as scents and flavoring agents: ethyl butyrate is pineapple-like, octyl acetate is citrusy, and ethyl propionate smells rather like rum—and there are thousands more. Isoamyl acetate smells like every banana-flavored sweet you've ever had.

In 1962, Canadian researchers Rolf Boch and Duncan Shearer found isoamyl acetate to be a pheromone of the honeybee. Pheromones are odor messages sent by living creatures to one another and are particularly important among insects. Ants, moths, beetles, and butterflies all use pheromones to find mates, mark trails, and spread alarms. It had been known since 1814 that the smell of a freshly used bee sting prompted other bees to attack, and isoamyl acetate is one of the main ingredients in that signal. If you spill it on yourself and walk outdoors within range of a beehive, your problems will not end with smelling like a large, overripe banana.

SEE ALSO Soap (c. 2800 BCE), Purification (c. 1200 BCE), Functional Groups (1832), Fougère Royale (1881), Diazomethane (1894), Maillard Reaction (1912), Dean-Stark Trap (1920), Cellular Respiration (1937)

Bees are extremely sensitive to isoamyl acetate because it signals to them that one of their own has just attacked an enemy.

1963

Ziegler-Natta Catalysis

Karl Waldemar Ziegler (1898–1973), **Giulio Natta** (1903–1979)

Unless you're a chemist, you may have never heard of Ziegler-Natta catalysts, but they're another one of the hidden underpinnings of the modern world: they're used industrially on a huge scale to produce many everyday materials.

German chemist Karl Waldemar Ziegler and his team were investigating ways of making **polyethylene** through aluminum-containing Lewis acid catalysts. One run produced an almost perfect yield of butene gas (the product of two ethylenes condensing with each other), and Ziegler's team set out to discover what made this happen so cleanly. A great deal of investigation narrowed it down to nickel contamination at a very low level—the combination of this and the aluminum species was somehow a tremendous catalyst. This led to a full-on effort to investigate other metal combinations, and soon ethylene was being turned into polymers in unprecedentedly pure forms, without being taken up to high pressure (which no one had thought possible, as high pressure had been necessary for all earlier polyethylene syntheses).

The chemical company Montecatini in Italy licensed Ziegler's patents, where Italian chemist Giulio Natta's lab began work in the area as well. Using propylene instead of ethylene (which leaves a series of methyl groups coming off the long polymer chain) gave a startling result: a polymer with its methyl groups lined up in such a perfectly regular pattern that it could be crystallized. Normally such a reaction would be expected to give all sorts of randomly aligned products, but these metal catalysts were something new.

The various Ziegler-Natta catalysts have been developed and refined by researchers all over the world. They're used to produce many of the most common plastic and rubber materials (tires and golf balls), as well as more exotic applications such as surgical instruments and parts for sonar systems. Ziegler and Natta received the Nobel Prize in 1963, and the enormous value of their discovery is undiminished. If there are limits to the number of compounds that can be successfully polymerized, the variations that can be produced on their structures, and ways the catalysts can be modified, they haven't been discovered yet.

SEE ALSO Polymers and Polymerization (1839), Bakelite (1907), Acids and Bases (1923), Polyethylene (1933), Nylon (1935), Teflon (1938), Cyanoacrylates (1942), Kevlar (1964), Gore-Tex (1969), Olefin Metathesis (2005)

Plastic wraps are made from a form of polyethylene that were made commercially possible by the Ziegler-Natta catalysts.

Merrifield Synthesis

Robert Bruce Merrifield (1921–2006)

Our cells synthesize thousands of proteins constantly, but human hands can't work so quickly. Stringing together **amino acids**, one after another, is tedious, "steppy" work as the organic chemists say. But amino acids thrown together and coupled all at once would give you the biggest mess possible (and would probably have to be chipped out of the flask with a chisel), so to successfully synthesize a protein, chemists have to add amino acids one at a time.

American biochemist Robert Bruce Merrifield had a brilliant insight about how to make the process easier. He took the first amino acid needed for a protein chain and masked its amine end with a removable "protecting group," then attached the free acid end to a solid support, such as the surface of a plastic bead. He then cleaved off the amine-protecting group and reacted it with the next amino acid needed, which was also in that "amine-masked acid-open" form. This gave a dipeptide attached to the bead. He cleaved off the protecting group and ran the cycle again, and again . . . each time washing away all the spare reagents, because the growing peptide chain stayed attached to the solid support. At the end of the sequence, he used a completely different set of conditions from the previous steps to cleave the whole chain off the support, allowing the new protein to float freely (and making sure it would stay attached until then).

As long as the average yield of each step (coupling the new amino acid, then deprotecting it) was high, one could make surprisingly large products. Merrifield and his colleagues demonstrated this on longer and longer proteins and six years later made an entire enzyme (ribonuclease A), which worked just like the "real" one. Enzymes were indeed chemicals, and they could be made by chemical means.

Solid-phase peptide synthesis became a sensation after it was published in 1963. The process later became automated, with the cycles being fed from premade reagent reservoirs according to the protein sequence spelled out at the beginning. The field of custom protein synthesis exploded.

SEE ALSO Amino Acids (1806), Polymers and Polymerization (1839)

A micrograph of resin beads used in a Merrifield-style peptide synthesizer, magnified two hundred times. The growing protein chain is attached to these during synthesis, in which anywhere between one hundred and several thousand amino acids are linked together.

Dipolar Cycloadditions

Rolf Huisgen (b. 1920)

The **Diels-Alder reaction** is the classic example of a cycloaddition (a chemical reaction that leads to a ring formation in a compound), but there are many others. Over the decades, several were discovered that made five-membered rings instead of the Diels-Alder's six-membered ones. One partner in the reaction was an alkene (an unsaturated hydrocarbon containing a double bond), as usual, but the other side could be an array of three-atom species. German chemist Rolf Huisgen brought all of these reaction types under one tent in a series of papers, which he summarized in a 1963 overview.

What they all had in common was that the three-atom partner could be thought of as a dipole (a species with a negatively charged end and a positively charged one). Some of these dipoles are relatively stable, but many of them form transiently and react quickly. Huisgen's work established that these cycloadditions really were taking place through a concerted (all-at-once) mechanism like the Diels-Alder reaction, rather than through stepwise intermediates or **free radicals** (as put forth by competing theories), and he provided a framework to predict the orientation of the products.

For example, if there are very polar intermediates involved, then the reaction should be sensitive to changes in the polarity of the solvent (which could strongly stabilize or destabilize the transition states of such species), but concerted cycloadditions are not. If the two new bonds are being formed at the same time, then the arrangement of substituents around the alkene (the "dipolarophile") should be preserved, since they don't have time to rotate and scramble. And if free radicals are involved, then those could be intercepted by flooding the reaction with compounds that react quickly with them, stopping the reaction in its tracks.

The dipolar cycloadditions make a wide variety of five-membered heterocycles that are useful in drug structures, agricultural chemicals, and more. The most famous use of the Huisgen reactions, though, is probably the azide/alkene **click chemistry** made famous by Barry Sharpless.

SEE ALSO Ozone (1840), Free Radicals (1900), Diels-Alder Reaction (1928), Transition State Theory (1935), Reaction Mechanisms (1937), Kinetic Isotope Effects (1947), Woodward-Hoffman Rules (1965), Click Triazoles (2001).

Trees of the genus Aspidosperma *are found throughout South America and contain a number of very complex alkaloids. Synthesis of these in the lab has often required dipolar cycloaddition as a key step, since the compounds' structures include several fused rings.*

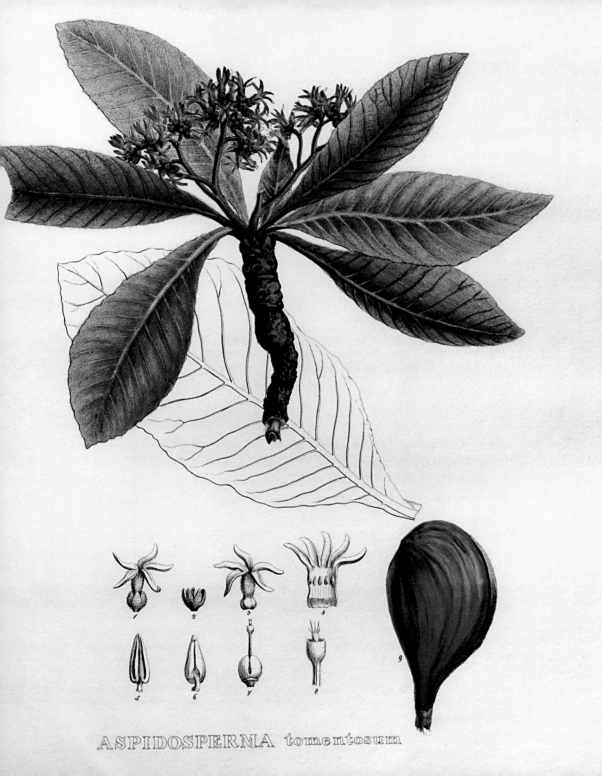
ASPIDOSPERMA tomentosum

Kevlar®

Stephanie Louise Kwolek (1923–2014)

Kevlar is a well-known polymer due to its high-profile uses in products such as bulletproof vests. It can be spun into very strong fibers, and, for its weight, it has several times the tensile strength of steel. These unusual properties come from some unusual behavior at the molecular level, and there was an unusual path to its discovery.

In 1964, American chemist Stephanie Louise Kwolek was working at DuPont on a project to discover strong and lightweight polymers for use in tires. The compounds she was working with showed **liquid-crystal** behavior, arranging in orderly bundles in solution. At the time, this was unexplored territory in polymer chemistry. The Kevlar monomer molecules are highly linear and full of aromatic rings, which makes them very rigid. The hope was that if they could be spun into fibers, the liquid-crystalline bundles would continue to line up in the fiber direction.

But there were problems at first. For one thing, the polymerization had to be done in very polar solvents, because otherwise the insoluble product crashed (precipitated) out of solution due to its highly ordered nature. A thick high-boiling liquid called HMPA, with one of the highest **dipole moments** of any industrially available solvent, worked well, but not many other solvents would. Even then, the solution didn't look promising, because it still didn't seem viscous enough to have fibers successfully spun from it. Kwolek persuaded her coworkers to test it anyway, and not only did it work, but the resulting fibers refused to break even under severe conditions.

Kwolek's managers at DuPont realized a good thing when they saw it, and Kevlar soon went into production. However, it gradually became clear that HMPA was highly toxic, and it was eventually classified as a carcinogen. The Dutch chemical firm Akzo had come up with a similar polymer recipe (Twaron) using a far less toxic solvent called NMP, and when DuPont switched solvents, Akzo filed a patent lawsuit. Eleven years of worldwide legal struggles ensued between the two companies until they finally licensed each other's patents in the late 1980s. Body armor, of course, is only the best known of the Kevlar products; it's used in everything from bicycle tires to smartphones.

SEE ALSO Polymers and Polymerization (1839), Liquid Crystals (1888), Bakelite (1907) , Dipole Moments (1912), Polyethylene (1933), Nylon (1935), Teflon (1938), Cyanoacrylates (1942), Ziegler-Natta Catalysis (1963), Gore-Tex (1969)

A U.S. infantryman fires an M-4 rifle during a training exercise. His helmet contains multiple layers of Kevlar, which is used in military and police supplies worldwide.

Protein Crystallography

John Desmond Bernal (1901–1971), **Dorothy Crowfoot Hodgkin** (1910–1994), **David Chilton Phillips** (1924–1999)

The enzyme lysozyme—which takes its name from the Greek word *lysis*, meaning "loosening"—had been known about for some time when British biophysicist David Chilton Phillips tried to solve its structure by **X-ray crystallography**. Alexander Fleming (of **penicillin** fame) named it in 1923, and it was known as the reason egg whites and tears showed antibacterial properties. That's the role it has in humans as well. As part of the innate immune system, it attacks compounds called *peptidoglycans*, which are found in the cell walls of bacteria. It's also found in saliva where it degrades starch, which is why a piece of popcorn disappears so rapidly when it touches your tongue.

Applying X-ray crystallography to proteins was a big step, since they were much larger and more complex than the usual molecules examined by this technique. British X-ray crystallographers John Desmond Bernal and Dorothy Crowfoot Hodgkin showed that protein diffraction was possible in 1934, but the first structure solved was myoglobin in 1958 by Phillips. His next target was an enzyme. It was thought that a three-dimensional structure might provide insights into how enzymes were able to speed up chemical reactions so dramatically.

Lysozyme had many things in its favor as a target—it's easily available, and it crystallizes well, which is a step that even today can be something of a black art in protein work. Solving the diffraction pattern was a huge effort with the computational resources available, but lysozyme's structure was determined, as were somewhat fuzzier structures of lysozyme with inhibitor molecules inside its active site. Crystallizing proteins in this way now provides countless insights into biochemistry and drug discovery (since most pharmaceuticals work by binding to specific protein sites). There are pitfalls, naturally. A crystal structure is a static shot of a mobile protein, a picture that may or may not reflect what goes on out in the real world where proteins can move and shift. But there's always **NMR** to fall back on, which can sometimes be used to solve protein structures in solution.

SEE ALSO Amino Acids (1806), X-Ray Crystallography (1912), Carbonic Anhydrase (1932), NMR (1961), Enzyme Stereochemistry (1975).

Lysozyme crystals as seen through a polarized-light microscope. These well-formed blocklike crystals are ideal for X-ray crystallographic studies, but few proteins form them as well as lysozyme does.

Cisplatin

Michele Peyrone (1813–1883), **Alfred Werner** (1866–1919), **Barnett Rosenberg** (1924–2009)

Cisplatin is one of the more unlikely looking drugs out there, and it was discovered by an unlikely route. American chemist Barnett Rosenberg had been struck by the way the chromosomes in dividing cells resembled metal filings in a magnetic field, and he wanted to see if there were any electromagnetic effects on the process of cell division. He and his coworkers investigated this by growing bacterial cells in a chamber with electrodes at each end.

The effect was dramatic: the bacteria (*E. coli*) are usually quite compact, but they began to take on a strange, stringy form that had never been seen before. They were continuing to grow but were unable to divide. After a good deal of work it was found that the electric current had nothing to do with the effect. It was the electrodes themselves. They were made of platinum, which was supposedly inert, but platinum compounds had diffused into the solution and were having profound effects on cell division. Testing other similar metal compounds directly on bacteria showed the same effects, and the most active form turned out to be the complex of platinum with two ammonia molecules on one side and two chlorines on the other (*cis*, in chemical nomenclature). Remarkably, if an ammonia and a chlorine were swapped, the resulting *trans* form (ammonias opposite from each other, and chlorines as well) showed no activity at all.

These sorts of complexes had actually been known since the 1840s, when Italian chemist Michele Peyrone first synthesized his Peyrone's chloride (in fact, cisplatin), and their structures (a source of much argument at the time) were a key part of Swiss chemist Alfred Werner's successful metal coordination theory in 1893. But no one thought that they could be drugs; such things seemed firmly rooted in the inorganic chemistry lab. Further work showed that cisplatin worked by sliding its flat, square structure into the double helix of DNA and disrupting its replication. And that immediately suggested a use in chemotherapy, since anything that stopped cell division would be expected to affect fast-growing cancer cells most of all. Cisplatin and related compounds are still in use as anticancer agents today.

SEE ALSO Toxicology (1538), Coordination Compounds (1893), Bari Raid (1943), Antifolates (1947), DNA Replication (1958), Thalidomide (1960), Rapamycin (1972), Taxol (1989).

Crystals of pure cisplatin, one of the more unlikely drug structures ever to be discovered.

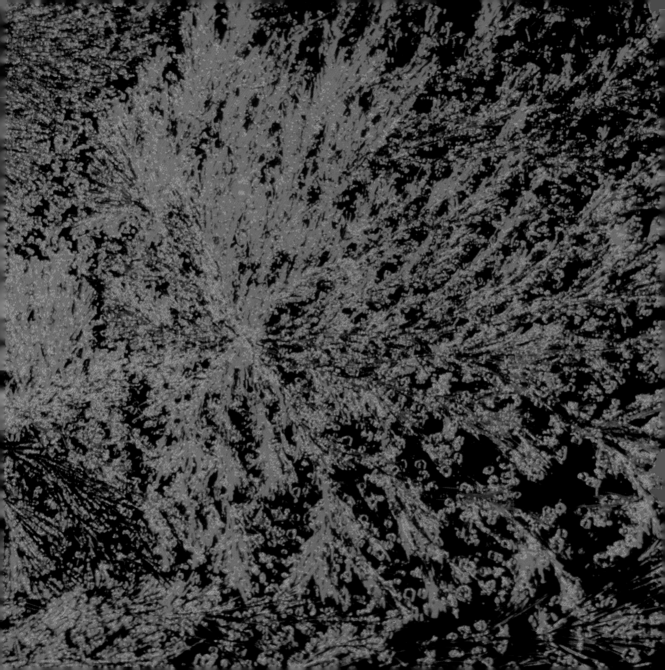

Lead Contamination

Clair Cameron Patterson (1922–1995)

Clair Cameron Patterson was an American geological chemist whose work involved establishing long-term radioisotope dating techniques (variations on the idea of carbon-14 dating, but on geological timescales). The decay of uranium **isotopes** down to lead isotopes, and the ratio of the different lead isotopes themselves, provides dates over billion-year timescales because of the long half-lives involved. Using this, one of his projects was to estimate the age of the Earth, which he did in 1956. It's about four and a half billion years old—an age scientists still agree is essentially correct.

It was while working on these analyses, which required painstaking measurement of lead levels in many different samples, that Patterson became aware of how much lead was being added to the environment—the atmosphere, the water, and the food chain. Ice cores from Greenland (which could be dated year by year) showed that the addition of **tetraethyl lead** to gasoline was the biggest contributor, but several other industrial processes were also releasing lead far and wide. His figures showed much larger increases over time than most others in the field could believe, leading to a long period of disagreement with both the lead industry and other analytical chemists. In 1965, he published a book on his findings, *Contaminated and Natural Lead Environments of Man*, which set off sharp debates among scientists—especially those employed by the lead industry.

They were debates that Patterson—and his data—eventually won. The U.S. Environmental Protection Agency announced a plan to start phasing out leaded gasoline in 1973, and lead was completely removed from automobile gas in the U.S. by 1986 (some aircraft fuel still uses a lead additive, and there are other countries that have not yet completed a phaseout). Similar reductions were made in water pipes, paints, glazes for food storage containers, and other manufactured goods. Lead levels in the blood of an average U.S. citizen began to drop immediately and have continued to decline ever since. Since no safe level of lead in the bloodstream has yet been identified, this is good news indeed, and we have Clair Patterson to thank for it.

SEE ALSO Toxicology (1538), Isotopes (1913), Tetraethyl Lead (1921), Donora Death Fog (1948), Thallium Poisoning (1952)

An ice core from Greenland is melted for trace-element analysis. Such studies showed that leaded gasoline had contaminated the environment in the twentieth century (and interestingly, that the Roman Empire had contributed significant amounts in its day as well).

Methane Hydrate

Humphry Davy (1778–1829), **Yuri F. Makogon** (b. 1930)

Methane hydrate is a substance that makes you realize how odd the natural world is. Methane (CH_4, the simplest of all the hydrocarbons) is a gas at room temperature and pressure. It's not particularly soluble in water—at room temperature, you can get about twenty milligrams of it into one kilo of water, only one part in fifty thousand by weight. But put water and methane together under cold, high-pressure conditions and the water will freeze so that it "cages" methane molecules, a form known as a *clathrate* (from the Latin word *clathratus*, meaning "furnished with a lattice"). This methane hydrate looks like cloudy ice, but it has about 13 percent methane in it by weight, which is enough for it to catch fire. The sight of a dripping ice cube with blue and orange flames coming from it is a bit difficult to process.

English chemist Humphry Davy first discovered this strange stuff in 1810, and in the 1940s it turned up clogging natural gas pipelines, where it was considered a major pain for the engineers (it still is). No one thought it could be a naturally occurring substance—not on this planet, anyway—until 1965, when it was reported in the sediments under a Siberian natural gas field by Ukrainian engineer Yuri F. Makogon. After that, vast amounts were found under the oceans, at the bottom of the comparatively shallow waters of the continental shelves. The amount of methane contained in these layers is hard to estimate but it is probably many times the amount known to exist in traditional deposits. Naturally enough, this has attracted attention in a world always looking for new sources of energy, and Japan is the first country that appears to be starting serious mining of its methane hydrate fields.

Careful study of the **isotopes** found in these deposits has indicated that the most abundant ones are formed from gas produced over the years by bacteria. Some other deposits, though, have a different isotopic signature, suggesting that their gas seeped up from deeper in the earth without a biochemical origin at all.

SEE ALSO Ideal Gas Law (1834), Isotopes (1913), Isotopic Distribution (2006)

A lump of flaming methane hydrate, a common substance in the Arctic and the ocean floor.

Woodward-Hoffman Rules

Robert Burns Woodward (1917–1979), **Kenichi Fukui** (1918–1998), **Roald Hoffman** (b. 1937)

As chemists learned more about stereochemistry (the spatial arrangement of atoms and groups in molecules), it became clear that some reactions behave in a very controlled fashion indeed. In general, the class of ring-breaking and ring-forming reactions whose mechanisms can be drawn simply by rearranging bonds (similar to the **Diels-Alder reaction**, with no free radicals and no new ions) tend to give products in set patterns. Particularly dramatic examples were found when the same reaction could be activated either by heat or by light. It turned out that the thermal reaction gave one stereochemical outcome, while the photochemical reaction always gave a different one, and very specifically.

Something important was at work there, clearly, and in 1965 American chemists Robert Burns Woodward and Roald Hoffman explained it all in terms of molecular orbitals. They set forth a few rules of thumb: if the number of electrons moving around is divisible by four, the reaction will go in one direction, if the number of electrons is a multiple of four plus two more, it will go in another. (All these reactions work by moving bonds around, two electrons at a time, so these patterns may also remind you of the electron-counting rules for aromaticity.) The rules also predicted which sorts of reactions would be favored, which explained why thermally accelerated reactions tended to be like the Diels-Alder and form six-membered rings, while photochemistry tended to produce four-membered rings. A huge variety of reactions fit into these schemes, and the Woodward-Hoffman rules applied to them all, making sense of a long list of experimental results. Reactions of this kind that have not even been tried yet can have their products predicted.

Hoffman and Japanese chemist Kenichi Fukui (who'd worked out a very similar scheme from a different molecular orbital approach) got the Nobel for this in 1981, and Woodward would surely have shared it (as his second Nobel Prize in Chemistry) if he had lived long enough. This work had a significant impact on organic chemistry, making every synthetic chemist pay attention to molecular orbital theory even if they'd felt safe ignoring it before.

SEE ALSO Photochemistry (1834), Diels-Alder Reaction (1928), Dipolar Cycloadditions (1963), B_{12} Synthesis (1973).

Roald Hoffman, after being awarded the American Institute of Chemists' gold medal in 2006. (For a picture of his partner, Robert Burns Woodward, see B_{12} Synthesis.)

Polywater

Sérgio Pereira da Silva Porto (1926–1979)

The polywater story is a strange one, highlighting how even scientists can fool themselves and, therefore, must take care to rigorously check every sensational new report. In the 1960s, scientists in the Soviet Union were working with water confined in very small quartz tubes. It had strange properties, such as a much lower freezing point and a much higher boiling point, and it was quite thick and viscous compared to ordinary water. The findings didn't garner much attention in the West, but later in the decade they were presented at conferences in other countries, and the wider world of physical chemistry took notice.

What happened next was worth watching. Some researchers were able to reproduce the original findings, while others could not. Some believed that a new form of water had been produced, while others very forcefully disagreed. The popular press picked up the story, producing all sorts of speculations about the supposedly world-ending properties of polywater (what if it transformed all the water in the world on contact?). The first thing every chemist wanted to find out, though, was whether or not that was actually pure water in the quartz capillaries. Impurities would be expected to lead to all the physical changes that had been noted, but many of the early polywater papers maintained that this possibility had been scrupulously checked out.

It hadn't been. Samples of "authentic" polywater seemed to be quite impure (American biophysicist Denis L. Rousseau and Brazilian physicist Sérgio Pereira da Silva Porto found them to be hard to distinguish from human sweat), which led to countercharges that the tested samples weren't the right ones. But no one seemed to be able to provide any "pure" polywater, and with more meticulous techniques, the substance could not be produced at all. Opinion began to turn strongly negative on polywater's existence, though for years there were still true believers who refused to give up on the idea. A great deal of time and effort were invested (or wasted) in this discovery and subsequent invalidation, but in the end, the scientific record did correct itself.

SEE ALSO Hydrogen Bonding (1920), Zone Refining (1952), Recrystallization and Polymorphs (1998)

The saga of polywater began in the city of Kostroma, Russia, where this magnificent Church of the Resurrection is located.

HPLC

Joseph Jack Kirkland (b. 1925), **Josef Huber** (1925–2000), **Csaba Horváth** (1930-2004), **John Calvin Giddings** (1930–1996)

HPLC, or high-performance liquid **chromatography**, is well named. It's probably the final stage in the evolution of column chromatography, which began back in 1901 with Mikhail Tsvet and his columns of powdered chalk. American chemist John Calvin Giddings, German chemist Josef Huber, and Hungarian-American chemical engineer Csaba Horváth worked out much of the theory and practice of using smaller particle sizes and controlled flow rates in the early 1960s, and American chemist Joseph Jack Kirkland discovered a range of new solid phases during the rest of the decade. These finer-grained packings had more surface area and separating power but made flow rates slow to a crawl. What good is a column if nothing ever comes out the other end? The solution was to add mechanical pumps to force the solutions through the columns, which then had to be made out of stronger materials.

Take all these ideas to the maximum, and you have modern HPLC, with Horváth building the first such instrument in 1967. The columns are made of thick-walled metal, and the solid phases in them are extremely fine powders whose particles have the most uniform shapes, sizes, and pores that manufacturing can provide. Samples move across them at very high pressure, forced along by expensive pump heads made of high-performance metals such as **titanium**. Most separations are done in reverse-phase mode (meaning the stationary phase is hydrophobic and the mobile phase is a mixture of solvent and water), which gives better separations on a wider variety of samples. HPLC systems were first built for the analytical scale, but larger columns and pumps now enable real preparative-size samples to be purified.

HPLC is everywhere in chemistry and biology, and the varieties of columns and applications seem to be without limit. Organic synthesis, drug research and testing, food science and environmental science all make extensive use of HPLC. If it hasn't been run down a high-pressure column yet, then someone's out there figuring out how to do it.

SEE ALSO Titanium (1791), Chromatography (1901), Gas Chromatography (1952), Resolution and Chiral Chromatography (1960), Reverse-Phase Chromatography (1971), Electrospray LC/MS (1984), Acetonitrile (2009)

The pumps of a modern HPLC machine are made of titanium and other alloys that can stand up to the high pressure and range of solvents.

BZ Reaction

Boris Belousov (1893–1970), **Alan Mathison Turing** (1912–1954), **Anatol Zhabotinsky** (1938–2008)

Most chemical reactions have a time course that's fairly easy to follow. In general, the starting material gradually gets consumed, the products gradually accumulate, and the reactions can be sped up or slowed down by changing the conditions—especially temperature. Not so with the BZ (Belousov-Zhabotinsky) reaction, the most famous of the class known as oscillating reactions, which swing out further than usual from chemical equilibrium and then swing back the other way, overshooting back and forth until they eventually run down. If you run the BZ reaction in a standard flask with stirring, the solution's color keeps changing periodically. If it is run in a shallow dish with no stirring, expanding waves and spirals of color move through the solution in patterns reminiscent of mineral cross-sections, bacterial colonies, and animal camouflage.

Soviet chemist Boris Belousov discovered the basic mixtures during the 1950s but had trouble publishing his results because they seemed hard to believe and were difficult to explain. Soviet chemist Anatol Zhabotinsky, a graduate student, rediscovered the reaction in 1961, but it was still unknown outside a few groups of Russian chemists until he published a short description of it at a 1968 conference. We know the basic ideas behind the reaction—bromine goes up and down through various **oxidation states** and species, with reactions running in either direction depending on local concentrations. There's always another reactant that gets gradually consumed, since the reaction certainly doesn't violate the laws of thermodynamics—it just takes a bumpier ride downhill than most. The detailed mechanism involves at least eighteen steps and twenty-one separate chemical species, so getting a complete picture of the reaction process is no small matter.

The resemblance to the patterns found in living creatures is no coincidence. In 1952, the brilliant British mathematician and computer scientist Alan Turing, well known for his instrumental role in breaking the Germans' Enigma code during World War II, worked out a theory of pattern formations in biology that also predicted the existence of oscillating reactions like the Belousov-Zhabotinsky. He would surely have been happy to see a real one in action.

SEE ALSO Oxidation States (1860), Gibbs Free Energy (1876), Le Châtelier's Principle (1885).

The classic BZ reaction in a shallow dish. The bright yellow bands move slowly through the solution, driven by a complex series of reaction steps.

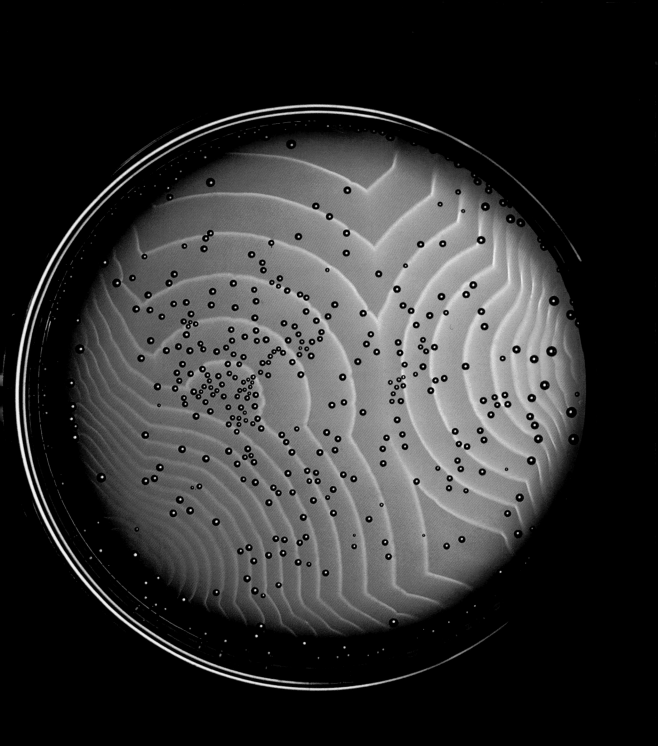

Murchison Meteorite

The organic chemistry of life on Earth is familiar—**amino acids**, DNA bases, carbohydrates, and so on. But what about organic chemistry in space? A large meteorite that broke into pieces in the skies over Murchison, Australia, in 1969 gives us one answer: outer space, it turns out, is full of these kinds of molecules.

The Murchison meteorite, pieces of which landed over five square miles, is a prime example of a meteorite class known as carbonaceous chondrites, which look like chunks of hardened asphalt. They're extremely primitive matter left over from the formation of the solar system. Some of them even have an oily or sulfurous smell when they've recently fallen, and the presence of those volatile compounds is vivid proof that such specimens have never undergone any prolonged heating over their billions of years of existence, since that heat would have driven off some of those compounds and caused further reactions.

What are things like amino acids doing in there, then? Are they signs of life? The answer is no, almost certainly not, because the chiral molecules found are invariably mixtures of both enantiomers (the right- and left-handed forms), which are not what any living system we know can use. But the compounds found in carbonaceous chondrites *are* signs of how life might have begun, since they are reminiscent of the mixtures formed in the **Miller-Urey experiment**. Lots of elements and simple organic molecules are floating around out there in space, and mixing and heating them apparently produces (among other things) the molecules of life.

But there's even more information hidden in these treasure troves disguised as asphalt. The most recent study of the Murchison material, using the latest liquid chromatography/mass spectrometry (LC/MS) technology, estimates that there could be tens of thousands of different organic molecules present in trace amounts, which means our quest to understand the chemistry of outer space has just begun. The more we look, the more we see.

SEE ALSO Amino Acids (1806), Chirality (1848), Miller-Urey Experiment (1952), Tholin (1979), Electrospray LC/MS (1984)

A fragment of the Murchison meteorite is on display at the Smithsonian in Washington, DC. This unimpressive dark gray mass is probably one of the most intensively studied rocks in the world.

Murchison meteorite
4600 million years old
Murchison, Australia

This is a fragment of a meteorite that landed in Murchison, Australia, in 1969. The Murchison meteorite and others like it, called **carbonaceous chondrites**, have been dated radiometrically and are thought to be remnants of the birth of the solar system.

Gore-Tex®

Robert W. Gore (b. 1937)

Polymer chemistry is somewhat unpredictable. Many of its greatest advances have come about by accident, or when people deliberately try unknown combinations just to see what happens (see the entries on **Rubber**, **Bakelite**, and **Cyanoacrylates**). As chemical understanding of polymers has increased, the study of polymer chemistry has gradually moved onto more solid ground, but there are always unexpected pits to step into.

Some of those pits turn out to be filled with gold. That's what American chemist Robert W. Gore stepped into while he was carrying on a family tradition of polymer research. His father had worked for DuPont before forming W. L. Gore & Associates, and Robert had already discovered how to coat electrical cables with polytetrafluoroethylene (PTFE), leading to a key Gore product. In 1969, Robert was working with rods of PTFE, trying to carefully heat and stretch them, but he wasn't getting very far. As he told the story later, after yet another unsuccessful run, he gave the next heated rod a furious yank in frustration, then watched in amazement as it expanded to nearly eight times its original length. The rod's diameter hardly changed, though, and experiments showed that "expanded PTFE" was like nothing anyone had made in the field before. It was about 70 percent air and formed of very fine fibers in a porous net.

And it turned out to be of great use in, for example, textiles (as the familiar Gore-Tex material), wire and cable insulation for electronics, medical devices including implants and sutures, and more. But as these applications were developed, a competing company contested Gore's patents, saying New Zealand inventor John W. Cropper had produced similar material three years earlier. But the company that used Cropper's invention kept it a trade secret, filing no patents. A court found for Gore, since he had disclosed his technique (in his patent applications) and had discovered it without knowing of Cropper's work. Industrial chemists, like other inventors, may keep their findings secret or file patents, but even a well-kept secret can be rediscovered.

SEE ALSO Polymers and Polymerization (1839), Rubber (1839), Bakelite (1907), Polyethylene (1933), Nylon (1935), Teflon (1938), Cyanoacrylates (1942), Ziegler-Natta Catalysis (1963), Kevlar (1964).

Water droplets beading on a piece of Gore-Tex fabric. The fluorinated surface repels liquid water and gives it many other useful properties as well.

Carbon Dioxide Scrubbing

Robert Edwin Smylie (b. 1929), **Jerry Woodfill** (b. 1945), **Gene Krantz** (b. 1933), **James Arthur Lovell Jr.** (b. 1928), **John Leonard Swigert Jr.** (1931–1982), **Fred Wallace Haise Jr.** (b. 1933)

Most familiar chemical reactions happen when the reaction partners are dissolved in a liquid. But that's not always the case. Gases can react with each other just as readily, and they can slip in and out of liquid solution quickly, too (as the bubbles in a fizzing glass of soda water show). It may be surprising, though, to find out that gases can react well with solids. For this to take place, it helps if the solid is very fine-grained, to expose the maximum surface area. Surface area has a profound effect on chemical reactions of all types, actually. Many reactions that are manageable with a coarsely ground reagent turn all too lively (even explosively so) when very fine powders are used.

Concern about the levels of **carbon dioxide** in the atmosphere has led researchers to explore a gas/solid reaction that would allow CO_2 to be absorbed into another substance (and out of the air we breathe). The alkali metal (or alkaline earth) hydroxide salt is the usual choice; it reacts with carbon dioxide to form the solid carbonate. A large-scale practical solution hasn't yet been found, but small-scale applications have proved vital. These carbon dioxide "scrubbers" are used whenever people have to survive in tightly sealed vessels, such as submarines or spaceships.

During the *Apollo 13* mission, in 1970, the well-known oxygen-tank explosion in the spacecraft's service module left the three astronauts on board—Jim Lovell, Jack Swigert, and Fred Haise—in grave danger in several ways, not least because of CO_2 buildup. The lunar module, which was being used as a lifeboat during most of the remaining trip back to Earth, did not have enough capacity to absorb the CO_2 from three astronauts for that long. Engineers on the ground (led by crew systems division chief Robert Edwin "Ed" Smylie and including Donald D. Arabian, Jerry Woodfill, and Gene Krantz) devised an ingenious fix involving duct tape (of course!), cardboard, and plastic bags that allowed the astronauts to adapt and use the lithium hydroxide canisters from the command module, whose shape was incompatible with the lunar module. It maintained the CO_2 scrubbing process and saved the astronauts' lives.

SEE ALSO Carbon Dioxide (1754), Solvay Process (1864), Greenhouse Effect (1896)

In a photograph taken by fellow crewmember Fred Haise, Apollo 13 *astronaut Jack Swigert holds the connection to the duct-taped command module scrubber apparatus that saved all of their lives.*

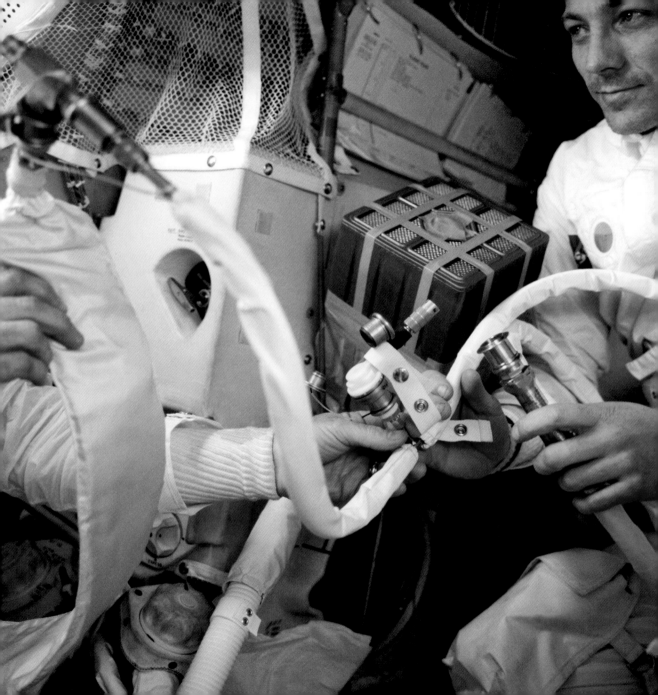

Computational Chemistry

Michael James Steuart Dewar (1918–1997), John Anthony Pople (1925–2004)

Chemists have been trying to tackle problems using computer models since computers were invented. Predicting the course of chemical reactions could save time and money, and such models also allow chemists to investigate many compounds that might be very difficult to isolate. But exact quantum mechanical solutions for any system larger than the hydrogen atom are just not possible, so chemists must figure out what approximations to use and what errors these might introduce, while still trying to solve real-world problems.

The year 1970 marked the introduction of the first off-the-shelf computational chemistry package, known as *Gaussian 70*. By today's standards, it's primitive—but it allowed chemists to apply computational techniques without having to program from scratch. Today there are many such packages (some free and open-source, others quite expensive), but Gaussian is still there, with more than forty years of updates and additions. In 1998, English theoretical chemist Sir John Anthony Pople was awarded a Nobel for his contribution to its creation. British-American theoretical chemist Michael James Steuart Dewar also developed many computational methods of his own based on quantum mechanical approximations during the 1970s and 1980s.

Giving chemists the ability to set up their own calculated models has been (from some viewpoints) a mixed blessing. As well-trained computational chemists understand, every model has built-in assumptions and limits. Even choosing the best computational approach for a given problem is a job for experienced hands. But without pioneering programs like Gaussian, there wouldn't be very many of those experienced hands.

The field has advanced tremendously since 1970 as hardware and software have become faster and more powerful, but the problems that chemists would like solved have grown as well. Figuring out how to approach them and how to model **hydrogen bonding**, molecular attraction, solvent effects, thermodynamics, and the other factors that go into chemical behavior advances both real-world experiments and the methods used to model them.

SEE ALSO Gibbs Free Energy (1876), Hydrogen Bonding (1920), Sigma and Pi Bonding (1931), *The Nature of the Chemical Bond* (1939), Engineered Enzymes (2010).

A man adjusts the random wiring network between the light sensors and association unit of scientist Frank Rosenblatt's Mark I Perceptron computer at the Cornell Aeronautical Laboratory, Buffalo, New York, c. 1960. Computational methods have advanced tremendously since this early attempt at a neural-net computer.

Glyphosate

John E. Franz (b. 1921)

Glyphosate (known to many under its original trade name, *Roundup*) is surely the most widely used weed killer in the world. First synthesized in the 1950s but rediscovered as an herbicide by organic chemist John E. Franz in 1970, it's an excellent example of the specialty that chemists know as "ag-chem." Glyphosate is a small molecule and a powerful inhibitor of an enzyme with a name that only a chemist could love: *5-enolpyruvylshikimate-3-phosphate synthase*. In living cells, this enzyme catalyzes a key step in the synthesis of three **amino acids**, so shutting it down is a tough obstacle for cells to overcome. The effect is strongest in growing plants, and because weeds are vigorous and spreading by definition, they are ideal targets. In recent years, some crops have been genetically modified to tolerate glyphosate, allowing farmers to use the chemical to kill other plants that show up in the fields without damaging them.

So why isn't glyphosate poisonous to everything it touches? The amino acids whose production it shuts down are synthesized only in microorganisms and plants—not in animals, including humans (we have to get them from our diet). Since we and all other animals lack this enzyme completely, targeting it should, in theory, provide a plant-specific poison. There are other things to consider, though. First is selectivity: an inhibitor that works well against its target may hit other closely related enzymes at the same time. Fortunately, glyphosate's target has no close relatives in higher organisms, but it still can hit related enzymes in bacteria, and considerable work has gone into finding out what that might lead to. As far as humans are concerned, some studies indicate that the surfactants and detergents in the commercial product might be more troublesome than the active ingredient itself.

But as with every other enzyme inhibitor, there's always the chance of a mutated form of its target enzyme showing up—one that's resistant to the poison. This happens again and again—viruses, bacteria, insects, plants, and cancer cells alike keep finding ways around our attempts to kill them off.

SEE ALSO Toxicology (1538), Amino Acids (1806)

Spraying a weed with glyphosate. The chemical doesn't kill just weeds; it will damage almost any plant that comes into contact with it, since weeds and plants all tend to use the same biochemical pathway for amino acid synthesis.

Reverse-Phase Chromatography

Archer John Porter Martin (1910–2002), **Joseph Jack Kirkland** (b. 1925), **Csaba Horváth** (1930–2004), **Sidney Pestka** (b. 1936)

Since Mikhail Tsvet's first experiment with **chromatography**, it has been used throughout chemistry and biology, as everything from low-tech disposable cartridges to highly optimized state-of-the-art systems that cost huge amounts (and had better deliver huge results). Until the 1970s, a chemist took a solid polar stationary phase, dissolved a sample in a relatively nonpolar solvent, and eluted it through a column of the stationary phase. The sample's more polar components would stick to the column more tightly and lag behind, while the less polar ones would continue to pass through and come out of the column first.

In 1950, English chemists G. A. Howard and Archer John Porter Martin reported on a technique using two liquid phases in which the desired compounds were carried in the polar phase—a reversal of the usual method. It took until the 1970s for chemists using solid-phase column chromatography to adapt this idea, coating the usual silica column-packing material with a nonpolar layer, and then using polar solvents (even water) to move compounds over it. In this world, the most polar components came out of the column first, and the less polar ones lagged behind.

This reverse-phase chromatography has several advantages. Almost everything can be washed off the column with different solvents, as opposed to classic silica columns, whose initial zones become hopelessly fouled with colorful polar contaminants. And the separating power of the technique was startlingly good, especially after 1971, when American chemist Joseph Jack Kirkland introduced columns where the nonpolar phases (there could be many) were covalently bonded onto the silica instead of just being layered there. Combining reversed-phase columns with **HPLC**—a technique pioneered by Hungarian American chemist Csaba Horváth and American biochemist Sidney Pestka for the purification of proteins—led to the powerful techniques that are now standard all over the world.

SEE ALSO Separatory Funnel (1854), Chromatography (1901), Resolution and Chiral Chromatography (1960), HPLC (1967), Electrospray LC/MS (1984), Acetonitrile (2009)

Archer J. P. Martin at the lab bench, 1952—the year he won a Nobel, with Richard L. M. Synge, for their invention of partition chromatography. Developing the new forms of chromatography was not a matter of working out a new theory, but more a process of relentless tinkering and experimentation.

Rapamycin

Suren Sehgal (1932–2003)

Natural products' complexity, variety, and range of biochemical activities have always made them an important part of organic chemistry. The year 1972 marks the discovery of one of the most remarkable of them all, both for its origin and for its benefits to human health.

Rapamycin is named for Rapa Nui, the native name for Easter Island. Indian-Canadian pharmacist Suren Sehgal discovered the compound in 1972 from soil samples obtained by a 1964 expedition to this Chilean island in the southeastern Pacific. Rapamycin was originally studied because of its complex structure and its antifungal activity, but tests in animals quickly showed side effects that were even more useful. The compound was a potent suppressor of the immune system, a relatively rare effect that's of great interest for treating autoimmune disorders and keeping transplanted organs from being rejected. Later tests showed that it had surprisingly good anticancer activity as well, and rapamycin analogs are now used in chemotherapy. Scientists' efforts to understand these effects required groundbreaking research into new biological processes, a beneficial side effect of unknown compounds like this. Eventually, a target protein named mammalian target of rapamycin (mTOR) was found. Rapamycin binds simultaneously to it and another protein called FKBP in a three-way complex, disrupting mTOR's functions.

Those functions are many. A great deal of biomedical research has been done on mTOR since its discovery, and it has been found to be involved in several crucial signaling pathways, including regulation of cell growth, cell proliferation, cell motility, and cell survival. For example, in a dramatic demonstration of rapamycin's effect on mTOR, it was found in 2009 to prolong the lives of laboratory mice when it was added to their diet. Further study showed that this effect was not a general slowdown of the aging process but was the result of cutting down the number of fatal cancers the mice developed. These findings have led to additional research in the application of rapamycin in Alzheimer's disease, muscular dystrophy, lupus, HIV, and kidney disease.

SEE ALSO Natural Products (c. 60 CE), Bari Raid (1943), Antifolates (1947), Thalidomide (1960), Cisplatin (1965)

Easter Island (Rapa Nui) is home to these mysterious monoliths and one of the most interesting natural-product molecules in the world.

B$_{12}$ Synthesis

Robert Burns Woodward (1917–1979), **Albert Eschenmoser** (b. 1925), **Dorothy Crowfoot Hodgkin** (1910–1994)

The chemical structures of the different vitamins are all over the map. There aren't many similarities, since they serve completely different purposes in biochemistry. The only thing that they have in common is that they are vital for human health and that our cells don't have the ability to make them.

Vitamin B$_{12}$ is used to treat cyanide poisoning and pernicious anemia, but you'll find it in your daily multivitamin, too. Its structure is an outlier even by the standards of the other vitamins, since it alone contains a metal—a cobalt atom in the middle of a large cyclic complex, broadly similar to the one that holds iron inside hemoglobin. This framework has nine chiral centers, a formidable synthetic challenge. Coming off this framework is a "tail" region, joined to the complex part of the molecule by an amide linker. British biochemist Dorothy Crowfoot Hodgkin, whose work was instrumental in understanding **penicillin** and in protein crystallography, identified these features in 1956 with yet another high-profile **X-ray crystallography** structure.

Then American chemist Robert Burns Woodward and Swiss chemist Albert Eschenmoser joined forces in a ten-year effort for one of the most difficult organic syntheses yet attempted. B$_{12}$ has four linked nitrogen-containing rings circling the cobalt atom, and all four are chiral. Rings A and B were made via "resolution" (physically separating the left- and right-hand isomers), while C and D were made from chiral starting materials. The A and D rings were joined and then coupled to the B-C piece through an intricate series of reactions. The cobalt atom was then introduced, and it helped hold the ends of the molecule close to each other for the final cyclization.

Seemingly every major reaction type of the day found its way into Woodward and Eschenmoser's lengthy route (seventy-two total steps), including the **Birch reduction**, the **Grignard reaction**, diazomethane, ozonolysis, and a **Diels-Alder reaction** whose details helped pave the way for the **Woodward-Hoffman rules**. The synthesis of vitamin B$_{12}$, reported in 1973, is a landmark in the field.

SEE ALSO Ozone (1840), Chirality (1848), Asymmetric Induction (1894), Diazomethane (1894), Grignard Reaction (1900), X-Ray Crystallography (1912), Diels-Alder Reaction (1928), Birch Reduction (1944), Woodward-Hoffman Rules (1965).

R. B. Woodward holds a model of a difficult section of the vitamin B$_{12}$ molecule, 1973. The light-colored atom in the center is the cobalt discussed above.

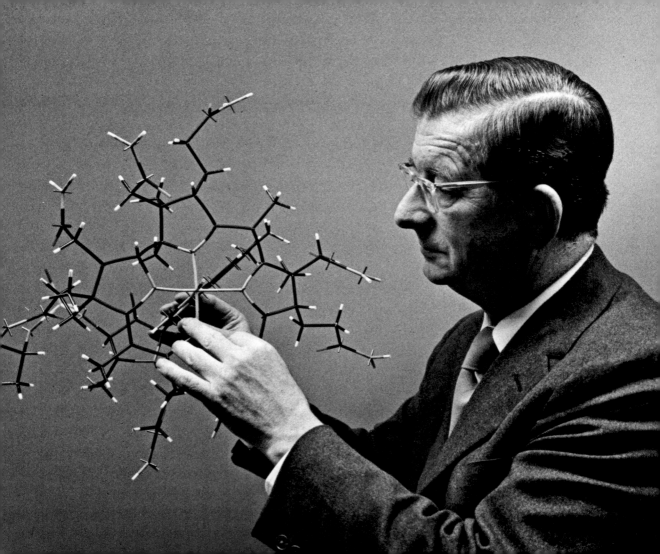

Virtually everyone has heard of the hole in the ozone, and some can remember how many uses for fluorocarbon gases (known by their trade name, *Freon*) were banned because they contribute to it. When American chemist Frank Sherwood Rowland and Mexican chemist Mario José Molina reported their research about the effect of CFCs on the ozone layer in 1974, though, it came as a surprise to almost everyone.

Ozone is a very reactive allotrope of oxygen—a molecule with three oxygen atoms instead of the usual two. It can be formed when the sun's strong ultraviolet (UV) light hits the upper atmosphere, but it can be broken down up there as well. The amount of high-level ozone stays fairly steady, then, because of this cycle of formation and destruction. The UV light the ozone absorbs, though, would be quite harmful to organisms on the ground (including humans) if it could get through the "ozone shield."

Free radicals also break down ozone, and although normally there are very few chlorine free radicals in the upper atmosphere, loose Freon-like compounds increase those levels dramatically, because these **chlorofluorocarbons** (CFCs) also break down when exposed to short-wavelength UV rays. Moreover, the free radicals can go through a number of cycles that destroy ozone and regenerate the radicals to repeat the process. This catalytic cycle of ozone destruction means that the amounts of gases that were released had a disproportionate effect. One chlorine radical could easily run through tens of thousands of ozone molecules!

The problem with the chlorofluorocarbons is that, unlike most sources of chlorine radicals, they aren't "rained out" of the lower atmosphere because they aren't very water soluble. Evidence quickly emerged that ozone depletion was going on in the stratosphere, and, eventually, the major industrialized nations passed a number of laws regulating and phasing out various halogen-carbon compounds. CFC levels have been falling since the 1990s, despite some atmospheric indications that some nations are cheating, and the ozone has started working its way back. And in 1995, Rowland and Molina were awarded the Nobel Prize in Chemistry.

SEE ALSO Photochemistry (1834), Ozone (1840), Free Radicals (1900), Chlorofluorocarbons (1930)

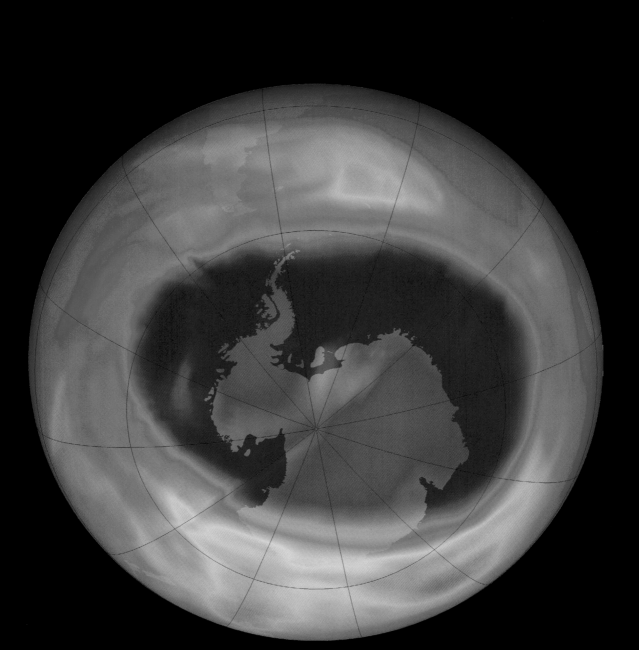

Enzyme Stereochemistry

Frank H. Westheimer (1912–2007), **John Cornforth** (1917–2013), **Hermann Eggerer** (1927–2006), **Duilio Arigoni** (b. 1928)

One of the many remarkable things about enzymes is their extraordinary sensitivity toward **chirality** (left- and right-handedness). If they use chiral substrates, they tend to show great selectivity for one form over the other, and if they start from a nonchiral molecule but generate chiral products, they generally make only one form of those as well. The **amino acids** that make up an enzyme all have chiral centers (except for the simplest one, glycine), giving them plenty of opportunity to have discriminating active sites and binding pockets.

Australian-British chemist Sir John Cornforth played a large role in discovering just how selective these reactions were by using one of the physical organic–chemist's most powerful techniques: labeling with **isotopes** (two or more of the same element with different atomic masses). Switching a specific hydrogen atom in a structure with a deuterium (or a radioactive tritium) and tracking where it ends up after a reaction, or watching for rate differences due to **kinetic isotope effects**, can illuminate reaction mechanisms like no other technique. Cornforth and his collaborators, including German biochemist Hermann Eggerer, synthesized labeled forms of the substrates for various enzymes—often no small problem in itself—and were able to determine biosynthetic pathways by tracking where the labels ended up. Cornforth's lab developed the stereochemical details of **cholesterol**'s synthesis from the starting material used by cells for all the steroids, a compound called mevalonic acid, which has six hydrogens that can be tracked. All six of them were eventually (and painstakingly) labeled in various combinations, and in 1975, Cornforth received the Nobel Prize in Chemistry for this work.

American chemist Frank H. Westheimer, Swiss chemist Duilio Arigoni, and many others also applied Cornforth's techniques, and, in general, the more enzyme mechanisms were investigated, the more impressive they looked. Organic chemists are still catching up and trying to engineer enzymes to their own specifications.

SEE ALSO Amino Acids (1806), Cholesterol (1815), Chirality (1848), Tetrahedral Carbon Atoms (1874), Asymmetric Induction (1894), Isotopes (1913), Radioactive Tracers (1923), Deuterium (1931), Carbonic Anhydrase (1932), Kinetic Isotope Effects (1947), Protein Crystallography (1965), Isotopic Distribution (2006), Engineered Enzymes (2010)

Strawberry wine fermentation, which involves the use of enzymes in yeast to convert carbohydrates into carbon dioxide and alcohol.

PET Imaging

Louis Sokoloff (1921–2015), **Alfred P. Wolf** (1923–1998), **Abass Alavi** (b. 1938), **Joanna Sigfred Fowler** (b. 1942)

Positron emission tomography, or PET, is a nuclear-medicine imaging procedure. Positrons are the antimatter equivalent of the electron, and they react immediately with normal matter in a distinctive burst of energy that can be mapped by detectors, allowing three-dimensional imaging if enough positrons are emitted from a sample. When the sample is a whole person, and the positron source is a radioactive drug or metabolite, a great deal can be learned about what's happening inside a working human body.

But preparing that labeled compound is organic chemistry's version of speed chess. The two **isotopes** used for almost all PET scans (carbon-11 and fluorine-18) have short half-lives (about twenty minutes for carbon-11, and a bit longer, one hundred ten minutes, for fluorine-18), which means that they have to be incorporated into the test molecule as quickly and cleanly as possible, and then that sample is rushed down the hall to the waiting patient. (This also means that the reagent delivering these isotopes has to be made freshly as well, which requires a small cyclotron to irradiate the precursors.)

Adding targeted fluorines to organic molecules is an important process, since fluorination can change a compound's properties so drastically. But PET fluorination is a subset all its own, with its practitioners always looking for new reactions to allow them to work under stopwatch conditions. American neuroscientists Louis Sokoloff and Martin Reivich showed, in the early 1970s, that radioactive glucose was an excellent tracer for brain activity, and American chemist Alfred P. Wolf suggested that the fluorine-18 version would be very useful. In the year 1976, Iranian-American neuroscientist Abass Alavi first gave this compound (synthesized by Wolf and American chemist Joanna Sigfred Fowler) to human volunteers and measured its distribution in a PET scanner. Since the brain uses only glucose for fuel, the regions that light up during a fluoroglucose scan are believed to be the ones with the greatest blood flow—that is, the parts that are working the hardest. Fluorinated drug molecules can be tracked the same way and provide a molecular-level look into living creatures, tracking biomolecules and drug candidates.

SEE ALSO Isolation of Fluorine (1886), Isotopes (1913)

A PET image of a patient's brain minutes after taking an injection of fluorine-18–labeled glucose. The red areas have the most signal and the blue the least, showing differences in brain activity.

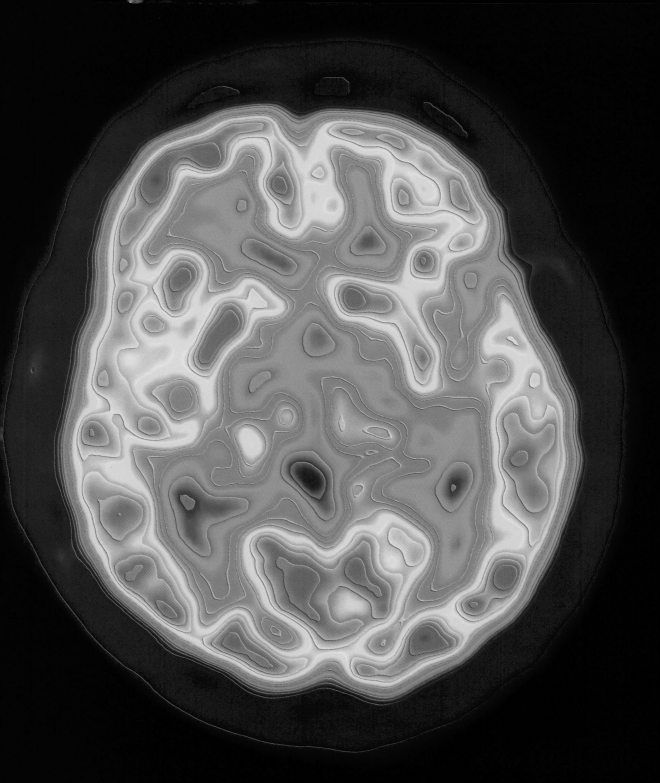

Nozaki Coupling

Hitosi Nozaki (b. 1922), **Yoshito Kishi** (b. 1937), **Tamejiro Hiyama** (b. 1946)

The Nozaki coupling reaction illustrates two key points in modern organometallic chemistry. One is that metal-catalyzed couplings are extremely useful, and the other is that figuring out just what's going on in a metal-catalyzed coupling can be extremely difficult.

The original Nozaki coupling reaction, reported in 1977 by Japanese chemists Tamejiro Hiyama and Hitosi Nozaki, was sort of a chromium variation of the magnesium-based **Grignard reaction**. Unlike the classic magnesium reagents, these were very selective toward reacting with aldehydes, and the chemistry was later extended to other sorts of functional groups that could form organochromium reagents. But as other chemists tried out these transformations, they discovered that sometimes these couplings would work, and sometimes they wouldn't, even on the same starting materials.

This was frustrating, because the reagents had potential to solve some very delicate synthetic problems. Japanese chemist Yoshito Kishi's group was involved at this time in trying to synthesize **palytoxin**, and these sorts of mild carbon-carbon-bond-forming reactions were just what they needed. But they ran into the same reproducibility issues, which were especially horrifying when applied to precious intermediates that had taken months to make.

Both the Kishi and Nozaki groups eventually found the problem: different batches of chromium chloride were giving the different results. And the difference between those batches? Tiny amounts of nickel. Ironically, the cleanest and most expensive sources of chromium chloride gave the worst results, since they were least likely to have the essential nickel contamination. Deliberately spiking these with a couple of hundredths of a percent of nickel chloride, though, made them fly right every time. What's known as the Nozaki-Hiyama-Kishi (NHK) reaction is still a valuable tool for bond formation, now that it's been tamed.

SEE ALSO Grignard Reaction (1900), Palytoxin (1994), Metal-Catalyzed Couplings (2010)

Spheres of pure nickel—the metal that caused all the difficulty in the Nozaki coupling procedure.

Tholin

Bishun Khare (1933–2013), Carl Sagan (1934–1996)

If you look through a telescope at Jupiter and Saturn, you'll notice a range of yellow, orange, red, and brown colors in their atmospheres. Saturn's largest moon, Titan, has colorful clouds and a yellow-orange surface, and there are colorful deposits on the surfaces of other icy moons. An organic chemist would find these shades familiar, though. Although most simple organic chemicals are colorless, most organic lab reactions aren't. Impurities and side reactions often turn reaction mixtures a light yellow shade, and that works its way down through the spectrum to the reddish brown stuff that sticks at the top of many **chromatography** columns. This material is difficult to separate into pure substances; it seems to be a mix of large polymeric molecules produced by organic chemistry gone astray.

And so it is in the outer solar system and beyond. There are huge numbers of small molecule building blocks out in space (**nitrogen**, water, ammonia, **cyanide**, **acetylene**, and so on), and they're being irradiated by hard ultraviolet rays from sunlight, heated and compressed in deep atmospheric currents, and shot through with gigantic lightning bolts—processes that have been going on for billions of years. In other words, there's a great deal of totally uncontrolled chemistry going on out there, and it seems to be producing similar kinds of colorful gunk wherever it has a chance. In 1979, American astronomer Carl Sagan and Indian chemist and physicist Bishun Khare named this mixture *tholin* (after the Greek word *tholós*, meaning "muddy") after they produced such mixtures while experimenting on a blend of gases that was thought to mimic Titan's atmosphere. The original category has now been divided into several classes (Titan tholin, Triton tholin, ice tholin, etc.) depending on where and how it is formed, and Pluto's reddish color indicates the presence of another likely class.

More recently, similar stuff has been found in the disks around young stars, and it can probably be found in most planetary systems. When it comes to chemistry, the universe apparently has decided that autumn shades are its preferred palette.

SEE ALSO Photochemistry (1834), Chromatography (1901), Miller-Urey Experiment (1952), Murchison Meteorite (1969)

One scheme for the formation of tholins on Titan. Ultraviolet light from the sun causes complex chemistry in the upper atmosphere, with the products gradually settling down to coat the moon's surface.

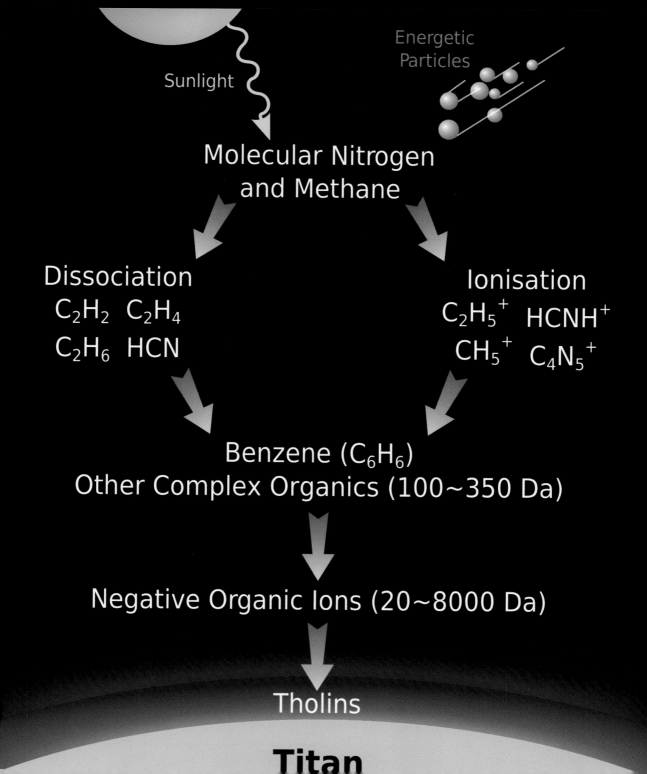

Iridium Impact Hypothesis

Luis Walter Alvarez (1911–1988), **Frank Asaro** (1927–2014), **Helen Vaughn Michel** (b. 1932), **Walter Alvarez** (b. 1940)

Iridium is not a common element near the surface of Earth (it dissolves readily in iron, so most of it is down in the core with all that molten iron). There's a lot more of it in several sorts of asteroids, though, along with other precious metals, which has led to some interesting schemes to bring one of these huge metal-rich space rocks close enough to Earth to be mined. But a strange layer of sediment found around the globe wherever Cretaceous-era rock layers are found strongly suggests that such an asteroid once came a lot closer to Earth than these would-be miners would want, and in a significantly less satisfying way for everything that lived here at the time.

The idea was that analyzing for iridium might help determine how many years the end-of-the-Cretaceous boundary layer represented, but the iridium numbers were so large that they looked like mistakes. Frank Asaro and Helen Vaughn Michel, American chemists specializing in radioisotope decay, had assigned dates and origins to many artifacts and geological samples before, but this was something new. In 1980, the Alvarezes, American geologist–son Walter and physicist-father Luis, advanced the bold hypothesis that this layer marked the worldwide debris from a gigantic asteroid impact, and that this caused the mass extinctions (dinosaurs and more) seen in the fossil record at this exact point.

It's hard to remember the doubt with which this idea was first received. Geologists prefer not to invoke gigantic one-time-only special events as explanations. If you allow too much of that sort of thing, you can hand-wave a justification for whatever theory you like. But the evidence has only strengthened over the years. Chromium **isotope** ratios, among other lines of evidence, also point to exactly the same theory. The amount of iridium in the global layer suggests the impact of an asteroid about the size of Manhattan, likely at the site of Chicxulub crater in southern Mexico, which would have catastrophically altered the entire planet's climate. We should not be surprised that the dinosaurs died—we should be surprised that anything else survived.

SEE ALSO Isotopes (1913)

Radar imaging of the Yucatán peninsula has helped to pin down the traces of an enormous impact crater whose age and size make it the leading candidate for the Cretaceous-ending extinction event.

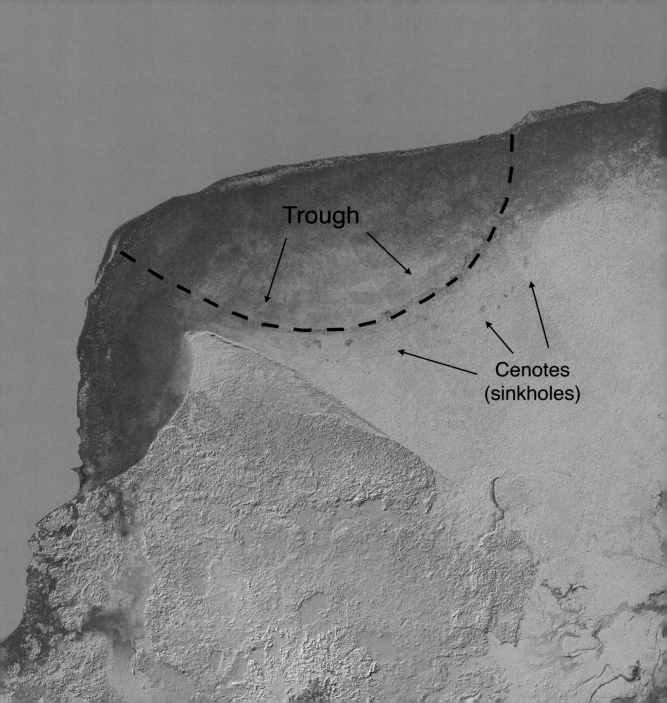

Unnatural Products

Leo Paquette (b. 1934), **Philip E. Eaton** (b. 1936)

The synthesis of **natural products** has been a big part of the story of organic chemistry (think **quinine**, indigo, **penicillin**, vitamin B_{12}), but what about the synthesis of *unnatural* products? That may sound odd, but there are many difficult or exotic molecules that (as far as we know) have never existed except in the imagination and then the lab. Chemists work on these to push the boundaries of chemical structure, of stability, and of organic synthetic techniques.

In that last category, the most well-known example is dodecahedrane. Mathematically, a dodecahedron is a platonic solid, made up of nothing but identical polygonal sides (in this case, twelve pentagons). Cubane, a smaller platonic molecule, was made in 1964 by American chemist Philip E. Eaton and coworkers, and its geometrically strained carbon-carbon bonds were quite unusual (see **Tetrahedral Carbon Atoms**). But no one had ever prepared dodecahedrane, although several talented chemists (Eaton, as well as the legendary chemist Robert Burns Woodward) had taken some swings at it.

Assembling all those five-membered rings was clearly a major challenge, but in 1982, American chemist Leo Paquette at Ohio State succeeded in around twenty-nine steps. Half the carbons in the final product come from cyclopentadiene molecules (the same five-membered ring compound, used for **ferrocene**), and four more come through **Diels-Alder reactions**. The rest are installed in a tricky series of reactions around the lip of the increasingly cup-shaped intermediates, with some bonds being formed by **photochemistry** reactions.

What about the other platonic solids? Tetrahedrane, the smallest, has never been made, although some substituted forms are known. It may not be stable. No one has ever prepared octahedrane or icosahedrane (the eight- and twenty-sided solids), and it won't be happening any time soon. Octahedrane would have no hydrogen atoms and would be a new allotrope of carbon, but it would need its bonds bent back like an umbrella in the wind, and icosahedrane would need five-bonded carbons, which (although they might exist in some exotic situations) usually means that someone who is just learning organic chemistry is about to lose points on an exam for giving carbon more than four!

SEE ALSO Natural Products (c. 60 CE), Quinine (1631), Photochemistry (1834), Tetrahedral Carbon Atoms (1874), Indigo Synthesis (1878), Diels-Alder Reaction (1928), *The Nature of the Chemical Bond* (1939), Penicillin (1945), Ferrocene, (1951), B_{12} Synthesis (1973), Fullerenes (1985)

A ball-and-stick model of tetranitrocubane, a derivative of cubane (note the block-shaped section in the middle). No cubane derivative has ever been found in nature, but the number of such compounds is, effectively, limitless.

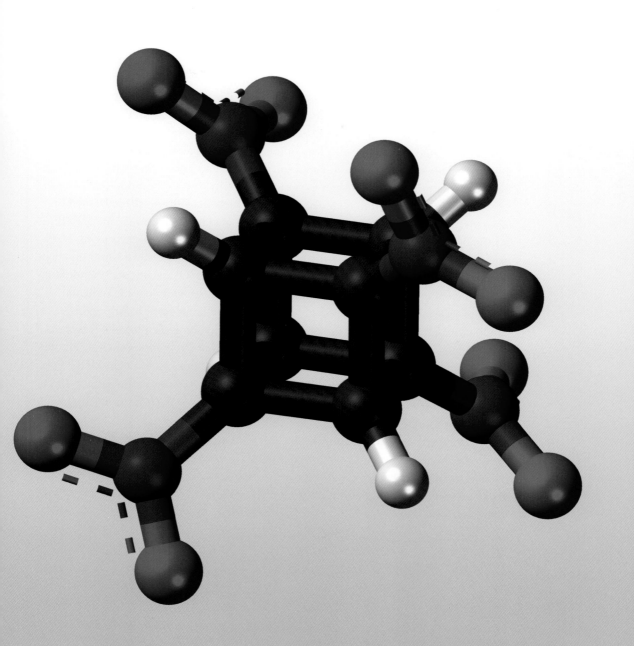

MPTP

MPTP is a terrifying compound. It has a simple structure, one that a medicinal chemist might make as a drug intermediate. Unfortunately, it's a contaminant produced during the careless synthesis of the street drug MPPP, a synthetic opioid with similar effects to heroin, and most people who are interested in making illegal opioids are not very picky about their analytical chemistry. In 1982, several young drug users turned up in emergency rooms near San Francisco with profound motor deficits that seemed to be permanent. Their diagnosis took a leap of imagination, because no one had ever seen anyone that young with advanced Parkinson's disease. Parkinson's normally presents in older patients and takes many years to get to the severity seen in these addicts.

A combination of criminal, medical, and chemical detective work established that all of them had been exposed to the same contaminated batch of homemade MPPP. It turned out that there had been a warning shot when, in 1976, a single drug user had exposed himself to the same mixture (his own product) and come down with immediate parkinsonian symptoms. But tests in rats had shown no MPTP toxicity, and the question had remained open. Later research (in the lab of American neurologist J. William Langston) showed that the compound crosses the blood-brain barrier easily and is a good fit for a protein that takes the neurotransmitter dopamine up into cells. The cells with the largest amounts of such dopamine transporters are in the substantia nigra, the very region in the brain that deteriorates in Parkinson's. Once inside a cell, MPTP gets metabolized to another species that (by chance) inhibits a key metabolic pathway, and that shutdown proves deadly. (Rats, it turns out, escape these effects by producing much less of the second toxic species.)

This tragedy opened up entirely new lines of Parkinson's research. The disease itself probably isn't caused by environmental MPTP (there isn't any), but (as one hypothesis) researchers are exploring whether some other compounds might exploit the same metabolic vulnerability, and whether some people may be more susceptible to those effects.

SEE ALSO Toxicology (1538), Modern Drug Discovery (1988)

Photomicrographs of substantia nigra brain tissue from a patient who had Parkinson's disease. The red-colored objects—shown at various magnifications—are "Lewy bodies," abnormal tangles of a protein called alpha-synuclein *that are one of the signs of the disease.*

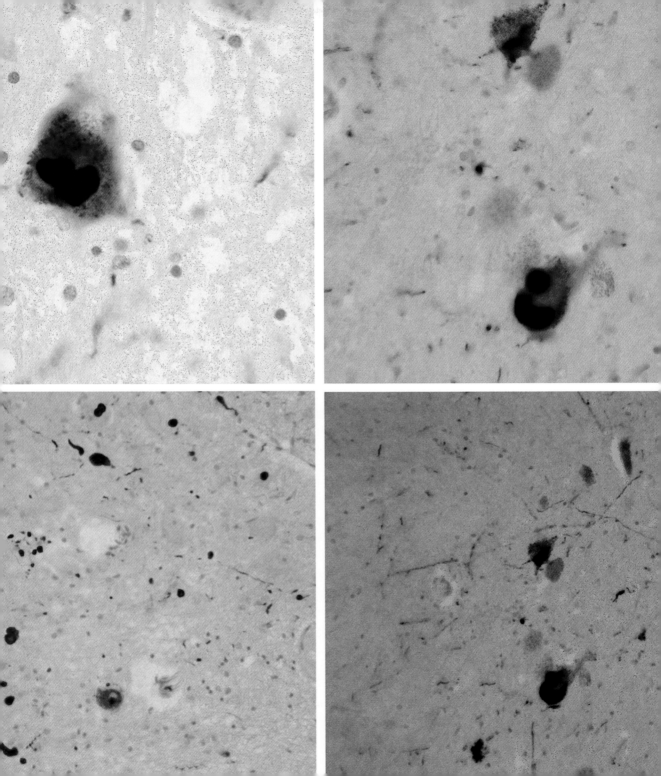

Polymerase Chain Reaction

Har Gobind Khorana (1922–2011), **Kjell Kleppe** (1934–1988), **Kary Banks Mullis** (b. 1944)

Like many great discoveries, the polymerase chain reaction (PCR)—the rapid synthesis of large quantities of a given DNA segment—seems simple, and like many great discoveries, it had been anticipated. DNA polymerase was a known enzyme that would copy an existing DNA segment. Indian-American chemist Har Gobind Khorana and Norwegian biochemist Kjell Kleppe had used it in the late 1960s to make several copies of a DNA sequence, proving that it could be done outside of a living cell. American biochemist Kary Banks Mullis added a key insight—if the reaction temperature was cycled up and down, it might be possible to amplify very small samples of DNA in an unheard-of fashion. Mullis was awarded the Nobel Prize in Chemistry in 1983 for his discovery.

Heating the DNA unravels the hydrogen bonds holding the double helix together, leaving it open for the polymerase. Once the solution cools, the enzyme can assemble a DNA sequence from individual nucleotide building blocks, but it needs "primers," short stretches of preassembled DNA that are complementary to the starting and ending regions of the whole sequence being copied. Far from being a complication, though, these primers allow a researcher to zero in on exactly the sequence of interest. The newly formed DNA sequences go on to serve as templates for still more copies, leading to an exponential increase of newly synthesized DNA.

Mullis had some trouble convincing his coworkers that he'd gotten his ideas to work, but his colleagues Randall Saiki and Henry Erlich were able to improve the technique. The original DNA polymerase was inactivated in each heating cycle and had to be added fresh, but a variant of the enzyme *Taq*, from an organism that lived in boiling hot springs, was rock-stable. *Taq* polymerase PCR became a huge success, the subject of both huge financial deals and huge patent lawsuits. Later versions are now indispensable in every field that touches on DNA—and they are numerous, with anthropology, archaeology, genetics, forensics, medicine, biotechnology, and molecular biology among the most prominent. PCR (and fast DNA sequencing) has remade the world.

SEE ALSO Hydrogen Bonding (1920), DNA's Structure (1953), DNA Replication (1958)

Morning Glory Pool in Yellowstone National Park, the sort of habitat where the Taq-*containing bacterium was found. Extreme environments breed extreme life-forms, which contain extreme enzymes.*

Electrospray LC/MS

John Bennett Fenn (1917–2010)

LC/MS stands for *Liquid **Chromatography**/**Mass Spectrometry***, and it is an amazingly versatile technique. It consists of an **HPLC** (high-performance liquid chromatography) system feeding into a mass spectrometer—a combination of two powerful machines that seems like a natural fit. But the first such instrument only arrived on the market in 1984, nearly twenty years after HPLC was invented, and there's good reason it took so long.

The two technologies have some fundamental incompatibilities. HPLC needs a solvent to push mixtures through its columns, but mass spectrometers need a vacuum. Switching from high-pressure liquid to outer-space conditions required some substantial engineering. Several methods were tried, but the "electrospray" technique really worked. It involves sampling the solvent stream from the HPLC column and spraying it as a fine mist into the mass spectrometer's vacuum chamber. The droplets naturally start to evaporate and get progressively smaller. Meanwhile, they're taken through a thin capillary tube with a very high voltage where they pick up electric charges. Those charges start to repel one another as the droplets evaporate, so they break apart into even smaller droplets, eventually producing charged "naked" molecules (and molecular fragments) flying along in a vacuum.

Electrospray allowed really large molecules to "fly" reliably in mass spectrometry experiments (as with matrix-assisted laser desorption and ionization, **MALDI**, developed around the same time). Any compound from an HPLC could now be identified by its exact molecular weight. Small-molecule chemists rushed to analyze their HPLC traces with the power of mass spectrometry, while protein and macromolecule chemists found themselves with a completely new tool. LC/MS-pioneer American chemist John Bennett Fenn (later involved in a patent dispute over the invention with his university, Yale) first reported the technique in 1984 and shared a well-deserved Nobel in 2002. Since its introduction, LC/MS has quickly moved from an exotic curiosity to something that no chemist wants to live without.

SEE ALSO Natural Products (c. 60 CE), Chromatography (1901), Mass Spectrometry (1913), Gas Chromatography (1952), HPLC (1967), Murchison Meteorite (1969), Reverse-Phase Chromatography (1971), MALDI (1985), Acetonitrile (2009).

A close-up of the electrospray unit of a mass spectrometer. Behind the glass of the apparatus is a harder vacuum than one would find in low Earth orbit.

AZT and Antiretrovirals

Jerome Horwitz (1919–2012), **Samuel Broder** (b. 1945), **Hiroaki Mitsuya** (b. 1950), **Robert Yarchoan** (b. 1950)

When the HIV/AIDS epidemic became a growing medical problem in the early 1980s, drug researchers began investigating existing compounds to see if they could be used as treatments. In 1984, a team at the National Cancer Institute (NCI)—including American oncologist Samuel Broder, Japanese virologist Hiroaki Mitsuya, and American doctor and medical researcher Robert Yarchoan—developed an assay that tested compounds' ability to keep T cells (cells active in the body's immune response) from being killed by HIV. They began by screening compounds that had already been tested as antivirals, and one provided to the NCI under a collaboration agreement with Burroughs Wellcome, a London-based firm with antiviral research experience, turned out to be a potent hit in the cell assays. This was the soon-to-be-famous azidothymidine, or AZT.

It had been made back in 1964 by American chemist Jerome Horwitz, as part of a long-running effort in medicinal chemistry to make analogs of the nucleic acids found in DNA and RNA. These analogs had the potential to block cell replication when they were mistaken by enzymes for the real thing. They'd been studied in antiviral, antibacterial, and chemotherapy research programs, all of which could find a use for disrupting **DNA replication**, but AZT had not been active enough in the 1960s to become a drug for Burroughs Wellcome. However, the firm maintained it for testing should a potential new use emerge. (This is standard practice in drug companies, where collections of hundreds of thousands—or even millions—of compounds are common.)

The company quickly moved AZT into clinical trials, where it showed life-extending effects in infected patients. It was approved by the FDA in 1987, in what was one of the shortest development periods since modern clinical trials were instituted. Manufacturing the drug took some care, since the azide group is an unusual one in drug structures and can produce toxic and explosive side products depending on the conditions. Researchers continued to develop antiretroviral drugs targeting a variety of mechanisms as HIV became better understood, and for those who have access to the drugs, they can turn the disease into a manageable, rather than fatal, condition.

SEE ALSO Salvarsan (1909), Sulfanilamide (1932), Streptomycin (1943), Penicillin (1945), DNA Replication (1958), Modern Drug Discovery (1988), Recrystallization and Polymorphs (1998), Click Triazoles (2001)

Crystalline azidothymidine as viewed under polarized light. AZT is still one of the only drugs ever to contain the azide functional group, partly because such compounds are not often tested for activity.

Quasicrystals

John Cahn (b. 1928), **Dan Shechtman** (b. 1941), **Paul Steinhardt** (b. 1952)

In 1912, **X-ray crystallography** had shown everyone what **crystals** were like: regular, repeating-unit cells that filled three-dimensional space. Symmetry was the key, and some kinds of symmetry were allowed, while other kinds weren't. In 1984, though, a startling report claimed that this picture was very incomplete. Israeli materials engineer Dan Shechtman presented electron-microscope data from an aluminum-magnesium alloy that showed fivefold symmetry—a structure thought to be impossible, for the same reason that you can't use a regular pentagon shape to tile a floor, while regular triangles, squares, and hexagons are fine for this purpose. The physical arrangement that led to this weird behavior was also a shock: it was a crystalline solid that had order but never repeated.

Shechtman said later that when he saw the first electron-diffraction data, he told himself, "There can be no such creature." It took him two years of follow-up work to get up the nerve to publish, and even then, his first manuscript (written with his Israeli colleague Ilan Blech, the first person to believe him) was flatly rejected. German-born American chemist John Cahn urged him to resend it to a more widely read journal, which attracted plenty of attention to these new "quasicrystals."

Unfortunately, much of that attention was negative, at least at first. The crystallography community was skeptical, and Linus Pauling—still a force to be reckoned with—was vocal in his disdain. "There are no quasicrystals," he said, "only quasi-scientists," but other groups were able to replicate Shechtman's work. American theoretical physicist Paul Steinhardt published a mathematical explanation that fit the data, and Pauling's position began to erode, especially after X-ray data on larger crystals confirmed the proposal. Pauling never accepted that he had been wrong, but in 2011 Shechtman won the Nobel for sparking a whole new field of research, since quasicrystalline materials have unusual mechanical and optical properties that could be useful in coatings and laser applications. Arguments are integral to science, but the field also provides the means for arguments to be settled—and Shechtman definitely won this one.

SEE ALSO Crystals (c. 500,000 BCE), X-Ray Crystallography (1912)

Dan Shechtman shows off a model of a quasicrystal to colleagues in 1985, not long after publishing his startling results.

1984

Bhopal Disaster

With more than half a million injured, as many as sixteen thousand dead, and thousands permanently disabled, the Bhopal disaster is probably the worst chemical catastrophe the world has ever seen. The causes of this tragedy are still a matter of argument, but these facts are indisputable: late on the night of December 2, 1984, some thirty metric tons of the reactive and toxic chemical methyl isocyanate (MIC) leaked from the Union Carbide India pesticide plant, spreading over twenty-five square miles in the central Indian city of Bhopal.

The devastating effects were no wonder, because MIC is extremely dangerous. It can be detected (as an eye irritant) down to about two parts per million in the air, and once it reaches over twenty parts per million, severe consequences follow from damage to the lungs. The compound was manufactured on site, as an intermediate to make the pesticide carbaryl, but a significant amount of the MIC had built up, unused. It's not a chemical that's typically stored for long periods—it's too reactive for that—and Union Carbide India had been warned of the possibility of a runaway reaction in the storage tanks. There had been numerous safety problems at the facility, with workers being exposed to both MIC and the toxic phosgene used to make it, among other hazards. It was found later that the safety systems that should have been installed on the MIC storage tanks were poorly maintained or not operating at all.

When water entered one of the tanks and began to react vigorously with the MIC, the safety equipment was totally inadequate to deal with the resulting breach. Some have alleged sabotage, while others cite appalling incompetence and negligence. Once the tank contents (MIC and its breakdown products) were released into the atmosphere in the middle of the night, however, the reasons did not matter to the half million innocent people, mostly impoverished, who lived around the plant. This accident stands as a horrible lesson for anyone working on industrial safety.

SEE ALSO Toxicology (1538), Donora Death Fog (1948)

Survivors of the Bhopal disaster who suffered ocular trauma sit in front of the Union Carbide factory. Most eye problems from the disaster were temporary, but there were also many cases of long-term damage.

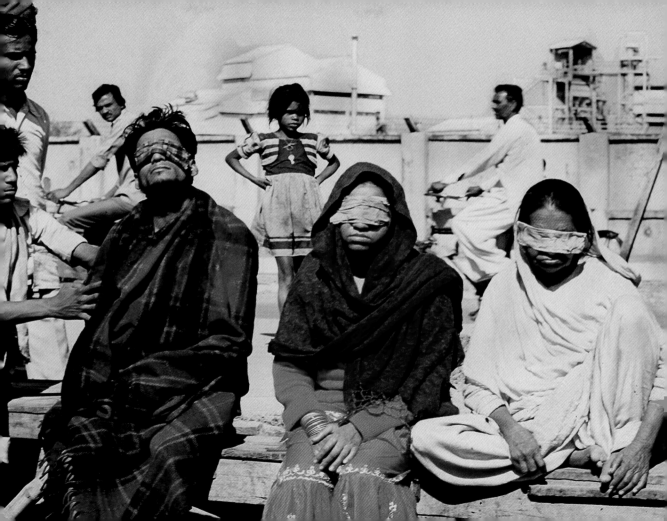

Fullerenes

Richard Buckminster Fuller (1895–1983), **Richard Smalley** (1943–2005), **Harold Walter Kroto** (b. 1939), **Robert Floyd Curl Jr.** (b. 1933)

If you'd told chemists in the early 1980s that there was a significant undiscovered allotrope of carbon, you would not have gotten much of a hearing. But one got discovered anyway, without anyone even looking for it.

English chemist Harold "Harry" Walter Kroto wanted to study long carbon chains that he thought might exist in interstellar space, and he talked American chemist Richard Smalley into trying out his specially made apparatus on the problem. Its laser blasted atoms from a sample into the gas phase, and as they cooled and began to clump together, these clusters were analyzed by **mass spectrometry**. When Kroto, Smalley, fellow American chemist Robert Floyd Curl Jr., and several graduate students began trying this technique on solid carbon in 1985, they saw one particular species forming over and over. It weighed in at exactly 60 carbons. Under some conditions, it was almost the only thing formed, but it was sometimes mixed with an equally mysterious C_{70} compound. Both species, whatever they were, were extremely stable, which suggested that they didn't have any spare bonding possibilities left over.

Some sort of curled-up structure seemed likely, but trying to assemble possible candidates from six-membered rings (like benzene) didn't form any believable C_{60} molecules. Smalley began experimenting with paper models that mixed in some five-membered rings and quickly found a round sixty-carbon structure with the same form as a classic soccer ball. This explained the stability perfectly, and later analysis confirmed the new carbon form. (It is so symmetrical that its **NMR** spectrum has only one peak, since it has only one kind of carbon atom!) In honor of the inventor of the geodesic domes it resembles, American polymath Richard Buckminster Fuller, the compound was named *buckminsterfullerene*—known to most chemists as a *buckyball*—and the whole class of enclosed ball-like structures are called *fullerene*s. Smalley's discovery immediately suggested an egg-shaped C_{70} molecule, and many others have now been discovered. Their unique properties have been the subject of research ever since, and their discovery led to a Nobel Prize for Smalley, Kroto, and Curl in 1996.

SEE ALSO Mass Spectrometry (1913), Synthetic Diamond (1953), NMR (1961), Unnatural Products (1982), Carbon Nanotubes (1991), Graphene (2004)

The C_{60} buckyball now looks like an obvious structure, and it has since turned up under many conditions (once you know to look for it), but until recently no one had apparently considered that it might really exist!

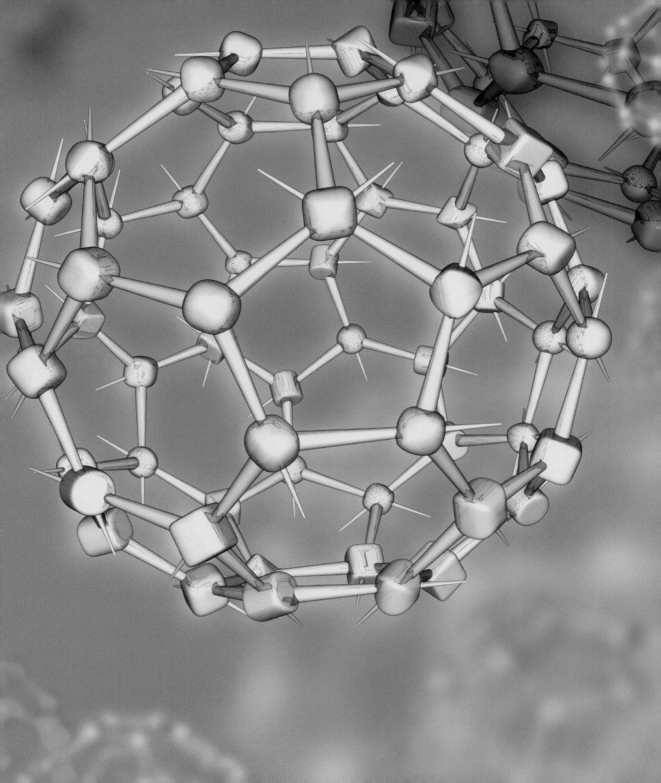

MALDI

Franz Hillenkamp (1936–2014), **Michael Karas** (b. 1952), **Koichi Tanaka** (b. 1959)

MALDI is short for *matrix-assisted laser desorption and ionization*, which sounds pretty formidable, but the process isn't hard to understand. German chemists Franz Hillenkamp and Michael Karas, who coined the term in 1985, were investigating pulses of laser light as a way to blast ions into the vacuum chamber of a mass spectrometer. A laser is very efficient at delivering a large amount of energy to a small sample, so the idea was sound. But because different compounds can have very different preferred wavelengths, depending on their chemical structures, it was thought at first that the wavelength of the laser light would have to be adjusted for each type of sample, so that the laser's energy could be absorbed by the compounds of interest. That could be a time-consuming complication.

What they found, though, was that the **amino acid** tryptophan would serve as a sort of energetic lever when it was irradiated, sending its excess energy into the molecules nearby. All sorts of ions could be sent out into the machine, even those of small proteins, as long as they were mixed together with tryptophan into a thin solid layer. The next big breakthrough came in 1987 from Japanese engineer Koichi Tanaka, whose group used very tiny cobalt particles and a thick liquid matrix (glycerol) to get molecules of unheard-of massive weights to show up in the mass spectrometer. Eventually, however, the solid-matrix technique became more popular, as better matrix molecules were discovered that worked well even with easy-to-obtain lasers. During the early 1990s, such machines came onto the commercial market, and biochemists, molecular biologists, and others who might not have paid much attention to **mass spectrometry** suddenly found themselves with a powerful new tool to analyze biomolecules and samples that had previously been very hard to characterize.

For large molecules, MALDI was particularly useful when hooked up to a time-of-flight–detecting mass spectrometer that measures how long it takes ions to show up at the other end of the vacuum chamber (heavier ones, naturally, are a bit slower). MALDI-TOF (pronounced *mal-dee-toff* is now a familiar phrase to chemists everywhere.

SEE ALSO Amino Acids (1806), Mass Spectrometry (1913)

In this depiction of an advanced MALDI technique, the crystalline matrix has actually been grown on the surface of a nerve cell, and the proteins from the cell's surface are then sent into the mass spectrometer by a laser beam.

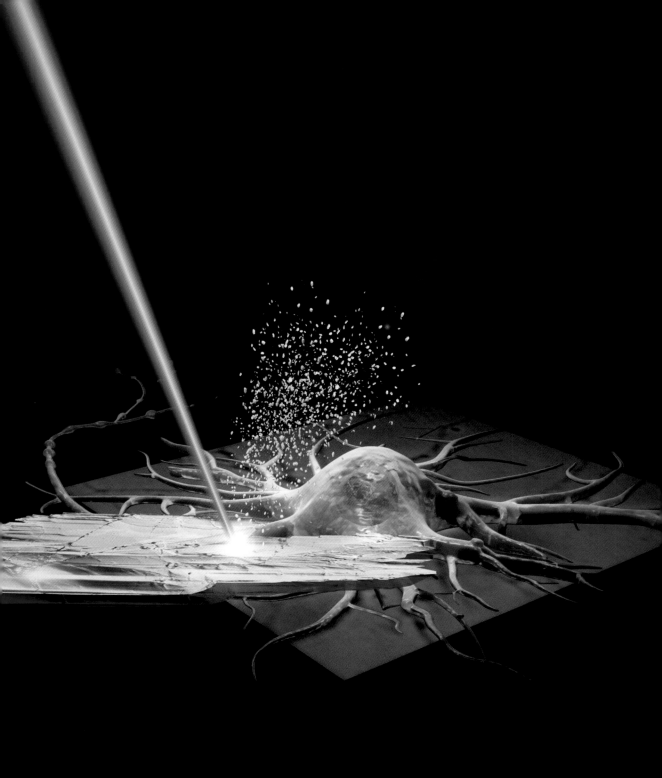

Modern Drug Discovery

George Herbert Hitchings (1905–1998), Gertrude Belle Elion (1918–1999), James Whyte Black (1924–2010)

In 1988, three of the biggest names in the history of medicinal chemistry were awarded the Nobel Prize in Physiology or Medicine "for their discoveries of important principles for drug treatment." American coworkers Gertrude Belle Elion and George Herbert Hitchings were involved in making purine derivatives, a class of compounds found in the structure of DNA and other significant biomolecules. This important role for the purine framework made it an excellent starting point for drug discovery. Their work yielded drugs for malaria, organ transplantation, bacterial infections, and cancer and formed the basis for many more projects. Elion and Hitchings were also responsible for pioneering research methods that led to the development of **AZT and antiretrovirals** (critical for the treatment of HIV/AIDS). Scotsman Sir James Whyte Black was instrumental in developing two compounds (propranolol for heart disease and cimetidine for ulcers) that became the best-selling drugs in the world.

The key to what its practitioners call *med chem* is finding compounds with the desired activity, even if they're weak, then varying their structures to make them more potent, more selective, and better tolerated by patients. Medicinal chemists must be nimble in response to new assay results, using any techniques they can think of to make new test molecules. "There are only two reaction yields," goes one saying, "enough and not enough."

Elion, Hitchings, and Black worked during the "classic age" of drug discovery, when medicinal chemists were learning how to optimize molecules toward specific targets. At the time, many compounds were tested directly in living cells or in rodent models of disease, since the practice of testing against pure cloned proteins had not yet been developed. But there's value in testing phenotypically, i.e., looking for the desired effect in a living system without necessarily knowing all the details of how it happens. After many years of "target-driven" drug discovery, in which drugs are designed to interact with a known biological target, modern forms of phenotypic screening, a descendant of the techniques used by the scientists who made some of the most important contributions to clinical medicine in the twentieth century, are making a comeback.

SEE ALSO Salvarsan (1909), Sulfanilamide (1932), Streptomycin (1943), Penicillin (1945), Antifolates (1947), Cortisone (1950), The Pill (1951), MPTP (1982), AZT and Antiretrovirals (1984), Taxol (1989)

George Hitchings and Gertrude Elion in an autographed photo from 1998.

PEPCON® Explosion

When the space shuttle *Challenger* was destroyed during liftoff in 1986, the whole NASA shuttle program came to a halt for more than two years while the deadly accident was investigated. One of the lesser-known effects of this hiatus was the accumulation of solid rocket propellants at their manufacturing sites. With no rocket boosters being made, several million pounds of ammonium perchlorate, a component of their solid fuel propellants, were stored at the PEPCON chemical company in the Nevada desert.

It's in everyone's best interest for any perchlorate facility to be located as far away from everything else as possible. Perchlorates feature chlorine atoms in their highest **oxidation state**, ready to oxidize anything they come in contact with. They are notoriously treacherous compounds, unforgiving of sloppy handling, and known to cause fires and explosions. The large amount of oxygen carried on the perchlorate anion means that once a fire starts, it doesn't have to rely on a supply of air. This feature has long made perchlorates a key ingredient of solid rocket fuel mixtures (and a common ingredient in fireworks manufacturing, which is a disturbingly similar field in many ways!).

On May 4, 1988, what may have been a welding accident started a fire at PEPCON, and the workers (fully aware of the potential for disaster) tried frantically to put it out. But the flames spread to a storage building packed with perchlorate-filled drums, causing the employees to flee for their lives. Local responding firemen parked a mile away, but all the windows of their trucks were blown out when the main stockpile of ammonium perchlorate exploded. A neighboring building and employees' vehicles were destroyed, and other buildings nearby suffered extensive damage. The gigantic blast threw a visible shock wave across the desert and was measured by distant seismographs at about 3.5 on the Richter scale. Only two people were killed, but hundreds of others, as distant as ten miles away, were injured by blown-in window glass and the effects of the shock wave. It was a dramatic demonstration of the unharnessed power of chemicals that humans may think they have under control.

SEE ALSO Gunpowder (c. 850), Nitroglycerine (1847), Oxidation States (1860)

The 1986 explosion of the Challenger *space shuttle brought a halt to manned space launches in the United States while the cause was tracked down, but the resulting oversupply of solid rocket fuel created its own problems.*

Taxol

Monroe Eliot Wall (1916–2002), **Mansukh C. Wani** (b. 1925), **Pierre Potier** (1934–2006), **Robert A. Holton** (b. 1944)

In the early 1960s, the National Cancer Institute (NCI) started a program to test extracts from as many different plants as possible, in a hunt for new chemotherapy agents. In 1964, an extract of the Pacific yew tree (*Taxus brevifolia*) was found to be toxic to cancer cells, and a team at RTI, then called the Research Triangle Institute, began working to isolate the active compound. American chemist Monroe Eliot Wall and Indian chemist Mansukh C. Wani reported a natural product they called *taxol* in 1966, and its complex structure was worked out by 1971.

Further studies required stripping thousands of pounds of bark from the yew trees. In 1979, the compound became even more interesting when it was found to have a mechanism that no one had ever seen—it bound to structures called microtubules, essential players in cell division. Taxol continued to be active in animal models, it passed toxicity testing, and the NCI started human clinical trials in 1984. However, it began to look as if turning it into a successful drug could wipe the Pacific yew tree off the face of the earth.

In 1989, the NCI turned to the drug company Bristol-Myers Squibb (BMS) to get the compound through clinical trials. The company (controversially) trademarked the name *Taxol* and changed the generic name to paclitaxel. Approval to treat ovarian cancer came in 1992. BMS and several academic groups worked to solve the supply problem; a key step was French chemist Pierre Potier's finding that a related yew species gave an advanced intermediate from its needles—a renewable source compared to stripping bark. American chemist Robert A. Holton's group had been working on the problem for years as well and was able to synthesize the drug in 1994. By 1995, no yew trees were being destroyed to supply the drug.

It was found, in the end, that the compound wasn't being made by the yew tree itself, but by a fungus colonizing its cells. Taxol is now made in fermentation tanks by growing these two organisms together and not by synthetic chemistry at all!

SEE ALSO Natural Products (c. 60 CE), Toxicology (1538), Bari Raid (1943), Antifolates (1947), Thalidomide (1960), Cisplatin (1965), Modern Drug Discovery (1988).

Yew tree bark is, fortunately, no longer needed to produce Taxol. Natural product–based drugs can be major challenges to synthetic organic chemistry.

Carbon Nanotubes

Sumio Iijima (b. 1939), Donald S. Bethune (b. 1948)

In November of 1991, Japanese physicist Sumio Iijima published a paper showing how to prepare tiny, hollow tubes of pure carbon atoms. These tubes were actually multilayered sheets of what we now call **graphene**, but with the edges bonded into a cylinder (called *multiwalled nanotubes*, or *MWNTs*). The electron microscope images were compelling, and the field took off in a flurry of publications from around the world. Carbon nanotubes could be used as probes, miniature electrodes and wires, and catalysts, and could eventually replace silicon in microprocessors.

A look through the scientific literature showed that this discovery had been made before, but its significance had not been recognized. Soviet researchers showed electron micrographs of such structures as far back as 1952, and further reports of MWNTs came from Japan, the U.S., and Russia during the 1970s and 1980s. One thing that seems to have really accelerated research in this area was the prediction that single-walled nanotubes would have very unusual physical properties. It wasn't long before Iijima and American physicist Donald S. Bethune each reported techniques to prepare these single-walled nanotubes (known as *SWNTs*) as well. Like **fullerenes**, nanotubes appear to have been hiding in plain sight—well, if you consider something that has to be looked at with an electron microscope to be in "plain sight" (and there are those who do). They can be found in the material sprayed off by carbon rods in electrical arcs, for example, and seem to be formed under many other conditions.

Controlled synthesis of these structures is now a huge field of research, because many of those predicted physical properties have turned out to be real. And a wide variety of nanotubes can be formed, with different widths and twists in their structures, each with different possible applications. Currently they are used in batteries, automotive and boat parts, sporting goods, and water filters. Proposed applications—some of which sound futuristic indeed—include artificial muscles made of wax-filled carbon nanotube yarn and oil-loving, absorbent, carbon nanotube sponges for use in oil spills—test samples can absorb more than a hundred times their weight in oil.

SEE ALSO Synthetic Diamond (1953), Fullerenes (1985), Graphene (2004)

Just a few of the many possible varieties of carbon nanotubes. Gardeners will find it impossible not to think of rolls of chicken wire, which is not a bad model!

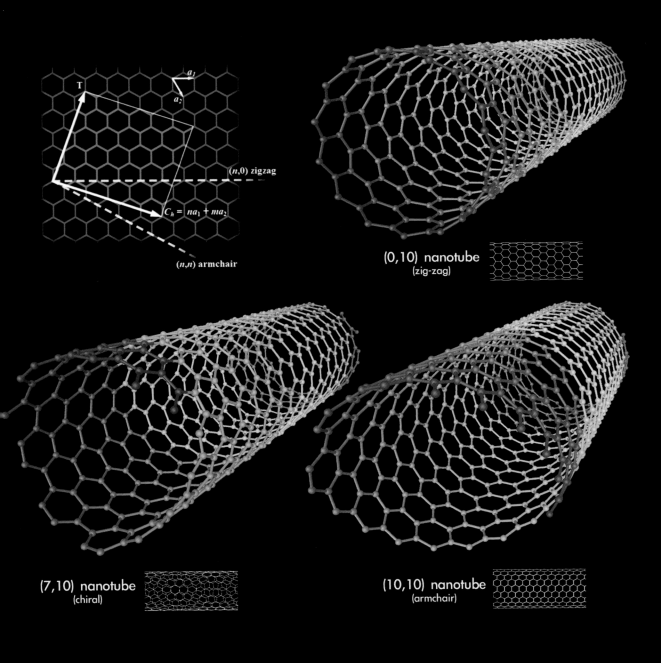

Palytoxin

Paul J. Scheuer (1915–2003), **Richard E. Moore** (1933–2007), **Yoshito Kishi** (b. 1937), **Daisuke Uemura** (b. 1945)

The complexity of some **natural products** boggles the mind. Case in point: palytoxin, isolated in 1971 by a research group at the University of Hawaii headed by American chemists Paul J. Scheuer and Richard E. Moore. The ancient Hawaiians had told stories of a particularly poisonous seaweed (actually a soft coral) that was supposed to have been cursed by a god. Japanese chemist Daisuke Uemura's group, who later worked out the palytoxin structure, and the later chemists who synthesized it were probably willing to believe the legend by the time they were through.

To begin with, the compound is hideously toxic. One billionth of a gram of palytoxin is sufficient to kill a mouse by injection, and many humans have been sickened or even died after being exposed. The most common way a person can encounter it is by eating contaminated fish from tropical waters, but a few owners of high-end saltwater aquariums have been killed by palytoxin-containing organisms (or just the water) from their own collections.

To get a handle on the molecule's complexity, consider that it has seventy-one chiral centers, which means it has 2^{71} possible isomers. This is a number that the mind cannot really grasp. Imagine a beach full of sand, and then try to imagine every beach and every grain of sand in the entire world. If every one of those sand grains were itself another Earth, full of its own beaches, and then if you counted every sand grain in every one of them, and then did that all a billion times over, it would be nothing compared to the number of possible palytoxins. Synthesizing just one of them is a true challenge.

It's been done, though, in what some have called the "Mount Everest of synthetic chemistry." The research group of Japanese chemist Yoshito Kishi at Harvard published their work in 1994, starting from six different chemical sources and converging on the product in some 140 separate synthetic steps. It was a massive and impressive effort, and it demonstrated that modern organic chemists are capable of making almost anything, given enough time and effort (and funding, of course!).

SEE ALSO Natural Products (c. 60 BCE), Toxicology (1538), Chirality (1848), Nozaki Coupling (1977).

Zoanthids very much like these can contain dangerous amounts of palytoxin, as some owners of exotic saltwater aquariums have found out by accident.

Coordination Frameworks

Makoto Fujita (b. 1957), **Omar Yaghi** (b. 1965)

The chemistry of metal **coordination compounds** has provided innumerable catalysts for the laboratory, various dyes for industry, and even platinum anticancer drugs and gadolinium MRI contrast agents for the medical field. In 1997, Jordanian-American chemist Omar Yaghi produced a paper investigating a new sort of metal complex. This type starts by making a rigid and symmetrical organic molecule, with groups at its ends that can easily complex metals. If these groups are reacted in solution with the appropriate metal ions, a crystalline solid can grow as the organic molecules arrange themselves around the metal atoms in a repeating three-dimensional lattice. There are many possible coordinating groups, which can be built onto all sorts of molecular scaffolds. And the various metals (as first shown by Swiss chemist Alfred Werner late in the nineteenth century) can take all sorts of geometries around themselves. This sounds like a recipe for a bewildering variety of different structures, and it certainly is. At present, every month brings reports of still more new materials, sometimes called *metal-organic frameworks* or *coordination polymers*.

The names vary because the **crystals** that form can be very odd structures with correspondingly odd properties. If the rigid organic units form large enough spacers, the resulting solid can have very large channels and pores—in fact, the crystals can be almost entirely empty space! These ordered cavities can be filled, in turn, by all sorts of other molecules, and the hope is that these materials might be used to store hydrogen, sequester **carbon dioxide**, serve as a matrix for new battery technologies, and more. A 2013 paper from the lab of Japanese chemist Makoto Fujita even reported that other small molecules can soak into some of these frameworks in such an orderly way that **X-ray crystallography** can be done on them, perhaps opening a convenient way to get crystal structures of things that otherwise might never crystallize at all.

SEE ALSO Crystals (c. 500,000 BCE), Carbon Dioxide (1754), Coordination Compounds (1893), X-Ray Crystallography (1912), Hydrogen Storage (2025)

The structure of the coordination framework MIL-53. The backbone is made up of benzene rings (with carbons as black spheres) that have carboxylic acid groups on each side (oxygens as red spheres). These coordinate around metal atoms (inside the blue polyhedrons), and the yellow spheres illustrate the amount of empty space in the resulting pores.

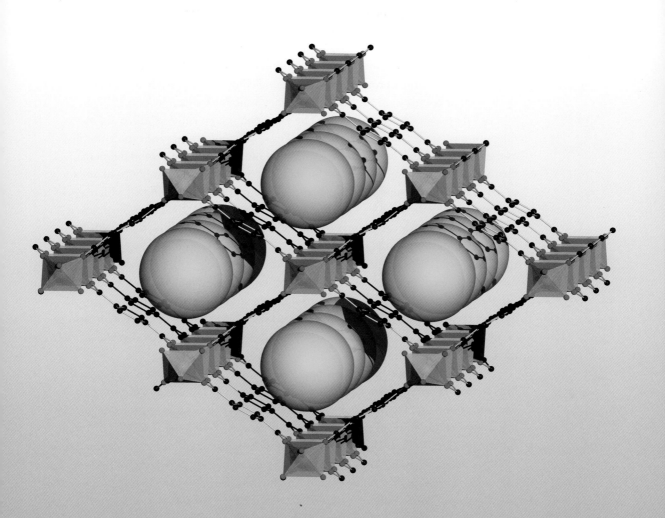

Recrystallization and Polymorphs

Eugene Sun (b. 1960)

Crystallization is a fundamental technique in chemistry and has often been used to purify compounds. A solvent that dissolves a compound while it's heated but doesn't do so when it cools down provides the opportunity to recrystallize. When the process is done right, the impurities remain in the solution, and the clean **crystals** can be filtered off. In the days before **chromatography** became common, recrystallization of solids and distillation of liquids were the main techniques available for purification.

But there can be complications. The same compound can crystallize out in different ways under different conditions. These different crystal forms (polymorphs, as they're known) can be laboratory curiosities, but they can also exhibit different characteristics, occasionally putting human lives and hundreds of millions of dollars at risk. That's what happened with an antiretroviral drug called *ritonavir*, which was approved by the FDA in 1996 for the treatment of HIV/AIDS. In 1998, a new polymorph appeared (called *form II*), with a lower-energy crystalline state (with better **hydrogen bonding**) than the form I found in the original capsules. Since it was more stable as a crystal, form II ritonavir did not dissolve nearly as well, making it ineffective as a drug. The compound's developer, Abbott Laboratories, had to pull the existing supplies from distribution, since they could not guarantee their crystalline form—an increasing number of sample capsules at the factory were failing to dissolve on testing, as outlined by American clinician Eugene Sun, speaking for the company at a series of press conferences.

A liquid suspension of form I ritonavir was available, but it was foul tasting and hard to tolerate. After much frantic work, the company's chemists and formulation scientists found that the only way to formulate ritonavir in capsule form was to package it as a thick solution in gel caps, requiring refrigeration. Later, the team (led by Indian-American chemist Sanjay Chemburkar) found methods to produce form I again under tightly controlled conditions, isolated from any traces of form II (which would have converted it back again!). Abbott is far from the only drug company that has experienced polymorph problems, but theirs may have been the most dramatic.

SEE ALSO Crystals (c. 500,000 BCE), Chromatography (1901), Hydrogen Bonding (1920), Polywater (1966), AZT and Antiretrovirals (1984)

Sodium acetate starting to crystallize out from a solution. Depending on the concentration, temperature, and solvent, even a simple compound, such as this one, can form several crystalline polymorphs.

Click Triazoles

Karl Barry Sharpless (b. 1941), **Carolyn Ruth Bertozzi** (b. 1966), **Valery Fokin** (b. 1971)

In 2001, American chemist Karl Barry Sharpless (already a Nobel laureate for his work in asymmetric synthesis, using one chiral center to help set another one) proposed an idea that no one had ever phrased quite so succinctly. He called for what he termed "click" reactions—two reacting partners, each with functional groups that don't cross-react with anything else, that would immediately bond with each other without the need for other reagents or catalysts. He also pointed out that there weren't any reactions yet that met all those criteria, but as a close approximation he suggested the Huisgen cycloaddition between **acetylene** derivatives and a class of nitrogen compounds called *azides*, a reaction that produces a triazole (a five-membered ring with three nitrogens).

That reaction can take place just by heating the reactants up, but not many scientists had tried this, since azides (especially small ones) have a reputation for being explosive. Sharpless and Russian chemist Valery Fokin found a method that used very small amounts of a copper catalyst (a "near click" reaction) that allowed the two compounds to react at room temperature, and this caught the attention of many chemists. Synthetic organic and medicinal chemists were probably the first to start making more of this relatively unfamiliar type of triazole, but click chemistry was taken up by materials science researchers, nanotechnologists, and many in the expanding field of chemical biology. The reaction was useful for attaching fluorescent tags to biomolecules, and American chemist Carolyn Ruth Bertozzi and others even found variations that could run inside living cells without the need for copper catalysts at all. Sharpless and his coworkers showed that some triazoles could assemble themselves inside enzymes' active sites, if the azide and acetylene starting materials fit well.

The original triazole linkage has been applied recently to DNA chemistry, inorganic complexes, organic semiconductors, and every field where a scientist needs to tie two molecules together. Other click-style reactions have since been discovered, but Sharpless set off the search in the first place by showing how useful a specific molecular-level "super glue" could be.

SEE ALSO Fluorescence (1852), Dipolar Cycloadditions (1963), AZT and Antiretrovirals (1984)

Tumor cells with their internal structures labeled by fluorescent dyes. Click chemistry provides a way to do such experiments with extreme selectivity, all the way down to individual classes of molecules.

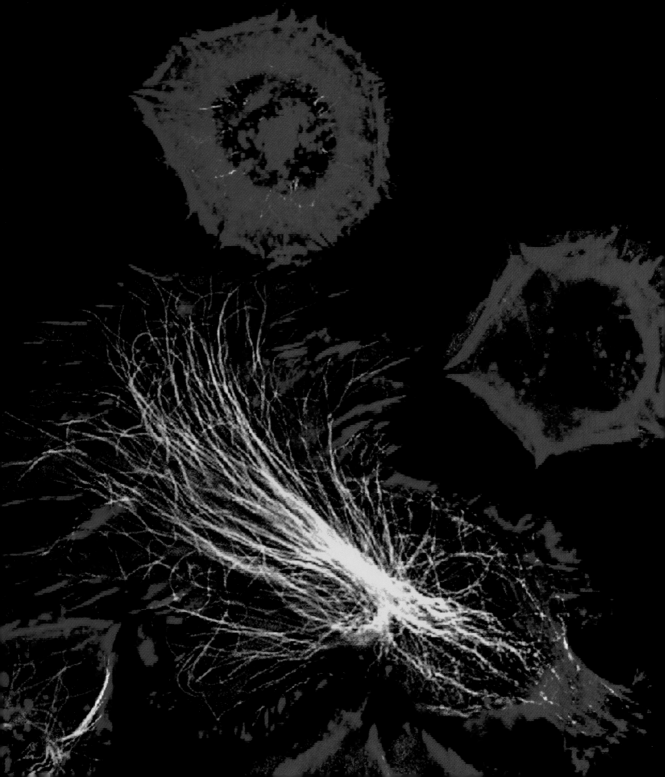

2004

Graphene

Andre Konstantin Geim (b. 1958), **Konstantin Novoselov** (b. 1974)

One would think that an element explored as thoroughly as carbon has been wouldn't have too many secrets left to give up, at least not about its pure allotropes, but just the opposite has happened. **Fullerenes** and **carbon nanotubes** were the first new carbon forms to show up at the nanotechnology party, and everyone was waiting to see when graphene would appear. That term describes a single layer of graphite, one carbon thick—basically, a plane of fused benzene rings stretching off in all directions.

Graphene was a rare case of a well-known compound that had never been isolated. It was well known because of the study of progressively thinner and thinner graphite layers and from the work on carbon nanotubes, which are basically cylinders of rolled-up graphene sheets. No one had convincingly demonstrated any flat material with single-atom thickness, though, and it was thought that it might spontaneously curl into nanotubes even if it were produced.

In 2004, Dutch-British physicist Andre Konstantin Geim and Russian-British physicist Konstantin Novoselov, though, delivered reproducible samples of inarguable graphene sheets through a startlingly simple technique: they used adhesive tape on graphite and peeled off the layers. There are several variations on the technique, but rubbing a tiny amount of graphite on a piece of tape, then repeatedly sticking together and pulling apart the tape is enough by itself. The challenge is detecting the graphene when it gets down to the thinnest layers—by then, it's transparent.

Because of its strength, transparency, and ability to conduct electricity, graphene holds tremendous interest for its optical, electrical, and mechanical properties, which are being explored at a pace that might best be described as frantic. Graphene layers deposited on other substrates are also a big field of study, as are sheets with deliberate defects and impurities doped into them, which might provide unusual semiconductors or solar cell materials. The same techniques are being applied to other sheet-layer substances (such as black **phosphorus**, which gives you two-dimensional phosphorene). Silicon and germanium sheets have been reported, and other elements will probably follow!

SEE ALSO Phosphorus (1669), Benzene and Aromaticity (1865), Synthetic Diamond (1953), Fullerenes (1985), Carbon Nanotubes (1991).

A molecular model of graphene, which is over two hundred times stronger than steel by weight. Such structures are the thinnest possible (only one layer of atoms) and have properties that are strangely different from those of ordinary bulk solids.

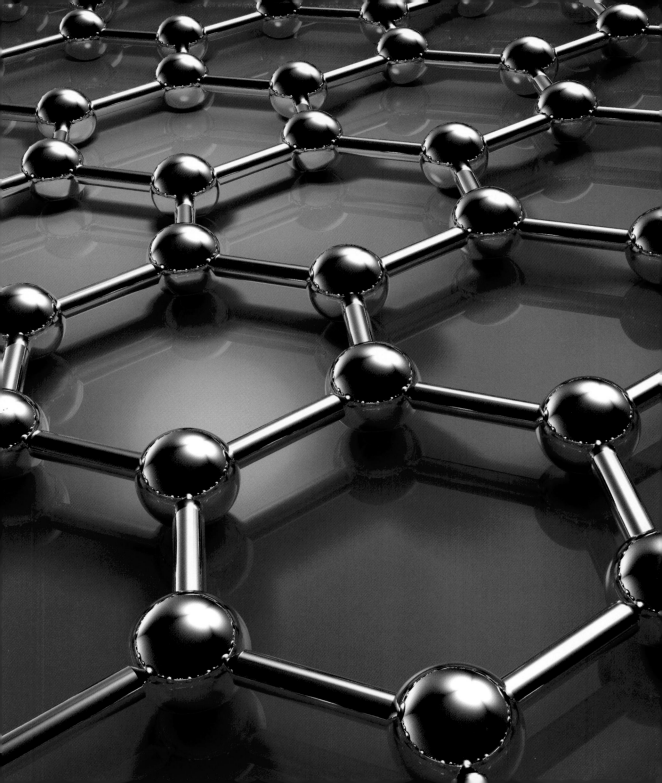

Shikimic Acid Shortage

John C. Rohloff (b. 1960)

Synthetic-organic chemists pride themselves on being able to make all sorts of complicated molecules. In the pharmaceutical industry, they take particular pride in being able to do so economically, reproducibly, and on a very large scale. That means not making any extra work for themselves. If there's a starting material available that gives the chemical route a head start, then it's the one to use.

That principle is seen in the synthesis of **Taxol** and the well-known drug for flu, oseltamivir (brand name *Tamiflu*). Oseltamivir is an inhibitor of the enzyme neuraminidase, which is involved in setting viral particles loose from their host cells. Its structure has three chiral carbons next to each other in a six-membered ring. Setting **chirality** correctly through synthesis is often an issue—you have to start from an existing chiral molecule, use a chiral reagent, or separate the enantiomers, which are the left- and right-handed forms (with the unappealing loss of half your material every time).

But there's a **natural product** called *shikimic acid* whose arrangements of chiral carbons make it an excellent starting material for oseltamivir. In 1998, a team led by American chemist John C. Rohloff at the drug company Gilead Sciences used the substance in its large-scale route to oseltamivir synthesis, which was later improved by two teams at Hoffmann-La Roche. Shikimic acid is found in most plants in small amounts, but a particularly good source is star anise, an aromatic spice. In 2005, worries about an avian flu pandemic sent the demand for Tamiflu—and for star anise—way up. Hoffmann-La Roche was caught short, and it appeared that most of the available star anise in the world was bought up as a starting material.

Many well-known synthetic chemists proposed new sources of shikimic acid or completely different routes of oseltamivir synthesis to get around its use. If another spike in demand comes, and if Tamiflu is effective (always a worry with constantly evolving viruses), the hope is that enough can be made without cleaning off the shelves of every grocery store in the world.

SEE ALSO Natural Products (c. 60 CE), Chirality (1848), Asymmetric Induction (1894), Resolution and Chiral Chromatography (1960), Taxol (1989)

Chinese star anise, used in five-spice powder, in "red-cooked" beef, and as a not-always-reliable source of shikimic acid.

Olefin Metathesis

Yves Chauvin (1930–2015), **Robert Howard Grubbs** (b. 1942), **Richard Royce Schrock** (b. 1945)

Metathesis, also called *double decomposition*, is a reaction between two compounds in which both compounds appear to be broken in half and switched to form two new combinations. In olefin (alkene) metathesis two carbon-carbon double bonds get reordered and rearranged, in a process that can unzip rings, form new ones, and stitch carbon chains together. A simple example is two molecules of propylene (the three-carbon alkene) going into the reaction, to emerge as a fifty-fifty mixture of two-carbon and four-carbon alkenes. The same number of carbons and double bonds are present, but in a new combination. The reaction provides a unique way to form carbon-carbon bonds, and that's a topic that will never fail to attract the attention of synthetic-organic chemists.

Metathesis has its origins in the **Ziegler-Natta catalysis**, since both chemists, Germany's Karl Waldemar Ziegler and Italy's Giulio Natta, separately observed reactions that we now know to have been part of this chemistry. Industrial groups at companies like Shell, Phillips, Goodyear, and DuPont also discovered new reactions with olefins, which began to fit into an overall pattern, and French chemist Yves Chauvin suggested in the early 1970s that these reactions involved a four-membered ring intermediate containing a metal atom from the catalyst. American chemist Robert Howard Grubbs proposed a five-membered metallocycle intermediate, but his own **isotope**-labeling experiments showed that Chauvin was probably right. American chemist Richard Royce Schrock developed a series of metal catalysts for the transformation containing tungsten and molybdenum, and Grubbs found easy-to-handle ruthenium complexes that helped make the reaction popular among the wider organic community. It's now being applied even to biomolecules, creating new cross-linked varieties with different biological activities.

Olefin metathesis went from a curiosity to an industrial processing route—where millions of tons of ethylene are turned into longer-chain compounds that are used as feedstocks for plastics and detergents—to a technique used in even the most delicate total synthesis reactions. As such, Chauvin, Grubbs, and Schrock were awarded the Nobel Prize in 2005.

SEE ALSO Isotopes (1913), Ziegler-Natta Catalysis (1963).

A crystal structure of a first-generation Grubbs catalyst. The blue atom in the middle is a ruthenium, with two green chlorines flanking it. It's also coordinated to two bulky substituted orange phosphorus atoms.

Flow Chemistry

Steven Victor Ley (b. 1945)

Most chemists are used to working in "batch mode": a reaction is started in a vessel, run until it's complete, and then taken out to be purified or used in the next step of a sequence. This chemistry is scaled up by using a bigger flask, by running the reaction again and again, or both.

But there's another way to run reactions, long used in industrial plants and now gaining favor at the lab bench. Continuous-flow reactors pump the starting materials through a reaction zone and if conditions are adjusted well, the reaction can be made to go nearly to completion in the time it takes to go through the system once. The process can be scaled up simply by using bigger flasks of reactants on one end and a bigger flask at the other to catch the products.

It's an appealing idea, but for many years flow chemistry was found mostly in large production plants and built around particular processes (such as **nitroglycerine** manufacturing). Smaller equipment, built for versatility and easy experimentation, brought new prominence to small-scale flow work. The "reaction zone" can be just a coiled metal tube long enough to keep the reactants together, ready to be heated as required. It can be an illuminated zone for **photochemistry** or a column filled with a catalyst. Different reagents (liquids or gases) can be introduced at various points in the flow, and the crude reaction can be run over solid reagents at the end to clean up the products.

British chemist Steven Victor Ley's group first demonstrated an entire natural-product synthesis using an array of flow reactors in 2006. Many chemists have used the technique for especially reactive or hazardous intermediates (**diazomethane**, for example), since there's only a small amount of these present in the reaction zone at any one time. Flow chemistry's benefits seem compelling: reactions can be superheated to speed up the process, products are cleaner, the process is safer, and variables in the reaction can be altered quickly. Traditional flasks now have competition.

SEE ALSO Photochemistry (1834), Nitroglycerine (1847), Thermal Cracking (1891), Diazomethane (1894), Hydrogenation (1897), Haber-Bosch Process (1909)

A small flow-chemistry apparatus. The channel (covered by a transparent film) is filled with tiny beads that have been layered with an enzyme catalyst, and the flow is adjusted so that the starting material is completely converted by the time it leaves the reactor.

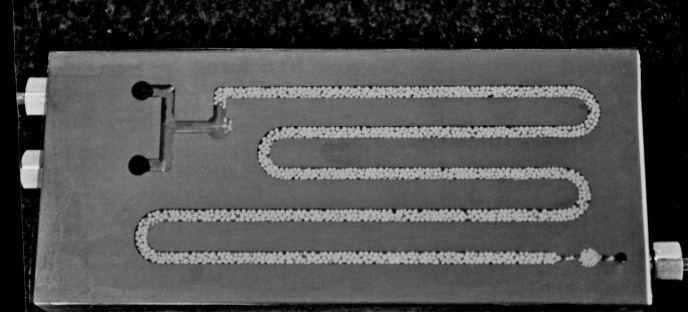

Isotopic Distribution

The **kinetic isotope effect**, or the change in a chemical reaction's rate when an atom in one of the reactants is substituted with one of its **isotopes**, has popped up in some unexpected circumstances (and for some people, rather disconcerting ones). The difference between the carbon-12 and carbon-13 isotopes, for example, isn't large, but because a cell's carbon compounds are constantly being processed, the ratio between the two in the various carbon atoms of a biomolecule (the isotopic distribution) is noticeably different in samples from living creatures: the lighter carbon-12 has been getting enriched for a billion years as plants grow on the remains of other plants, as animals eat them, and so on. In fact, **mass spectrometry** techniques can even tell whether a given molecule was produced from the biochemistry of tropical plants or from that of temperate-zone ones.

This aspect of analytical chemistry made headlines in 2006 when American cyclist Floyd Landis, original winner of that year's Tour de France, was accused of having taken testosterone supplements to enhance his performance. Landis claimed that his body naturally tended to produce more testosterone and that he had not boosted his biochemistry. But mass spectrometry told another story.

A testosterone molecule has nineteen carbon atoms. A bit over 1 percent of all carbon atoms are carbon-13. So in the absence of isotope effects, statistics would say how many testosterone molecules in a sample would contain two carbon-13s (about one in ten thousand), how many would have three (about one in a million), and so on. But steroid molecules from natural sources have more carbon-12 than you would predict, and the pattern of this enrichment depends on the biochemical pathways that formed them.

Landis's testosterone did not show the human pattern, but one expected from warm-climate plants. And since synthetic testosterone is produced from warm-climate plant sterols (yams and soy), it has a completely different profile from the human variety. In 2010, Landis finally admitted to taking testosterone and other performance-enhancing drugs (but maintained that the 2006 results were the result of lab errors).

SEE ALSO Cholesterol (1815), Mass Spectrometry (1913), Isotopes (1913), Cellular Respiration (1937), Steroid Chemistry (1942), Kinetic Isotope Effects (1947), Photosynthesis (1947), Methane Hydrate (1965), Enzyme Stereochemistry (1975).

Floyd Landis racing in a time trial in California, 2009. A long list of professional bicycle racers have had their careers tarnished by evidence of performance-enhancing drug use, thanks to modern analytical chemistry.

Acetonitrile

Analytical chemistry labs in drug companies, academic research groups, and forensic science labs had a rough time of it in 2009, although it's safe to say that the rest of the world didn't notice. One of the most common solvents for **reverse-phase chromatography** is acetonitrile (a two-carbon compound consisting of a methyl group attached to a carbon-nitrogen triple bond). Its popularity is understandable: it mixes with water in any concentration; it dissolves a wide range of less polar organic molecules; and it's unreactive, relatively nontoxic, and easy to evaporate. But these advantages are all for nothing if you can't get enough acetonitrile, and for several months in late 2008 and well into 2009 there was an unprecedented shortage. Prices began creeping up, then skyrocketed, and then suppliers began calling their customers to say that many orders were just not going to be filled. Methanol could be used as a substitute in some cases but not all.

Several factors caused this situation. Factories in China that produced the solvent were shut down to improve the air quality for the Beijing Summer Olympics, and this put a crimp in the worldwide supply. A U.S. facility on the Gulf Coast was affected by Hurricane Ike. But the general problem was the global economic slowdown. Most of the world's acetonitrile comes as a by-product of the process used to make another chemical, acrylonitrile, which is used in many industrial polymers. Reduced demand for auto parts and other high-volume items caused many acrylonitrile plants to slow or shut down production, and the acetonitrile supply accordingly began to dry up as a side effect.

There are several other ways to make the solvent on an industrial scale, but no one had ever bothered, since recovering it from acrylonitrile plants was always cheaper. A few companies sprang up to try out these alternative routes during the shortage, but the gradual recovery of the acrylonitrile market (and the accompanying drop in acetonitrile prices) seems to have put these methods back on the shelf. Unless, of course, another banking crisis makes the **HPLC** systems stop running again . . .

SEE ALSO Polymers and Polymerization (1839), Chromatography (1901), Rotary Evaporator (1950), HPLC (1967), Reverse-Phase Chromatography (1971), Electrospray LC/MS (1984)

Heavy fog and haze at Tiananmen Square in Beijing, February 2013. The Chinese government's effort to avoid such scenes during the 2008 Summer Olympics was one of the factors in the sudden and unexpected short supply of acetonitrile.

2010

Engineered Enzymes

Jacob M. Janey (b. 1976)

Many organic chemists suffer from enzyme envy, because enzymes make many reactions run more quickly, more cleanly, and under much milder conditions than any chemist can manage. Harnessing these powers has been a long-term goal, and the year 2010 brought some progress.

The drug company Merck was looking for a new route to a key intermediate in the synthesis of sitagliptan, a diabetes drug. Sitagliptan contains a chiral amine, which had been made through asymmetric synthesis (using one chiral center to help set another one), but that route was hard to run on a large scale, and the product kept being contaminated with traces of the metal **hydrogenation** catalyst. To find another process, the company enlisted a group of scientists at the enzyme-engineering company Codexis. A class of enzymes called transaminases can make chiral amines, but none of the known varieties worked well on Merck's starting material—it was too large, and no natural transaminase had a binding pocket that could recognize it. So the team—including American chemists Chris Savile at Codexis and Jacob Janey at Merck—began modifying enzymes whose structures looked most accommodating, using computational models of their structures to change the size and shape of their binding pockets both randomly and deliberately.

Random choices are necessary because the effects of small changes on a larger protein structure (and on its activity) can be well beyond any human prediction. After each change, the new variant was quickly checked for its activity and selectivity on the Merck reaction, and the results were fed back into the design process. By the end, the team had made and screened over thirty-six thousand variations and processed an untold number of modeling possibilities that didn't make the cut. The final version of the engineered enzyme had twenty-seven **amino acids** changed from its starting point and could produce industrial quantities of the Merck intermediate with complete selectivity.

New approaches to enzyme engineering are being developed, but it's still expensive and time-consuming, and it's still in its infancy. Identifying faster, more cost effective, and more reliable methods, though, could fundamentally change the world of chemistry.

SEE ALSO Amino Acids (1806), Chirality (1848), Asymmetric Induction (1894), Zymase Fermentation (1897), Hydrogenation (1897), Carbonic Anhydrase (1932), Photosynthesis (1947), Computational Chemistry (1970), Enzyme Stereochemistry (1975).

A three-dimensional "ribbon structure" of the enzyme purine nucleoside phosphorylase, which is being engineered to synthesize antiviral and antitumor agents.

Metal-Catalyzed Couplings

Akira Suzuki (b. 1930), **Richard Fred Heck** (b. 1931), **Ei-ichi Negishi** (b. 1935)

In 2010, the Nobel Prize in Chemistry recognized a set of reactions whose fame had been rising for decades. American chemist Richard Fred Heck had discovered reactions between alkenes and palladium compounds in the late 1960s, leading to the formation of carbon-carbon bonds, which is an essential step in the production of new organic chemicals. He even found a way to run the reaction with only small catalytic amounts of palladium, which is important since the metal is so expensive. Later work from his group showed that other classes of compounds could form palladium intermediates as well, and he proposed what have turned out to be the correct mechanisms for these reactions.

There had been many metal-catalyzed coupling reactions reported in the literature, with copper, nickel, cobalt, iron, magnesium and other elements involved. In the 1970s and 1980s, Japanese chemist Ei-ichi Negishi and coworkers found that zinc reagents formed new carbon-carbon bonds under very mild conditions when palladium catalysts were used. Palladium chemistry was extended to the least reactive metal species yet in 1979 when Japanese chemist Akira Suzuki reported that carbon-linked boronic acids also participated in palladium couplings. These compounds, previously obscure, began to attract attention through the 1980s as more and more synthetic chemists realized how reliable such bond-forming reactions seemed to be.

The palladium wave grew and grew. The chemical catalogs filled up with new boronic acids, while other elements were brought in as reacting partners. A huge variety of palladium catalysts were investigated to allow even more classes of compounds to participate. "Suzukis" and the other palladium reactions became indispensable parts of the organic chemistry toolkit, allowing aromatic rings and other structures to be joined up in ways that once would have been thought nearly impossible. Some drug companies began to worry about the huge numbers of very similar palladium-made structures beginning to fill their compound files, since some rather un-"druglike" compounds were now so easy to make. When people begin to get worried that your reaction is too easy to run and too popular, your chemistry has indeed arrived.

SEE ALSO Grignard Reaction (1900), Nozaki Coupling (1977).

An image from a scanning electron microscope of pure crystalline palladium metal, one of the most valuable catalytic metals in the world.

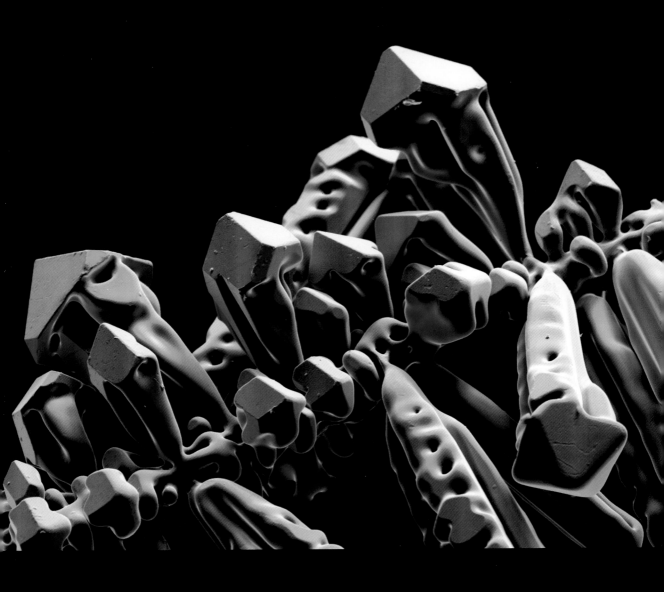

Single-Molecule Images

Gerhard Meyer (b. 1956), **Michael Crommie** (b. 1961), **Leo Gross** (b. 1973), **Felix R. Fischer** (b. 1980)

The 1980s saw the invention of a remarkable instrument called the atomic force microscope (AFM). This is essentially a very fine needle that's lowered extremely closely to a polished surface. The needle's tip is a single atom, and the distance from it to the surface is so small that the tip can interact with the other atoms it comes close to—it's like running a (very!) small fingertip over the surface. Depending on how the tip of the needle is engineered and what readouts are detected, atomic-level repulsive or attractive forces and other quantum mechanical effects can be identified. The inventors demonstrated the device by using the tip to drag individual atoms across a smooth surface, spelling out the name of their employer, IBM, as if with pool balls on a table.

The AFM and its variations have become important in **surface chemistry** (pioneers in that field such as American chemist Irving Langmuir and American physicist Katherine Blodgett would surely have loved to have seen one in operation). The technique has improved enough that species that would be otherwise difficult to detect can be imaged, among them the molecules of organic chemistry. For these, the probe's tip is a single carbon monoxide molecule, a technique discovered by German physicists Leo Gross and Gerhard Meyer at IBM. Working with the molecule's oxygen atom pointing down allows the tip to react to electron density, with the electrons on the oxygen atom being pushed away from the electrons on the target molecule.

At Berkeley, American chemists Felix R. Fischer and Michael Crommie and their coworkers used the AFM in 2013 to capture the first-ever images of a molecule changing in the course of a reaction, showing the internal cyclization of a molecule with several carbon-carbon triple bonds. The resolution in these experiments is startling, and many organic chemists find themselves fascinated and somewhat unnerved to see the structures that they draw on their whiteboards made real. A new way of determining chemical structures is at hand, and it will illuminate many complex molecules that are beyond the reach of **mass spectrometry** or **NMR**.

SEE ALSO Acetylene (1892), Surface Chemistry (1917), Sigma and Pi Bonding (1931), *The Nature of the Chemical Bond* (1939)

These amazingly detailed images show a starting material that forms three new cyclic products on heating. Their complex structures would be hard to tell apart using other instruments, but their actual carbon-bond frameworks can be scanned directly by the atomic force microscope.

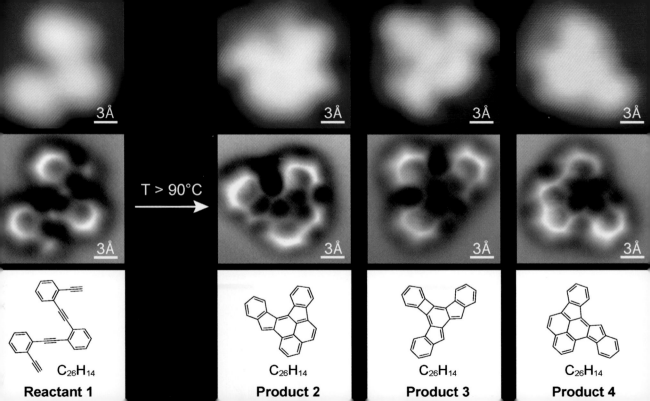

Hydrogen Storage

There's been talk since the 1970s about a coming "**hydrogen** economy," but there are many technical hurdles left to jump before such a thing becomes possible (let alone desirable). Burning hydrogen produces no emissions except water vapor, and given the emissions created by most existing energy sources, most people would pick hydrogen as their fuel of choice, if that were the only variable. But there's no such thing as a hydrogen mine. Rather than being an energy source, hydrogen is a substance that energy can be converted into. **Electrochemical reduction** will reverse the combustion reaction and produce hydrogen from water again, but electricity is required for that reaction, and the hydrogen produced then has to be stored for transport and later use — and that is a real challenge.

Hydrogen molecules are so small that they will soak into the solid structure of many metals. This is a key step in the use of **hydrogenation** catalysts, but it certainly complicates its use as a fuel. Its molecular weight and very low density also mean that hydrogen has to be highly compressed to be handled on a practical scale, and that adds to the energy cost of using it. Since hydrogen is also explosively flammable, any storage and transport system will have to be very well engineered for safety.

Hydrogen storage is currently a major area of research, with different groups trying new materials to try to find a workable system. **Coordination-framework** crystals might be one option, and compounds of hydrogen with various metals are also being investigated. Any viable system will have to be easily reversible (to make the gas accessible for use), capable of running for a great number of fueling and discharge cycles, able to store high densities of hydrogen, and (ideally) much safer than dealing with the gas itself. This is a tall order, but given the advances that chemistry, physics, and materials science have already made, there's no reason that it should be an impossible one.

SEE ALSO Hydrogen (1766), Electrochemical Reduction (1807), Hydrogenation (1897), Coordination Frameworks (1997), Artificial Photosynthesis (2030).

A coordination framework of carbon and zinc stores hydrogen atoms (in green) in an even denser packing than solid hydrogen would have.

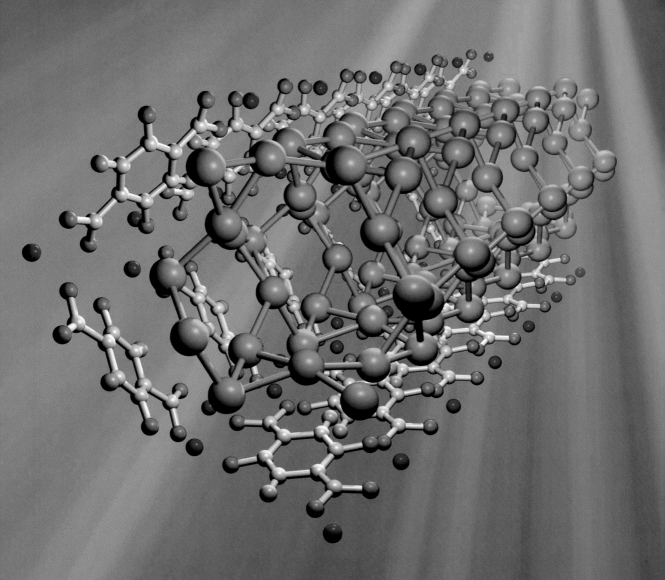

Artificial Photosynthesis

Akira Fujishima (b. 1942), **Daniel George Nocera** (b. 1957), **Andrew B. Bocarsly** (b. 1954)

Given the critical role **photosynthesis** plays in life on Earth, researchers have explored the process in detail, studying its efficiency, whether evolution has found the best system possible, and if humans can improve on it. These aren't academic questions. There are two sides to the photosynthesis reaction, and large-scale artificial methods for either could change the world. The **carbon dioxide** fixation reaction (the conversion of CO_2 to sugar) could provide an instant new food source as well as a renewable fuel source and industrial chemical feedstock. This process would pull carbon dioxide from the air and turn it back into small organic compounds, essentially reversing the combustion that put it into the air in the first place. In 2008, American chemist Andrew B. Bocarsly demonstrated a conversion of carbon dioxide to methanol, and his "Liquid Light" technology aims to recycle CO_2 into useful industrial hydrocarbons. Given concerns about land use, food supplies, and the climate, this could be a world-changing advance.

The other half of the natural photosynthesis reaction is also being intensively studied: splitting water back into hydrogen and oxygen. This can be done with electricity, but accomplishing this conversion without hooking it up to the power lines is something else again. Japanese chemist Akira Fujishima's 1967 discovery of **titanium** dioxide's ability to catalyze this reaction has led to a great deal of research into its possibilities and those of other semiconductors (American engineer William Ayers demonstrated this with a silicon wafer cell in 1983, and a much more efficient version was invented by American chemist Daniel Nocera and coworkers in 2011). Efficient water splitting and **hydrogen storage** technology would be the key steps to using the gas as a nonpolluting fuel.

It's too early to say which of the many competing technologies will be best. It seems safe to predict, though, that both processes could be driven by metal-containing catalysts. The challenge, and it's a big one, will be to find catalytic systems that use metals that aren't too exotic, have long lifetimes and high activity, and can be manufactured on a large scale. The task is enormous, but so are the benefits.

SEE ALSO Carbon Dioxide (1754), Titanium (1791), Electrochemical Reduction (1807), Chlor-Alkali Process (1892), Greenhouse Effect (1896), Photosynthesis (1947), Hydrogen Storage (2025).

Vast amounts of carbon dioxide are fixed by plants. No one knows yet whether this process can be duplicated artificially, but it's a prize being sought by research groups all over the world.

Notes and Further Reading

In addition to books and magazine articles, I have included a number of websites and sources below (be aware that these, of course, may disappear or the address may change). Chemistry articles on *Wikipedia* tend to be quite useful and have had attention from professional chemists (including, once in a while, me). Other broadly useful websites for chemistry include U.C. Davis's ChemWiki (chemwiki.ucdavis.edu), the Chemical Heritage Foundation (www.chemheritage.org), the Organic Chemistry Portal (www.organic-chemistry.org), Theodore Gray's PeriodicTable.com, and the American Chemical Society's home page (www.acs.org). The Periodic Table of Videos (www.periodicvideos.com) is also very worthwhile.

General Reading
Aldersey-Williams, H. *Periodic Tales*. New York: Ecco, 2011.
Coffey, P. *Cathedrals of Science*. New York: Oxford Univ. Press, 2008.
Gray, T. *The Elements*. New York: Black Dog and Leventhal, 2009.
Greenberg, A. *Chemistry Decade by Decade*. New York: Facts on File, 2007.
— — —*A Chemical History Tour*. New York: Wiley, 2000.
Kean, S. *The Disappearing Spoon*. New York: Little, Brown, 2010.
Le Couteur, P., and J. Burreson. *Napoleon's Buttons*. New York: Jeremy P. Tarcher/Penguin, 2003.
Levere, T. H. *Transforming Matter*. Baltimore: Johns Hopkins Univ. Press, 2001.

c. 500,000 BCE Crystals
Naica Caves official website, *www.naica.mx.com/english/*.
Shea, N. "Cavern of Crystal Giants." *National Geographic*, November 2008, ngm.nationalgeographic.com/2008/11/crystal-giants/shea-text.

c. 3300 BCE Bronze
Ekserdjian, D., ed. *Bronze*. London: Royal Academy of Arts, 2012.
Radivojević et al. "Tainted Ores and the Rise of Tin Bronzes in Eurasia." *Antiquity* 87 (2013): 1030.
Sherby, O. D., and J. Wadsworth. "Ancient Blacksmiths, the Iron Age, Damascus Steels, and Modern Metallurgy." U.S. Department of Energy, September 11, 2011, https://e-reports-ext.llnl.gov/pdf/238547.pdf.

c. 2800 BCE Soap
Hedge, R. W. *www.butser.org.uk/iafsoap_hcc.html*.
Verbeek, H. "Historical Review" in *Surfactants in Consumer Products*, 1–4. Berlin: Springer-Verlag, 1987.

c. 1300 BCE Iron Smelting
Hosford, W. G. *Iron and Steel*. New York: Cambridge Univ. Press, 2012.
Sherby, O. D., and J. Wadsworth. "Ancient Blacksmiths, the Iron Age, Damascus Steels, and Modern Metallurgy." U.S. Department of Energy, September 11, 2011, https://e-reports-ext.llnl.gov/pdf/238547.pdf.

c. 1200 BCE Purification
Rayner-Canham, M., and R. Rayner-Canham. *Women in Chemistry: From Alchemy to Acceptance*. Wash., D.C.: American Chemical Society, 1998.

c. 550 BCE Gold Refining
Heilbrunn Timeline of Ancient History, "Sardis," *www.metmuseum.org/toah/hd/srds/hd_srds.htm*.
Tassel, J., "The Search for Sardis." *Harvard Magazine*, March–April 1998, harvardmagazine.com/1998/03/sardis.html.

c. 450 BCE The Four Elements
See 48b in Plato's *Timaeus* at the Perseus Digital Library, Tufts Univ., *www.perseus.tufts.edu/hopper/text?doc=Plat.+Tim.+48b&redirect=true*.
Stanford Encyclopedia of Philosophy, "Empedocles," plato.stanford.edu/entries/empedocles/.

c. 400 BCE Atomism
Stanford Encyclopedia of Philosophy, "Democritus," plato.stanford.edu/archives/fall2008/entries/democritus/.

210 BCE Mercury
Emsley, J. *The Elements of Murder*. Oxford: Oxford Univ. Press, 2005.
Moskowitz, C. "The Secret Tomb of China's First Emperor." *livescience*, August 17, 2012, www.*livescience.com/22454-ancient-chinese-tomb-terracotta-warriors.html*.
Portal, J., *Terra Cotta Warriors*, Wash., D.C.: National Geographic, 2008.

c. 60 CE Natural Products
Firn, R. *Nature's Chemicals*. Oxford: Oxford Univ. Press, 2010.
Nicolaou, K. C., and T. Montagnon, *Molecules That Changed the World*. Weinheim, DE: Wiley-VCH, 2008.

c. 126 Roman Concrete
Brandon, C. J., et. al. *Building for Eternity*. Oxford: Oxbow Books, 2014.
Pruitt, S. "The Secret of Ancient Roman Concrete." *History in the Headlines* (blog), June 21, 2013, *www.history.com/news/the-secrets-of-ancient-roman-concrete*.

c. 200 Porcelain
Finlay, R. *The Pilgrim Art*. Berkeley: Univ. of California Press, 2010.

c. 672 Greek Fire
The classic work is J. R. Partington's *A History of Greek Fire and Gunpowder* (Cambridge: W. Heffer, 1960), which is available in various editions.

c. 800 The Philosopher's Stone
Principe, L. M. *The Secrets of Alchemy*. Chicago: Univ. of Chicago Press, 2013.

c. 800 Viking Steel
Hosford, W. G. *Iron and Steel*. New York: Cambridge Univ. Press, 2012.
PBS *Nova*, "Secrets of the Viking Sword," www.pbs.org/wgbh/nova/ancient/secrets-viking-sword.html
Sherby, O. D., and J. Wadsworth. "Ancient Blacksmiths, the Iron Age, Damascus Steels, and Modern Metallurgy." U.S. Department of Energy, September 11, 2011, https://e-reports-ext.llnl.gov/pdf/238547.pdf.

c. 850 Gunpowder
Kelly, J. *Gunpowder: Alchemy, Bombards, and Pyrotechnics*. New York: Basic Books, 2004.
Partington, J. R. *A History of Greek Fire and Gunpowder*. Cambridge: W. Heffer, 1960.

c. 900 Alchemy
Greenberg, A. *From Alchemy to Chemistry in Picture and Story*. Hoboken, NJ: Wiley-Interscience, 2007.
Principe, L. M. *The Secrets of Alchemy*. Chicago: Univ. of Chicago Press, 2013.

c. 1280 Aqua Regia
See Princeton Univ.'s online lab-safety manual, https://ehs.princeton.edu/laboratory-research/chemical-safety/chemical-specific-protocols/aqua-regia. Don't mess with the stuff!

c. 1280 Fractional Distillation
Books on distillation tend to be industrial chemical engineering handbooks or guides for homebrewed spirits. For a general overview, your best bet is, in fact, *Wikipedia*: en.wikipedia.org/wiki/Distillation.

1538 Toxicology
A definitive textbook on the subject is *Casarett and Doull's Toxicology* (8th ed.) by Curtis Klaassen (New York: McGraw-Hill, 2013). A shorter and less technical work is *The Dose Makes the Poison* by Patricia Frank and M. Alice Ottoboni (Hoboken, NJ: Wiley, 2011).

1540 Diethyl Ether
The history of diethyl ether can be found mostly in various anesthesiology textbooks. Also see *Wikipedia*: en.wikipedia.org/wiki/Diethyl_ether.

1556 *De Re Metallica*
Project Gutenberg has the entire text (with the woodcut illustrations) online for free at www.gutenberg.org/files/38015/38015-h/38015-h.htm. Interestingly, this English translation is by former U.S. president Herbert Hoover.

1605 The Advancement of Learning
Project Gutenberg, www.gutenberg.org/ebooks/5500. For several different translations of the *Novum Organum*, see en.wikisource.org/wiki/Novum_Organum. More on Francis Bacon himself can be found at the *Internet Encyclopedia of Philosophy*, www.iep.utm.edu/bacon/.

1607 Yorkshire Alum
Balston, J. *The Whatmans and Wove Paper*. West Farleigh: 1998, www.wovepaper.co.uk/alumessay2.html.
National Trust. "Yorkshire Coast," www.nationaltrust.org.uk/yorkshire-coast/history/view-page/item634280/.

1631 Quinine
Firn, R. *Nature's Chemicals*. Oxford: Oxford Univ. Press, 2010.
Nicolaou, K. C., and T. Montagnon. *Molecules That Changed the World*. Weinheim, DE: Wiley-VCH, 2008.
Rocco, F. *Quinine*. New York: Harper Perennial, 2004.

1661 *The Sceptical Chymist*
Boyle, R. *The Sceptical Chymist*. Project Gutenberg, www.gutenberg.org/ebooks/22914.
Hunter, M. *Boyle*. New Haven: Yale Univ. Press, 2009.

1667 Phlogiston
National Historic Chemical Landmarks program of the American Chemical Society. "Joseph Priestley and the Discovery of Oxygen," 2004, www.acs.org/content/acs/en/education/whatischemistry/landmarks/josephpriestleyoxygen.html.
Donovan, A. *Antoine Lavoisier*. Cambridge, MA: Cambridge University Press, 1996.
Johnson, S. *The Invention of Air*. New York: Riverhead Books, 2008.

1669 Phosphorus
Emsley, J. *The 13th Element*. New York: Wiley, 2000.

1700 Hydrogen Sulfide
As noted on *Wikipedia*, Isaac Asimov called Scheele "Hard-luck Scheele" because he probably made several discoveries that he is not given full credit for.
Smith, R. P. "A Short History of Hydrogen Sulfide" *American Scientist*, 98 (January–February 2010): 6. http://www.americanscientist.org/issues/num2/a-short-history-of-hydrogen-sulfide/4.

c. 1706 Prussian Blue
Kraft, A. "On the Discovery and History of Prussian Blue." *Bulletin for the History of Chemistry*, 33 (2008): 61. www.scs.illinois.edu/~mainzv/HIST/bulletin_open_access/v33-2/v33-2%20p61-67.pdf.
Senthilingam, M. "Prussian Blue." *Chemistry in Its Element* (podcast), *Chemistry World Magazine*, January 30, 2013, www.rsc.org/chemistryworld/2013/04/prussian-blue-podcast.

1746 Sulfuric Acid
Kiefer, D. "Sulfuric Acid: Pumping up the Volume." *Today's Chemist at Work*, pubs.acs.org/subscribe/archive/tcaw/10/i09/html/09chemch.html.

1752 Hydrogen Cyanide
If you need convincing not to encounter this compound, then the Centers for Disease Control and Prevention (CDC) should be able to give you some: www.cdc.gov/niosh/ershdb/EmergencyResponseCard_29750038.html.

1754 Carbon Dioxide
West, J. B. "Joseph Black, Carbon Dioxide, Latent Heat, and the Beginnings of the Discovery of the Respiratory Gases."

American Journal of Physiology - Lung Cellular and Molecular Physiology 306 (March 2014), L1057. *ajplung. physiology.org/content/early/2014/03/25/ ajplung.00020.2014.*

1758 Cadet's Fuming Liquid
Seyferth, D. "Cadet's Fuming Arsenical Liquid and the Cacodyl Compounds of Bunsen." *Organometallics* 20 (2001): 1488. *pubs.acs.org/doi/pdf/10.1021/om0101947.*

1766 Hydrogen
There are plenty of videos on YouTube of people entertaining themselves with hydrogen fires—*de gustibus non est disputandum.*
Rigden, J. S. *Hydrogen.* Cambridge, MA: Harvard Univ. Press, 2002.

1774 Oxygen
Johnson, S. *The Invention of Air.* New York: Riverhead Books, 2008.
National Historic Chemical Landmarks program of the American Chemical Society. "Joseph Priestley and the Discovery of Oxygen," 2004, *www.acs.org/content/acs/ en/education/whatischemistry/landmarks/ josephpriestleyoxygen.html.*

1789 Conservation of Mass
Donovan, A. *Antoine Lavoisier.* Cambridge, MA: Cambridge University Press, 1996.

1791 Titanium
Housley, K. L. *Black Sand.* Hartford, CT: Metal Management Aerospace, 2007.
Titanium Industries, Inc. "History of Titanium," *titanium.com/technical-data/ history-of-titanium/.*

1792 Ytterby
A detailed monograph is *Episodes from the History of the Rare Earth Elements* by C. H. Evans (Boston: Kluwer Academic Pub., 1996). Also see *RareMetalsMatter.com* and "Separation of Rare Earth Elements by Charles James" at the American Chemical Society, *www.acs.org/content/acs/ en/education/whatischemistry/landmarks/ earthelements.html.*

1804 Morphine
Booth, M. *Opium.* New York: St. Martin's Press, 1998.

1806 Amino Acids
Tanford, C., and J. Reynolds. *Nature's Robots.* Oxford: Oxford Univ. Press, 2001.

1807 Electrochemical Reduction
Davy's own presentation of these results, from the *Philosophical Transactions of the Royal Society,* can be found here: *www.chemteam. info/Chem-History/Davy-Na&K-1808.html.*
Knight, D. *Humphry Davy.* Cambridge: Cambridge Univ. Press, 1992.

1808 Dalton's Atomic Theory
Summaries of Dalton's theories can be found on *Wikipedia,* at General Chemistry Online (*antoine.frostburg.edu/chem/ senese/101/atoms/dalton.shtml*), and at the Chemical Heritage Foundation (*www. chemheritage.org/discover/online-resources/ chemistry-in-history/themes/the-path-to-the- periodic-table/dalton.aspx*).

1811 Avogadro's Hypothesis
Morselli, M., *Amedeo Avogadro.* Dordrecht, NL: Springer 1984.

1813 Chemical Notation
Melhado, E. M., and T. Frängsmyr, eds. *Enlightenment Science in the Romantic Era.* Cambridge: Cambridge Univ. Press, 2002.

1814 Paris Green
Meharg, A. *Venomous Earth.* New York: Macmillan, 2005.
University of Aberdeen. "Arsenic and the World's Worst Mass Poisoning," January 12, 2005, *www.abdn.ac.uk/mediareleases/release. php?id=104.*

1815 Cholesterol
Wikipedia is a good place to start online for the chemical story of cholesterol.
National Historic Chemical Landmarks program, American Chemical Society. "Russell Marker and the Mexican Steroid Hormone Industry," 1999, *www.acs.org/ content/acs/en/education/whatischemistry/ landmarks/progesteronesynthesis.html.*
UC Davis ChemWiki. "Steroids," *chemwiki. ucdavis.edu/Biological_Chemistry/Lipids/ Steroids.*

1819 Caffeine
Weinberg, B. A., and B. K. Bealer. *The World of Caffeine.* New York: Routledge, 2001.

1822 Supercritical Fluids
See the *Wikipedia* and UC Davis ChemWiki entries on the subject for an introduction. An excellent video demonstration of the phenomenon is at *www.youtube.com/ watch?v=GEr3NxsPTOA.*

1828 Wöhler's Urea Synthesis
Wöhler's letter to Berzelius is found here: *classes.yale.edu/01-02/chem125a/125/ history99/4RadicalsTypes/UreaLetter1828.html.*

1832 Functional Groups
Brock, W. B. *Justus von Liebig.* Cambridge: Cambridge Univ. Press, 1997.
Chemical Heritage Foundation. "Justus von Liebig and Friedrich Wöhler," *www. chemheritage.org/discover/online-resources/ chemistry-in-history/themes/molecular- synthesis-structure-and-bonding/liebig-and- wohler.aspx.*

1834 Ideal Gas Law
Book-length studies are, of necessity, technical. *Wikipedia* and UC Davis ChemWiki are better for an accessible overview.

1834 Photochemistry
A summary of photochemistry's history can be found at *turroserver.chem.columbia.edu/ PDF_db/History/intro.pdf.*
Natarajan et al. "The Photoarrangement of -Santonin Is a Single-Crystal-to-Single- Crystal Reaction," *Journal of the American Chemical Society* 129, 32 (2007): 9846. *http://pubs.acs.org/doi/abs/10.1021/ ja073189o?journalCode=jacsat.*
Roth, H. D. "The Beginnings of Organic Photochemistry." *Angewandte Chemie International Edition (English)* 28, 9 (1989): 1193.

1839 Polymers and Polymerization
Walton, D., and P. Lorimer. *Polymers*. Oxford: Oxford Univ. Press, 2000.

1839 Daguerreotype
Daguerreian Society. "About the Daguerreian Society," *daguerre.org/index.php*.

Wooters, D., and T. Mulligan, eds. *A History of Photography: The George Eastman House Collection*. London: Taschen, 2005.

1839 Rubber
Goodyear Tire & Rubber Company. "The Charles Goodyear Story," *www.goodyear.com/corporate/history/history_story.html*.

Korman, R. *The Goodyear Story*. San Francisco: Encounter Books, 2002.

1840 Ozone
A teaching resource about atmospheric ozone is found here: *www.ucar.edu/learn/1_5_1.htm*. Ignore the huge pile of "ozone therapy" books that are available.

1842 Phosphate fertilizer
McDaniel, C. N. *Paradise for Sale*. Berkeley: Univ. of California Press, 2000.

1847 Nitroglycerine
An extraordinary series of anecdotes about nitroglycerine's use in the oil fields is here: *www.logwell.com/tales/menu/index.html*. If it makes you want to experience it yourself, there's clearly no hope for you.

1848 Chirality
This is a deep, extremely important topic in chemistry, physics, and mathematics. There are many types of chirality that I have no space to mention. (Consider, for example, a curling screw-shaped molecule that can exist in right-hand and left-hand thread . . .) Surprisingly, someone has taken up the challenge of writing an introductory book on the topic: *Mirror-Image Asymmetry: An Introduction to the Origin and Consequences of Chirality* by James P. Riehl (Hoboken, NJ: Wiley 2010).

1852 Fluorescence
Technical works are beyond counting, as befits a phenomenon that touches on so many areas. On the inorganic side, see the Fluorescent Mineral Society (*uvminerals.org/fms/minerals*) or *users.ece.gatech.edu/~hamblen/uvminerals/*. On the biochemical side, fluorescent tags and proteins are used intensively in cell biology and microscopy. See *micro.magnet.fsu.edu/primer/techniques/fluorescence/fluorescenceintro.html* for a technical overview.

1854 Separatory Funnel
There are a variety of YouTube videos showing a sep funnel in action.

1856 Perkin's Mauve
Chemical Heritage Foundation. "William Henry Perkin," *www.chemheritage.org/discover/online-resources/chemistry-in-history/themes/molecular-synthesis-structure-and-bonding/perkin.aspx*.

Garfield, S. *Mauve*. New York: W. W. Norton, 2001.

1856 Mirror Silvering
A recipe for demonstrating this reaction can be found at the Royal Society of Chemistry: *www.rsc.org/Education/EiC/issues/2007Jan/ExhibitionChemistry.asp*. Just don't leave the solution around once you're finished!

1859 Flame Spectroscopy
Chemical Heritage Foundation. "Robert Bunsen and Gustav Kirchhoff," *www.chemheritage.org/discover/online-resources/chemistry-in-history/themes/the-path-to-the-periodic-table/bunsen-and-kirchhoff.aspx*.

1860 Cannizzaro at Karlsruhe
Nye, M. J., ed. *The Cambridge History of Science* (Vol. 5). Cambridge: Cambridge Univ. Press, 2002.

1860 Oxidation States
UC Davis ChemWiki illustrates the rules that have to be followed to make things consistent: *chemwiki.ucdavis.edu/Analytical_Chemistry/Electrochemistry/Redox_Chemistry/Oxidation_State*.

1861 Erlenmeyer Flask
Sella, A. "Classic Kit: Erlenmeyer Flask." *Chemistry World*, July 2008, *www.rsc.org/chemistryworld/issues/2008/july/erlenmeyerflask.asp*.

1861 Structural Formula
Wikipedia's introduction illustrates the basic kinds of chemical drawings, with some of the rules for producing them: *en.wikipedia.org/wiki/Structural_formula*.

1864 Solvay Process
Here's a teaching resource on the technology, with plenty of details: *www.hsc.csu.edu.au/chemistry/options/industrial/2765/Ch956.htm*. No new Solvay plants appear to have been built in years, but there are still dozens operating around the world.

1865 Benzene and Aromaticity
Rocke, A. J. *Image and Reality*. Chicago: Univ. of Chicago Press, 2010.

1868 Helium
Probably the most detailed account of this discovery is at the American Chemical Society's website: *www.acs.org/content/acs/en/education/whatischemistry/landmarks/heliumnaturalgas.html*.

1874 Tetrahedral Carbon Atoms
Chemical Heritage Foundation. "Jacobus Henricus van 't Hoff," *www.chemheritage.org/discover/online-resources/chemistry-in-history/themes/molecular-synthesis-structure-and-bonding/vant-hoff.aspx*.

Nobelprize.org. "Jacobus H. van 't Hoff - Biographical," *www.nobelprize.org/nobel_prizes/chemistry/laureates/1901/hoff-bio.html*.

1876 Gibbs Free Energy
A nontechnical treatment of this (and thermo-dynamics in general) is a tall order, because sooner or later, it's going to be Math or Nothing.

American Physical Society. "J. Willard Gibbs," www.aps.org/programs/outreach/history/historicsites/gibbs.cfm.
Set Laboratories, Inc. "Thermal Cracking," www.setlaboratories.com/therm/tabid/107/Default.aspx.
Wikipedia, "Josiah Willard Gibbs," en.wikipedia.org/wiki/Josiah_Willard_Gibbs.

1877 Maxwell-Boltzmann Distribution
Lindley, D. *Boltzmann's Atom*. New York: The Free Press, 2001.

1877 Friedel-Crafts Reaction
No nontechnical book exists. I suggest *Wikipedia* (en.wikipedia.org/wiki/Friedel-Crafts_reaction) for a nice overview, but any organic-chemistry textbook will have a section on this reaction as well.

1878 Indigo Synthesis
Glowacki et al. "Indigo and Tyrian Purple – From Ancient Natural Dyes to Modern Organic Semiconductors." *Israel Journal of Chemistry* 52, (2012): 1. https://www.jku.at/JKU_Site/JKU/ipc/content/e166717/e166907/e174991/e175004/2012-08.pdf.

1879 Soxhlet Extractor
Sella, A. "Classic Kit: Soxhlet extractor." *Chemistry World*, September 2007, www.rsc.org/chemistryworld/Issues/2007/September/ClassicKitSoxhletExtractor.asp.

1881 Fougère Royale
Turin, L. *The Secret of Scent*. New York: Ecco, 2006.

1883 Claus Process
The best overview I've seen for people who are not chemical engineers is at *Wikipedia*: en.wikipedia.org/wiki/Claus_process.

1883 Liquid nitrogen
A search through YouTube will yield examples of almost every strange liquid nitrogen demonstration that anyone can think up (as well as recipes for liquid nitrogen ice cream and other culinary creations).

1884 Fischer and Sugars
Kunz, H. "Emil Fischer—Unequalled Classicist, Master of Organic Chemistry Research, and Inspired Trailblazer of Biological Chemistry." *Angewandte International Edition (English)* 41, 23 (November 2002): 4439.

1885 Le Châtelier's Principle
Clark, Jim. "Le Chatelier's Principle," UC Davis ChemWiki, http://chemwiki.ucdavis.edu/Physical_Chemistry/Equilibria/A._Chemical_Equilibria/2._Le_Chatelier's_Principle.

1886 Isolation of Fluorine
When doing any fluorine-related searches, beware of the masses of crank literature on water fluoridation.
Wikipedia, "History of Fluorine," en.wikipedia.org/wiki/History_of_fluorine.

1886 Aluminum
National Historic Chemical Landmarks program of the American Chemical Society. "Production of Aluminum: The Hall-Héroult Process," 1997, www.acs.org/content/acs/en/education/whatischemistry/landmarks/aluminumprocess.html.

1887 Cyanide Gold Extraction
International Cyanide Management Code. "Use in Mining," www.cyanidecode.org/cyanide-facts/use-mining.

1888 Liquid Crystals
Collings, P. J. *Liquid Crystals*. Princeton, NJ: Princeton Univ. Press, 2002.
Gross, Benjamin. "How RCA Lost the LCD." *IEEE Spectrum*, November 1, 2012, http://spectrum.ieee.org/consumer-electronics/audiovideo/how-rca-lost-the-lcd.

1891 Thermal Cracking
Leffler, W. L. *Petroleum Refining in Nontechnical Language* (4th ed.). Tulsa, OK: PennWell, 2008.
Set Laboratories, Inc. "Thermal Cracking," www.setlaboratories.com/therm/tabid/107/Default.aspx.

1892 Chlor-Alkali Process
The entire chapter on the history of the chlor-alkali process from the *Handbook of Chlor-Alkali Technology* (New York: Springer, 2005) can be downloaded at rd.springer.com/chapter/10.1007%2F0-306-48624-5_2#page-1.

1892 Acetylene
National Historic Chemical Landmarks program of the American Chemical Society. "Commercial Process for Producing Calcium Carbide and Acetylene, 1998, www.acs.org/content/acs/en/education/whatischemistry/landmarks/calciumcarbideacetylene.html.

1893 Thermite
Wikipedia is a very good source on this topic. (The rest of the web is full of conspiracy-theory bizarreness about secret uses of thermite.) YouTube has a variety of pyrotechnic videos from home experimenters—watching them is a lot safer than trying it yourself.

1893 Borosilicate Glass
Watch Theodore Gray point out that not all heat-resistant glass these days is borosilicate, which can have some unfortunate consequences: www.popsci.com/science/article/2011-03/gray-matter-cant-take-heat.
SCHOTT Company. "SCHOTT Milestones," www.us.schott.com/english/company/corporate_history/milestones.html.

1893 Coordination Compounds
Kaufmann, G. "A Stereochemical Achievement of the First Order." *Bulletin for the History of Chemistry* 20 (1997): 50. www.scs.illinois.edu/~mainzv/HIST/bulletin_open_access/num20/num20%20p50-59.pdf.

1894 The Mole
June 2 (6/02) is celebrated as Mole Day every year, which you may find endearingly nerdy or alarmingly nerdy, depending on your disposition.

1894 Diazomethane
Here's a technical fact sheet from Sigma-Aldrich, one of the world's largest laboratory chemical suppliers, detailing the preparation of diazomethane and precautions that need to be taken: www.sigmaaldrich.com/content/dam/sigma-aldrich/docs/Aldrich/Bulletin/al_techbull_al180.pdf.

Mastronardi et al., "Continuous Flow Generation and Reactions of Anhydrous Diazomethane Using a Teflon AF-2400 Tube-in-Tube Reactor." *Organic Letters* 15, 21 (2013): 5590. pubs.acs.org/doi/abs/10.1021/ol4027914.

1895 Liquid Air
Johns, W. E. "Notes on Liquefying Air," www.gizmology.net/liquid_air.htm.

1896 Greenhouse Effect
The issue is, of course, soaked through with politics. Carbon dioxide, beyond doubt, is a greenhouse gas, and humans have, beyond doubt, added a great deal of it to the atmosphere. At that point, the arguing starts.

1897 Aspirin
Jeffreys, D. *Aspirin*. New York: Bloomsbury, 2004.

1897 Zymase Fermentation
Cornish-Bowden, A., ed. *New Beer in an Old Bottle*. Valencia, ES: Univ. of Valencia, 1998.

1898 Neon
Fisher, D. E. *Much Ado about (Practically) Nothing*. New York: Oxford Univ. Press, 2010.

1900 Grignard Reaction
Kagan, H. B. "Victor Grignard and Paul Sabatier." *Angewandte International Edition (English)* 51, 30 (2012): 7376. onlinelibrary.wiley.com/doi/10.1002/anie.201201849/abstract.

Nobelprize.org. "Victor Grignard - Biographical," www.nobelprize.org/nobel_prizes/chemistry/laureates/1912/grignard-bio.html.

1900 Free Radicals
American Chemical Society, www.acs.org/content/acs/en/education/whatischemistry/landmarks/freeradicals.html.

1900 Silicones
Dow Corning, www.dowcorning.com/content/discover/discoverchem/?wt.svl=FS_readmore_home_CORN.

European Silicones Centre, www.silicones-science.eu/.

1901 Chromatography
Wixom, R. L., and C. W. Gehrke, eds. *Chromatography: A Science of Discovery*. Hoboken, NJ: Wiley, 2010.

Wikipedia, "Chromatography," en.wikipedia.org/wiki/Chromatography.

1902 Polonium and Radium
Curie, E. *Madame Curie: A Biography*. New York: Da Capo Press, 2001.

Goldsmith, B. *Obsessive Genius*. New York: W. W. Norton, 2005.

1905 Infrared Spectroscopy
Rupawalla et. al. "Infrared Spectroscopy," UC Davis ChemWiki, chemwiki.ucdavis.edu/Physical_Chemistry/Spectroscopy/Vibrational_Spectroscopy/Infrared_Spectroscopy.

Wikipedia. "Infrared Spectroscopy," en.wikipedia.org/wiki/Infrared_spectroscopy.

1907 Bakelite®
Meikle, J. *American Plastic*. New Brunswick, NJ: Rutgers Univ. Press, 1995.

National Historic Chemical Landmarks program of the American Chemical Society. "Moses Gomberg and the Discovery of Organic Free Radicals," 2000, www.acs.org/content/acs/en/education/whatischemistry/landmarks/bakelite.html.

Sumitomo Bakelite Co. "Amsterdam Bakelite Collection," www.amsterdambakelitecollection.com.

1907 Spider Silk
Brunetta, L., and C. L. Craig. *Spider Silk*. New Haven, CT: Yale Univ. Press, 2010.

1909 pH and Indicators
A large table of indicator color changes can be found here: w3.shorecrest.org/~Erich_Schneider/tweb/Chemweb/datatables/indicators.jpg.

1909 Haber-Bosch Process
Hager, T. *The Alchemy of Air*. New York: Broadway Books, 2008.

1909 Salvarsan
Modern work with salvarsan and its chemistry (Waikato University) is found here: researchcommons.waikato.ac.nz/bitstream/handle/10289/188/content.pdf?sequence=1.

Hayden, D. *Pox*. New York: Basic Books, 2003.

1912 X-Ray Crystallography
Jenkin, J. *William and Lawrence Bragg, Father and Son*. New York: Oxford Univ. Press, 2008.

Kazantsev, R., and M. Towles. "X-Ray Crystallography," UC Davis ChemWiki, chemwiki.ucdavis.edu/Analytical_Chemistry/Instrumental_Analysis/Diffraction/X-ray_Crystallography.

University of Leeds. "William Thomas Astbury," arts.leeds.ac.uk/museum-of-hstm/research/william-thomas-astbury/.

1912 Maillard Reaction
McGee, H. *The Curious Cook*. San Francisco: North Point Books, 1990.

1912 Stainless Steel
Cobb, H. M. *The History of Stainless Steel*. Materials Park, OH: ASM Int., 2010.

1912 Boranes and the Vacuum-Line Technique
Wiberg, E. "Alfred Stock and the Renaissance of Inorganic Chemistry." *Pure and Applied Chemistry* 49 (1977): 691. pac.iupac.org/publications/pac/pdf/1977/pdf/4906x0691.pdf.

1912 Dipole Moments
Ball, P. "Letters Defend Nobel Laureate Against Nazi Charges." *Nature*, December 9, 2010, www.nature.com/news/2010/101209/full/news.2010.656.html.

1913 Mass Spectrometry
Griffiths, J. "A Brief History of Mass Spectrometry." *Analytical Chemistry* 80 (2000): 5676. pubs.acs.org/doi/pdf/10.1021/ac8013065.

1913 Isotopes
The printed literature on this topic is scattered between histories of physics, geology, chemistry, and medicine (which tells you what an important topic it is).

1915 Chemical Warfare
If you can find a copy, the eminent scientist J.B.S. Haldane wrote a vigorous defense of the entire idea of chemical warfare, titled *Callinicus*, in 1925.
Harris, R., and J. Paxman. *A Higher Form of Killing*. New York: Hill and Wang, 1982.

1917 Surface Chemistry
Coffey, P. *Cathedrals of Science*. New York: Oxford Univ. Press, 2008.

1918 Radithor
The Oak Ridge Assoc. Universities site has a terrifying online museum of radioactive quack cures (*www.orau.org/ptp/collection/quackcures/quackcures.htm*). An article with evidence that Eben Byers's remains were hot enough to expose film when the EPA reworked his grave site, is "The Great Radium Scandal" by Roger Macklis (August 1993 issue of *Scientific American*).

1920 Dean-Stark Trap
Sella, A. "Classic Kit: Dean-Stark Apparatus." *Chemistry World*, June 2010, www.rsc.org/chemistryworld/Issues/2010/June/DeanStarkApparatus.asp.

1920 Hydrogen Bonding
Wikipedia, "Hydrogen Bond," en.wikipedia.org/wiki/Hydrogen_bond.

1921 Tetraethyl Lead
Midgley, T. *From the Periodic Table to Production*. Corona, CA: Stargazer Publishing, 2001.
Warren, C. *Brush with Death*. Baltimore, MD: Johns Hopkins Univ. Press, 2000.

1928 Diels-Alder Reaction
Wikipedia and the Organic Chemistry Portal (*www.organic-chemistry.org/namedreactions/diels-alder-reaction.shtm*) are good places to start, but you'll rapidly find yourself looking at a lot of organic-chemistry reactions. The original Diels-Alder paper is here: dx.doi.org/10.1002%2Fjlac.19284600106.

1928 Reppe chemistry
ColorantsHistory.org. "Walter Reppe: Pioneer in Acetylene Chemistry," updated June 21, 2009, www.colorantshistory.org/ReppeChemistry.html.
Travis, A. "Unintended Technology Transfer: Acetylene Chemistry in the United States." *Bulletin for the History of Chemistry* 32, 1 (2007): 27. www.scs.illinois.edu/~mainzv/HIST/bulletin_open_access/v32-1/v32-1%20p27-34.pdf.

1930 Chlorofluorocarbons
Meiers, P. "Fluorocarbons - Charles Kettering, and 'Dental Caries,'" www.fluoride-history.de/p-freon.htm.
Midgley, T., *From the Periodic Table to Production*. Corona, CA: Stargazer Publishing, 2001.

1931 Deuterium
Dahl, P. F. *Heavy Water and the Wartime Race for Nuclear Energy*. Bristol, UK: Institute of Physics, 1999.
Mathez, A., ed. *Earth*. New York: New Press, 2000. www.amnh.org/education/resources/rfl/web/essaybooks/earth/p_urey.html.

1932 Carbonic Anhydrase
Kornberg, A. *For the Love of Enzymes*. Cambridge, MA: Harvard Univ. Press, 1989.

1932 Vitamin C
There is a lot of crank literature on this subject, thanks to Pauling and others.
Brown, S. R. *Scurvy*. New York: Thomas Dunne Books, 2003.
Le Couteur, P., and J. Burreson. *Napoleon's Buttons*. New York: Jeremy P. Tarcher/Penguin, 2003.
National Historic Chemical Landmarks program of the American Chemical Society, "Albert Szent-Györgyi's Discovery of Vitamin C," 2002, www.acs.org/content/acs/en/education/whatischemistry/landmarks/szentgyorgyi.html.

1932 Sulfanilamide
Hager, T. *The Demon Under the Microscope*. New York: Harmony Books, 2006.

1933 Polyethylene
Walton, D., and P. Lorimer. *Polymers*. Oxford: Oxford Univ. Press, 2000.

1934 Superoxide
This is a tough subject to research on a nonspecialist level, because any mention of oxygen or ROS sets off a massive flux of crank medical books and websites. And this is still a very active area of research, so opinions are changing constantly.

1934 The Fume Hood
The best introduction to this topic is on *Wikipedia* (en.wikipedia.org/wiki/Fume_hood).

1935 Nylon
National Historic Chemical Landmarks program of the American Chemical Society. "Wallace Carothers and the Development of Nylon," 2005, www.acs.org/content/acs/en/education/whatischemistry/landmarks/carotherspolymers.html.
Walton, D., and P. Lorimer. *Polymers*. Oxford: Oxford Univ. Press, 2000.

1936 Nerve Gas
Tucker, J. B. *War of Nerves*. New York: Pantheon Books, 2006.

1937 Elixir Sulfanilamide
Martin, B. J. *Elixir*. Lancaster, PA: Barkerry Press, 2014.

1938 Catalytic Cracking
Leffler, W. L. *Petroleum Refining in Nontechnical Language* (4th ed.). Tulsa, OK: PennWell, 2008.
National Historic Chemical Landmarks program of the American Chemical Society. "Houdry Process for Catalytic

Cracking," 1996, *www.acs.org/content/acs/ en/education/whatischemistry/landmarks/ houdry.html*.
Set Laboratories, Inc. "Thermal Cracking," *www.setlaboratories.com/therm/tabid/107/ Default.aspx*.

1939 The Last Element in Nature
A number of videos on the web claim to illustrate the testing of a "Francium bomb," but there is no such thing.

1939 *The Nature of the Chemical Bond*
Oregon State Univ. "Linus Pauling: The Nature of the Chemical Bond: A Documentary History," *scarc.library. oregonstate.edu/coll/pauling/bond/*.
Pauling, L. *The Nature of the Chemical Bond*. Ithaca, NY: Cornell Univ. Press, 1960.

1939 DDT
National Historic Chemical Landmarks program of the American Chemical Society. "Legacy of Rachel Carson's Silent Spring," 2012, *www.acs.org/content/acs/ en/education/whatischemistry/landmarks/ rachel-carson-silent-spring.html*.

1942 Steroid Chemistry
Wikipedia and UC Davis ChemWiki (*chemwiki.ucdavis.edu/Biological_Chemistry/ Lipids/Steroids*) have quick introductions to steroid chemistry. An excellent book about Russell Marker and the early days of the field is waiting to be written.
National Historic Chemical Landmarks program of the American Chemical Society. "Russell Marker and the Mexican Steroid Hormone Industry," 1999, *www.acs.org/content/acs/en/ education/whatischemistry/landmarks/ progesteronesynthesis.html*.

1942 Cyanoacrylates
Walton, D., and P. Lorimer. *Polymers*. Oxford: Oxford Univ. Press, 2000.

1943 LSD
Hofmann, A. *LSD My Problem Child*. Santa Cruz, CA: MAPS, 2009.

1943 Streptomycin
Chemical Heritage Foundation. "Selman Abraham Waksman," *www.chemheritage. org/discover/online-resources/chemistry-in-history/themes/pharmaceuticals/preventing-and-treating-infectious-diseases/waksman. aspx*.
National Historic Chemical Landmarks program of the American Chemical Society. "Selman Waksman and Antibiotics," 2005, *www.acs.org/content/acs/ en/education/whatischemistry/landmarks/ selmanwaksman.html*.

1943 Bari Raid
Mukherjee, S. *The Emperor of All Maladies*. New York: Scribner, 2010.

1945 Penicillin
The penicillin story has been told many times, but (as mentioned in this entry), not always correctly. More background can be found at the Nobel Prize Foundation's website (*www.nobelprize.org/nobel_prizes/ chemistry/laureates/1964/perspectives.html*).

1945 Glove Boxes
Mentions of the early Manhattan Project glove boxes can be found in an interview with Cyril Smith here: *www.manhattanprojectvoices.org/ oral-histories/cyril-s-smiths-interview*.

1947 Antifolates
Mukherjee, S. *The Emperor of All Maladies*. New York: Scribner, 2010.
Visentin, M., et al. "The Antifolates." *Visentin M, Zhao R, Goldman ID. The Antifolates. Hematology/Oncology Clinics of North America* 26, 3 (2012): 629. *www.ncbi.nlm. nih.gov/pmc/articles/PMC3777421/*.

1947 Kinetic Isotope Effects
UC Davis ChemWiki. "Kinetic Isotope Effects," *chemwiki.ucdavis.edu/Physical_ Chemistry/Quantum_Mechanics/Kinetic_ Isotope_Effect*.

1947 Photosynthesis
Baillie-Gerritsen, V. "The Plant Kingdom's Sloth." *Protein Spotlight* 38 (September 2003). *web.expasy.org/spotlight/back_ issues/038/*.

1948 Donora Death Fog
Murray, A. "Smog Deaths in 1948 Led to Clean Air Laws" *All Things Considered*, NPR, April 22, 2009, *www.npr.org/templates/story/story. php?storyId=103359330*.
Pennsylvania Historical & Museum Commission. "The Donora Smog Disaster October 30–31, 1948," *www. portal.state.pa.us/portal/server.pt/ community/documents_from_1946_-_ present/20426/donora_smog_ disaster?qid=63050470&rank=1*.
Peterman, E. "A Cloud with a Silver Lining: The Killer Smog in Donora, 1948," Pennsylvania Center for the Book, Spring 2009, *pabook.libraries.psu.edu/palitmap/ DonoraSmog.html*.

1949 Catalytic Reforming
Gembicki, S. "Vladimir Haensel 1914– 2002." *National Academy of Sciences Biographical Memoirs* 88 (2006). *www. nasonline.org/publications/biographical-memoirs/memoir-pdfs/haensel-vladimir.pdf*.
Leffler, W. L. *Petroleum Refining in Nontechnical Language* (4th ed.). Tulsa, OK: PennWell, 2008.
Set Laboratories, Inc. "Thermal Cracking," *www.setlaboratories.com/therm/tabid/107/ Default.aspx*.

1949 Nonclassical Ion Controversy
Peplow, M. "The Nonclassical Cation: A Classic Case of Conflict." *Chemistry World*, July 10, 2013. *www.rsc.org/ chemistryworld/2013/07/norbornyl-nonclassical-cation-brown-winstein-olah*.

1950 Conformational Analysis
Hermann Sachse tried several times to show that the rings could not be planar, but expressed himself in such an impenetrable fashion (to his fellow chemists) that he made little headway. See *https://webspace.yale.edu/ chem125_oyc/125/history99/6Stereochemistry/ Baeyer/Sachse.html*.

1950 Cortisone
National Historic Chemical Landmarks program of the American Chemical Society. "Russell Marker and the Mexican Steroid Hormone Industry," 1999, www.acs.org/content/acs/en/education/whatischemistry/landmarks/progesteronesynthesis.html.
Ophardt, C. "Steroids," UC Davis ChemWiki, chemwiki.ucdavis.edu/Biological_Chemistry/Lipids/Steroids.

1951 Sanger Sequencing
Streton, A. "The First Sequence: Fred Sanger and Insulin." *Genetics Society of America* 162, 2 (October 1, 2002): 527. www.genetics.org/content/162/2/527.full.

1951 The Pill
National Historic Chemical Landmarks program of the American Chemical Society, "Russell Marker and the Mexican Steroid Hormone Industry," 1999, www.acs.org/content/acs/en/education/whatischemistry/landmarks/progesteronesynthesis.html.
Ophardt, C. "Steroids," UC Davis ChemWiki, chemwiki.ucdavis.edu/Biological_Chemistry/Lipids/Steroids.

1951 Alpha-Helix and Beta-Sheet
University of Leeds. "William Thomas Astbury," arts.leeds.ac.uk/museum-of-hstm/research/william-thomas-astbury/.

1951 Ferrocene
An episode of the podcast *Chemistry in Its Element* from the Royal Society of Chemistry is devoted to this: www.rsc.org/chemistryworld/2013/05/ferrocene-podcast.

1951 Transuranic Elements
Chemical Heritage Foundation. "Glenn Theodore Seaborg," www.chemheritage.org/discover/online-resources/chemistry-in-history/themes/atomic-and-nuclear-structure/seaborg.aspx.

1952 Miller-Urey Experiment
The original Miller-Urey experiment's idea of a primitive atmosphere was probably wrong, but complex biochemicals can be formed under many other conditions. This takes us right into origin-of-life books, which are many and various (and often contain political or religious/antireligious agendas of their own).

1952 Zone Refining
Many of the accounts of the development of zone refining are found in the history of computer hardware, due to its use in purifying silicon.
McKetta, J. J. *Encyclopedia of Chemical Processing and Design* (vol. 68). New York: Dekker, 1999.

1952 Thallium Poisoning
Frank, P., and M. A. Ottoboni. *The Dose Makes the Poison*. Hoboken, NJ: Wiley, 2011.
Klaassen, C. D. *Casarett and Doull's Toxicology* (8th ed.). New York: McGraw-Hill, 2013.

1953 DNA's Structure
Crick, F. *What Mad Pursuit*. New York: Basic Books, 1988.
Watson, J. D. *The Double Helix*. New York: Atheneum, 1968.

1955 Electrophoresis
Rutty, C. J. "Sifting Proteins." *Conntact* (December 1995): 10. www.healthheritageresearch.com/CONNTACT9512-Smithies-StarchGel.pdf.
Vesterberg, O. "History of Electrophoretic Methods." *Journal of Chromatography* 480 (1989): 3.
Westermeier, R. *Electrophoresis in Practice*. Weinheim, DE: Wiley-VCH, 2005.

1956 The Hottest Flame
The original account of the combustion of dicyanoacetylene (*Journal of the American Chemical Society* 78 [1956]: 2020) can be read at pubs.acs.org/doi/abs/10.1021/ja01590a075.

1957 Luciferin
Pieribone, V., D. F. Gruber. *Aglow in the Dark*. Cambridge, MA: Belknap Press, 2005.

1960 Thalidomide
This story is another that has been told many times, and not always accurately.
Chemical Heritage Foundation. "Frances Oldham Kelsey," www.chemheritage.org/discover/online-resources/chemistry-in-history/themes/public-and-environmental-health/food-and-drug-safety/kelsey.aspx.

1960 Resolution and Chiral Chromatography
Chromatography Online. "The Evolution of Chiral Chromatography, www.chromatographyonline.com/lcgc/Column%3A+History+of+Chromatography/The-Evolution-of-Chiral-Chromatography/ArticleStandard/Article/detail/750627.

1961 NMR
The history of NMR, especially the development of imaging for medical use, is tangled. When the Nobel Prize was awarded for MRI, one disgruntled inventor took out full-page ads in major newspapers to protest being left out! A good account of the early days is at www.ray-freeman.org/nmr-history.html.

1962 Green Fluorescent Protein
NobelPrize.org press release, October 8, 2008, www.nobelprize.org/nobel_prizes/chemistry/laureates/2008/press.html.
Pieribone, V., and D. F. Gruber. *Aglow in the Dark*. Cambridge, MA: Belknap Press, 2005.
Zimmer, M. *Glowing Genes*. Amherst, NY: Prometheus Books, 2005.

1962 Noble Gas Compounds
National Historic Chemical Landmarks program of the American Chemical Society. "Neil Bartlett and the Reactive Noble Gases," 2006, www.acs.org/content/acs/en/education/whatischemistry/landmarks/bartlettnoblegases.html.

1962 Isoamyl Acetate and Esters
For an entertaining look at the use of ester compounds in perfumery, see *The Secret of Scent* by Luca Turin (New York: Harper Perennial, 2007), which also includes a case for a new theory of how the protein receptors in the nose detect aromas.

1963 Ziegler-Natta Catalysis
A fifty-year retrospective look at the Ziegler-Natta after the Nobel can be found at *onlinedigeditions.com/display_article.php?id=1340848*.
Walton, D., and P. Lorimer. *Polymers*. Oxford: Oxford Univ. Press, 2000.

1963 Merrifield Synthesis
Mitchell, A. R. "Bruce Merrifield and Solid-Phase Peptide Synthesis." *Peptide Science* 90, 3 (2008): 175.

1963 Dipolar Cycloadditions
Organic Chemistry Portal. "Huisgen Cycloaddition: 1,3-Dipolar Cycloaddition," *www.organic-chemistry.org/namedreactions/huisgen-1,3-dipolar-cycloaddition.shtm*.

1964 Kevlar®
Walton, D., and P. Lorimer. *Polymers*. Oxford: Oxford Univ. Press, 2000.

1965 Lead Contamination
Midgley, T. *From the Periodic Table to Production*. Corona, CA: Stargazer Publishing, 2001.
Warren, C. *Brush with Death*. Baltimore, MD: Johns Hopkins Univ. Press, 2000.

1966 Polywater
Franks, F. *Polywater*. Cambridge, MA: MIT Press, 1981.

1967 HPLC
Henry, R. "The Early Days of HPLC at DuPont," Chromatography Online, February 2, 2009, *www.chromatographyonline.com/lcgc/Column%3A+History+of+Chromatography/The-Early-Days-of-HPLC-at-DuPont*.

1969 Gore-Tex®
Chemical Heritage Foundation. "Robert W. Gore," *www.chemheritage.org/discover/online-resources/chemistry-in-history/themes/petrochemistry-and-synthetic-polymers/synthetic-polymers/gore.aspx*.
Walton, D., and P. Lorimer. *Polymers*. Oxford: Oxford Univ. Press, 2000.

1970 Carbon Dioxide Scrubbing
There are many accounts of the *Apollo 13* mission, the canonical one being *Lost Moon* (later renamed *Apollo 13*) by James Lovell and Jeffrey Kluger (Boston: Houghton Mifflin, 1993).

1970 Computational Chemistry
The literature on this subject is overwhelmingly technical, even for me. Introductory texts say things like "the reader will need some understanding of introductory quantum mechanics, linear algebra, and vector, differential and integral calculus." A good overview is this one by David Young: *www.ccl.net/cca/documents/dyoung/topics-orig/compchem.html*.

1970 Glyphosate
Many of the discussions of glyphosate are ax-grinding (and not by just one side of the debate, either). The EPA's fact sheet is at *www.epa.gov/safewater/pdfs/factsheets/soc/tech/glyphosa.pdf*, and Monsanto's own collection of history and background material is at *www.monsanto.com/products/pages/roundup-safety-background-materials.aspx*. There is, of course, a great deal of work in the primary literature. On the web and in the popular literature, the signal-to-noise ratio on this subject is very poor.

1971 Reverse-Phase Chromatography
Majors et. al. "New Horizons in Reversed-Phase Chromatography," Chromatography Online, June 1, 2010, *www.chromatographyonline.com/lcgc/Column%3A+Column+Watch/New-Horizons-in-Reversed-Phase-Chromatography/ArticleStandard/Article/detail/676044*.
Wikipedia. "Chromatography," *en.wikipedia.org/wiki/Chromatography*.
Wixom, R. L., and C. W. Gehrke, eds. *Chromatography*. Hoboken, NJ: Wiley, 2010.

1972 Rapamycin
Jenkin, J. *William and Lawrence Bragg, Father and Son*. New York: Oxford Univ. Press, 2008.
Sehgal, S. N. "Sirolimus: Its Discovery, Biological Properties, and Mechanism of Action." *Transplantation Proceedings* 35, 3, supplement (2003): S7. *dx.doi.org/10.1016/S0041-1345(03)00211-2*.

1973 B_{12} Synthesis
Woodward himself can be heard lecturing on the subject at *www.chem.umn.edu/groups/hoye/links/*.
Chemical Heritage Foundation. "Robert Burns Woodward," *www.chemheritage.org/discover/online-resources/chemistry-in-history/themes/molecular-synthesis-structure-and-bonding/woodward.aspx*.
Garg, N. "Vitamin B_{12}: An Epic Adventure in Total Synthesis," The Stoltz Group, California Institute of Technology, January 29, 2002, *stoltz.caltech.edu/litmtg/2002/garg-lit-1_29_02.pdf*.

1974 CFCs and the Ozone Layer
EPA. "Environmental Indicators: Ozone Depletion," August 19, 2010, *www.epa.gov/Ozone/science/indicat/*.

1979 Tholin
Sagan, C., and B. N. Khare. "Tholins." *Nature* 277, (1979): 102. *www.nature.com/nature/journal/v277/n5692/abs/277102a0.html*.
Waite et al. "The Process of Tholin Formation in Titan's Upper Atmosphere." *Science* 316, 5826 (May 2007): 870. *www.sciencemag.org/content/316/5826/870*.

1980 Iridium Impact Hypothesis
Chemical Heritage Foundation. "Helen Vaughn Michel," *www.chemheritage.org/discover/online-resources/chemistry-in-history/themes/atomic-and-nuclear-structure/michel.aspx*.
Lewis, J. S. *Rain of Iron and Ice*. Reading, MA: Addison-Wesley, 1997.

1982 Unnatural Products
Paquette's synthesis is annotated at *www.synarchive.com/syn/15*, and Paquette himself talked about the field in *Proceedings of the National Academy of Sciences*, available at *www.ncbi.nlm.nih.gov/pmc/articles/PMC346698/*.

1982 MPTP
Langston, J. W., and J. Palfreman. *The Case of the Frozen Addicts*. Amsterdam: IOS Press, 2014.
Wolf, L. K. "The Pesticide Connection." *Chemical and Engineering News* 91, 47, (November 25, 2013): 11. *cen.acs.org/articles/91/i47/Pesticide-Connection.html*.

1983 Polymerase Chain Reaction
Mullis, K. B. *Dancing Naked in the Mind Field*. New York: Pantheon Books, 1998.
Rabinow, P. *Making PCR*. Chicago: Univ. of Chicago Press, 1996.

1984 Quasicrystals
The book to read if you're already a materials scientist or crystallographer is *Quasicrystals: A Primer* by Christian Janot (New York: Oxford Univ. Press, 2012). If you're not, see *www.nobelprize.org/nobel_prizes/chemistry/laureates/2011/press.html*. An interview with Dan Shechtman about the difficulties of getting his proposals accepted is here: *www.theguardian.com/science/2013/jan/06/dan-shechtman-nobel-prize-chemistry-interview*.

1984 Bhopal Disaster
A review of the health impact of the disaster was published in *Environmental Health* and is available at *www.ncbi.nlm.nih.gov/pmc/articles/PMC1142333/*. The legal aspects are summarized here: *www.princeton.edu/~achaney/tmve/wiki100k/docs/Bhopal_disaster.html*.

1985 Fullerenes
Aldersey-Williams, H. *The Most Beautiful Molecule*. New York: Wiley, 1995.
National Historic Chemical Landmarks program of the American Chemical Society. "Discovery of Fullerenes," 2010, *www.acs.org/content/acs/en/education/whatischemistry/landmarks/fullerenes.html*.

1985 MALDI
Syed, B. "MALDI-TOF," UC Davis ChemWiki, *chemwiki.ucdavis.edu/Analytical_Chemistry/Instrumental_Analysis/Mass_Spectrometry/MALDI-TOF*.

1988 Modern Drug Discovery
Ravina, E., and H. Kubinyi. *The Evolution of Drug Discovery*. Weinheim, DE: Wiley-VCH, 2011.

1988 PEPCON® Explosion
A case study of the incident, prepared for NASA, can be found at *nsc.nasa.gov/SFCS/SystemFailureCaseStudyFile/Download/290*. There are also many copies of the film taken of the explosion on YouTube.

1989 Taxol®
Goodman, J., and V. Walsh. *The Story of Taxol*. Cambridge: Cambridge Univ. Press, 2001.

1991 Carbon Nanotubes
Iijima, S. "Synthesis of Carbon Nanotubes." *Nature* 354 (1991): 56. *www.nature.com/physics/looking-back/iijima/index.html*.
Monthioux, M., and V. L. Kuznetsov. *Carbon* 44 (2006): 1621. *nanotube.msu.edu/HSS/2006/1/2006-1.pdf*.

1994 Palytoxin
An alarming first-person account of palytoxin poisoning can be found at *www.advancedaquarist.com/blog/personal-experiences-with-palytoxin-poisoning-almost-killed-myself-wife-and-dogs*. Yoshito Kishi discussed the synthesis in *Pure and Applied Chemistry* (*media.iupac.org/publications/pac/1989/pdf/6103x0313.pdf*) and in many journal articles.

1997 Coordination Frameworks
This editorial at the Royal Society of Chemistry's *Chemistry World* blog is useful: *prospect.rsc.org/blogs/cw/2013/04/24/a-metal-organic-framework-for-progress/*. Also see "Taking the Crystals out of X-Ray Crystallography" by Ewen Callaway at *Nature*'s news site: *www.nature.com/news/taking-the-crystals-out-of-x-ray-crystallography-1.12699*.

1998 Recrystallization and Polymorphs
For an account written at the time, see *www.natap.org/1998/norvirupdate.html*.
Bauer et al., "Ritonavir: An Extraordinary Example of Conformational Polymorphism." *Pharmaceutical Research* 18, 6 (June 2001): 859. *rd.springer.com/article/10.1023%2FA%3A1011052932607*.
Chemburkar et al., "Dealing with the Impact of Ritonavir Polymorphs on the Late Stages of Bulk Drug Process Development." *Organic Process Research & Development* 4 (June 21, 2000): 413. *pubs.acs.org/doi/abs/10.1021/op000023y*.

2005 Shikimic Acid Shortage
Werner et al. "Several Generations of Chemoenzymatic Synthesis of Oseltamivir (Tamiflu)." *Journal of Organic Chemistry* 76, 24 (2011): 10,050.

2009 Acetonitrile
National Historic Chemical Landmarks program of the American Chemical Society. "Sohio Acrylonitrile Process," 2007, *www.acs.org/content/acs/en/education/whatischemistry/landmarks/acrylonitrile.html*.

2010 Engineered Enzymes
Bornscheuer et al. "Engineering the Third Wave of Biocatalysis." *Nature* 485 (May 10, 2012): 185. *www.nature.com/nature/journal/v485/n7397/full/nature11117.html*.

2010 Metal-Catalyzed Couplings
NobelPrize.org. "The Nobel Prize in Chemistry 2010," 2014, *www.nobelprize.org/nobel_prizes/chemistry/laureates/2010/*.

2013 Single-Molecule Images
IBM Zürich reported its pentacene images here: *www.zurich.ibm.com/st/atomic_manipulation/pentacene.html*.

Index

Abelson, Philip, 372
Abraham, Edward P., 340
Acetate, isoamyl, and esters, 404–405
Acetonitrile, 502–503
Acetylene, 198–199
Acids and bases, 272–273
Adams, Roger, 220
Agricola, Georgius, 56
Air, liquid, 212–213
Alavi, Abass, 450
Alchemy, 26, 46
Alder, Kurt, 278
Alderotti, Taddeo, 50
Alexander, Stewart F., 334
Allotropes, about, 68, 104, 130
Alpha-helix and beta-sheet, 368–369
al-Razi, Abu Bakr Muhammad ibn Zakariya', 46
Aluminum, 188–189, 200
Alum, Yorkshire, 60–61
Alvarez, Luis W., 456
Alvarez, Walter, 456
Amino acids, 96–97
Ampère, André-Marie, 186
Antifolates, 344–345
Aqua regia, 48–49
Arabian, Donald D., 434
Arigoni, Duilio, 448
Aristotle, 26
Arnold, James T., 398
Aromaticity/benzene, 158–159
Arrhenius, Svante, 214, 240
Arsenic-laden wallpaper, 106
Asaro, Frank, 456
Aspirin, 216–217
Astbury, William, 368
Aston, Francis W., 256, 258
Asymmetric induction, 208–209
Atomic theory, 100–101
Atomism, 28–29
Avogadro, Amedeo, 102, 148, 206
Avogadro's hypothesis, 102
AZT and antiretrovirals, 466–467, 476

Bacon, Francis, 58–59
Baekeland, Leo Hendrik, 236
Baeyer, Adolf von, 172
Bakelite, 236–237
Barbier, Philippe Antoine, 224
Bari raid, 334
Bartlett, Neil, 402
Barton, Derek H. R., 358
Bases and acids, 272–273
Bassham, James A., 348
Becher, Johann Joachim, 66
Becquerel, Antoine-Henri, 232
Belousov, Boris, 428
Benson, Andrew A., 348
Benzene/aromaticity, 158–159
Berg, Otto, 306
Bernal, John D., 414
Bertheim, Alfred, 244
Berthier, Pierre, 250
Berthollet, Claude-Louis, 76
Bertozzi, Carolyn R., 490
Beryllium, 114–115
Berzelius, Jöns Jacob, 104, 116
Beta-sheet, 368

Bethune, Donald S., 482
Bhopal disaster, 470–471
Bigeleisen, Jacob, 346
Birch, John, 336
Birch reduction, 336–337
Birth-control pill, 366–367
Black, James W., 476–477
Black, Joseph, 78
Blodgett, Katherine, 262, 508
Bocarsly, Andrew B., 512
Bolton, Elmer K., 302
Boltzmann, Ludwig Eduard, 168
Bonds, chemical, nature of, 320–321.
  See also Hydrogen bonding; Sigma/pi bonding
Boranes and vacuum-line technique, 252–253
Borosilicate glass, 202–203
Bosch, Carl, 242
Böttger, Johann Friedrich, 36
Bovet, Daniel, 292
Boyer, Paul Delos, 308
Boyle, Robert, 64–65, 66, 82, 86, 100, 120
Boyle's Law, 64
Bragg, William Henry, 246
Bragg, William Lawrence, 246
Brand, Hennig, 68–69
Branson, Herman, 368
Brearley, Harry, 250
Brickwedde, Ferdinand, 286
Broder, Samuel, 466
Brønsted, Johannes N., 272
Bronze, 16–17, 20
Brown, Herbert C., 252, 356
Brugnatelli, Luigi, 94
Buchner, Eduard, 218
Buchner, Hans E. A., 218
Buchner, Johann Andreas, 216
Bunsen, Robert, 80, 146, 160
Burton, William Merriam, 194
Bussy, Antoine, 114
BZ reaction, 428–429

Cadet, Louis Claude, 80
Cadet's fuming liquid, 80–81
Cady, Hamilton Perkins, 160
Caffeine, 110–111
Cahn, John, 468
Callow, Kenneth, 360
Calvin, Melvin, 348
Campbell, Walter, 310
Cannizzaro, Stanislao, 148, 206
Carbon dioxide: carbonic anhydrase, 288; fermentation and, 218; greenhouse effect and, 214; oxygen and, 84; photosynthesis and, 512; properties of, 78–79; scrubbing, 434–435; as supercritical fluid, 112, 396; supercritical phase, 112
Carbonic anhydrase, 288–289
Carbon nanotubes, 482–483
Carbon, tetrahedral atoms, 164–165
Cárdenas, Luis E. M., 366
Carothers, Wallace H., 302
Castner, Hamilton Young, 196
Catalysis, Ziegler-Natta, 406–407
Catalytic cracking, 314–315
Catalytic reforming, 352–353

Cavendish, Henry, 78, 82, 84
Cellular respiration, 308–309
Ceramics, porcelain, 36–37
CFCs. See Chlorofluorocarbons
Chadwick, James, 258, 286
Chain, Ernst B., 340
Chalfie, Martin, 400
Chaloner, Thomas, 60
Charles, Jacques, 120
Chauvin, Yves, 496
Chemical notation, 104–105
Chemical warfare, 260–261. See also Nerve gas
Chevreul, Michel-Eugène, 108
Chiral chromatography, resolution and, 396–397
Chirality, 136–137, 164, 208
Chlor-alkali process, 196–197
Chlorofluorocarbons (CFCs), 282–283, 316, 352, 446–447
Cholesterol, 108–109
Chromatography, 230–231; chiral, 396–397; gas, 374–375; HPLC, 426–427; LC/MS, 464–465; reverse-phase, 440–441
Ciamician, Giacomo, 122
Cisplatin, 416–417
Clapeyron, Benoît P. E., 120
Claus process, 178–179
Click triazoles, 490–491
Coblentz, William Weber, 234
Colton, Frank B., 366
Computational chemistry, 436–437
Concrete, Roman, 34–35
Condy, Henry Bollmann, 150
Conformational analysis, 358
Conservation of mass, 86–87
Contact process, 74
Coordination compounds, 204–205
Coordination frameworks, 486–487
Coover, Harry W., Jr., 328
Cordus, Valerius, 54
Corey, Robert, 368
Cornforth, John W., 360, 448
Cortisone, 360–361
Crafts, James Mason, 170
Craig, Lyman C., 362
Crick, Francis H. C., 382, 392
Criegee, Rudolf, 130
Croesus, 24
Crommie, Michael, 508
Crystallography, X-ray, 246–247
Crystals/crystallization: Cave of Crystals, 14–15; chirality, 136–137, 164, 208; factors affecting, 14; formation of, 14; gypsum, 14; liquid, 108, 192–193; protein crystallography, 414–415; quasicrystals, 468–469; recrystallization and polymorphs, 488–489; transition state theory and, 300–301
Curie, Marie and Pierre, 232
Curl, Robert F., Jr., 472
Cyanide: gold extraction, 190–191; hydrogen, 72, 76–77, 190
Cyanoacrylates, 328–329
Cycloadditions, dipolar, 410–411

Daguerreotype, 126–127
Dalton, John, 100–101, 102, 104, 148

Davankov, Vadim, 396
Davy, Humphry, 78, 98, 186, 336, 420
DDT, 332–333
Dean, Ernest Woodward, 266
Dean-Stark trap, 184, 266–267
Debye, Peter, 254–255
De Hevesy, George C., 48, 274
Democritus, 28–29, 100
Dempster, Arthur J., 256
De Re Metallica (Agricola), 56–57
Deuterium, 286–287
Dewar, Michael J. S., 436
Diamond, synthetic, 384–385
Diazomethane, 210–211
Diels-Alder reaction, 278–279, 410, 422, 444, 458
Diels, Otto P. H., 108, 278
Diethyl ether, 54–55
Diffusion, gaseous, 324–325
Dioscorides, Pedanius, 32
Dipolar cycloadditions, 410–411
Dipole moments, 254–255
Distillation, fractional, 50–51
Djerassi, Carl, 366
DNA: amino acids and, 96–97; antifolates and, 344; PCR, 462; replication, 392–393; structure, 382–383
Doering, William von Eggers, 62
Domagk, Gerhard, 292
Donora death fog, 350–351
Draper, John, 122
Drug discovery, modern, 476–477

Eaton, Philip E., 458
Eggerer, Hermann, 448
Ehrlich, Paul, 244
Eichengrün, Arthur, 216
Electrochemical reduction, 98–99, 510
Electrophoresis, 386–387
Electroplating, 94–95
Electrospray LC/MS, 464–465
Elements. See also specific elements: alchemy and, 26, 46–47; four classical, 26–27; last in nature, 318–319; periodic table, 26, 150, 162–163, 222, 258; *The Sceptical Chymist* and, 26; term origin, 26; transuranic, 372–373
Elion, Gertrude B., 476–477
Elixir sulfanilamide, 310–311
Empedocles, 26
Energy, Gibbs free, 166–167
Engineered enzymes, 504–505
Enzymes, engineered, 504–505
Enzyme stereochemistry, 448–449
Erlenmeyer flask, 152–153
Eschenmoser, Albert, 444
Esters, isoamyl acetate and, 404–405
Ether (diethyl ether), 54–55
Evans, Meredith Gwynne, 300
Ewald, Paul Peter, 246
Eyring, Henry, 300

Faraday, Michael, 78, 158
Farber, Sidney, 344
Fawcett, Eric, 294
Fenn, John B., 464
Fermentation, zymase, 218–219
Ferrocene, 370–371

INDEX 525

Fertilizer, phosphate, 132–133
Fire, essence of. See Phlogiston
Fire, Greek, 38–39
Fischer, Emil Hermann, 96, 182–183, 208, 238
Fischer, Ernst Otto, 370
Fischer, Felix R., 508
Fischer, Franz, 276
Fischer-Tropsch process, 276–277
Flame, hottest, 388–389
Flame spectroscopy, 146–147
Flask, Erlenmeyer, 152–153
Fleming, Alexander, 340, 414
Florey, Howard W., 340
Flow chemistry, 498–499
Fluorescence, 138–139
Fluorescent protein, 400–401
Fluorine, isolation of, 186–187
Fokin, Valery, 490
Formulas, chemical notation and, 104–105
Formulas, structural, 154–155
Fougère Royale, 176–177
Fourneau, Ernest, 292
Fowler, Joanna S., 450
Fractional distillation, 50–51
Francium, 318–319
Franck, James, 48
Franklin, Rosalind, 382
Franz, John E., 438
Fraunhofer, Joseph von, 146
Free radicals, 226–227
Fridovich, Irwin, 296
Friedel, Charles, 170
Friedel-Crafts reaction, 170–171, 312
Friedel, Georges, 192
Friedenthal, Hans, 240
Fujishima, Akira, 512
Fujita, Makoto, 486
Fukui, Kenichi, 422
Fullerenes, 472–473
Fuller, R. Buckminster, 472
Fume hood, 298–299
Functional groups, 118–119
Funnel, separatory/sep, 140–141

Gadolin, Johan, 90
Gas chromatography, 374–375
Gaseous diffusion, 324–325
Gas law, ideal, 120–121
Gas (noble) compounds, 402–403
Gay-Lussac, Joseph-Louis, 76, 102, 120, 148, 186
Geim, Andre Konstantin, 492
Gerhardt, Charles-Frédéric, 216
Gershman, Rebecca, 296
Gibbs free energy, 166–167
Gibson, Reginald O., 294
Giddings, John C., 426
Gilman, Alfred, Sr., 334
Glass, borosilicate, 202–203
Glauber, Johann Rudolf, 150
Glove boxes, 342–343
Glues (cyanoacrylates), 328–329
Glycerin, 18
Glyphosate, 438–439
Gold: alchemy and, 46, 52; cyanide extraction, 190–191; dissolving, 48; electroplating, 94; refining, 24–25
Goldschmidt, Hans, 200
Gomberg, Moses, 226

Goodman, Louis S., 334
Goodyear, Charles, 128
Gore-Tex (Robert W. Gore), 432–433
Graham, Thomas, 324
Graphene, 492–493
Gray, George William, 192
Greek fire, 38–39
Green fluorescent protein (GFP), 400–401
Greenhouse effect, 214–215
Green, Paris, 106–107
Gregor, William, 88
Grignard reaction, 224–225
Grignard, Victor, 224–225, 260
Grosse, Aristid V., 388
Grotthuss, Theodor, 122
Grubbs, Robert H., 496
Gunpowder, 44–45

Haber-Bosch process, 242–243
Haber, Fritz, 242, 260
Haensel, Vladimir, 352–353
Haise, Fred W, Jr., 434–435
Hall, Charles Martin, 188
Hall-Héroult process, 188
Hall, Howard T., 384
Hancock, Thomas, 128
Harries, Carl Dietrich, 130
Hassel, Odd, 358
Hata, Sahachiro, 244
Haynes, Elwood, 250
Heatley, Norman G., 340
Heck, Richard F., 506
Helium, 160–161
Helmont, Jan Baptist, 78
Hench, Philip S., 360
Henne, Albert Leon, 282
Héroult, Paul-Louis-Toussaint, 188
High-performance liquid chromatography, 426–427
Hillenkamp, Franz, 474
Hill, Julian Werner, 302
Hitchings, George H., 476–477
Hiyama, Tamejiro, 452
Hodge, John Edward, 248
Hodgkin, Dorothy C., 340, 414, 444
Hoff, Jacobus Henricus van't, 136, 164, 182
Hoffman, Felix, 216
Hoffman, Roald, 422–423
Hofmann, Albert, 330
Hofmeister, Franz, 96
Holton, Robert A., 480
Hook, Robert, 64
Horváth, Csaba, 426, 440
Horwitz, Jerome, 466
Hottest flame, 388–389
Houdry, Eugene J., 314
HPLC, 426–427
Huber, Josef, 426
Hückel, Erich, 284
Huggins, Maurice Loyal, 268
Hughes, Edward D., 312
Hydrated calcium aluminosilicate, 34
Hydrogen: chlor-alkali process and, 196; discovery and properties, 82–83; heavy (deuterium), 286; hottest flame and, 388; NMR and, 198; oxygen, water and, 82, 102; pH, acids, bases and, 240, 272; storage, 510–511
Hydrogenation, 220–221

Hydrogen bonding, 70, 268–269, 320, 346, 368
Hydrogen cyanide, 72, 76–77, 190
Hydrogen sulfide, 70–71, 84, 178, 352

Ibn Ḥayyān, Abū Mūsā Jābir, 26, 40
Ideal gas law, 120–121
Iijima, Sumio, 482
Indigo synthesis, 172–173
Induction, asymmetric, 208–209
Infrared spectroscopy, 234–235
Ingold Christopher K., 312
Ingram, Vernon, 354
Ion controversy, nonclassical, 356–357
Iridium impact hypothesis, 456–457
Iron smelting, 20–21
Isoamyl acetate and esters, 404–405
Isotope effects, kinetic, 346–347
Isotopes, 258–259
Isotropic distribution, 500–501
Itano, Harvey A., 354

James, Anthony T., 374
Janey, Jacob M., 504
Janssen, Pierre Jules César, 160
Jewett, Frank Fanning, 188
Johnson, Erling, 132
Joule, James Prescott, 212
Joyner, Fred, 328
Julian, Percy L., 360

Karas, Michael, 474
Karlsruh, Cannizzaro at, 148–149
Kealy, Thomas J., 370
Kekulé, Friedrich August, 158
Kellner, Karl, 196
Kelsey, Frances O., 310, 394–395
Kendall, Edward C., 360
Kettering, Charles F., 270, 282
Kevlar, 412–413
Khare, Bishun, 454
Khorana, Har Gobind, 462
Kinetic isotope effects, 346–347. See also Isotropic distribution
King, Victor L., 204
Kipping, Frederick S., 228
Kirchhoff, Gustav, 146, 160
Kirkland, Joseph J., 426, 440
Kishi, Yoshito, 452, 484
Klaproth, Martin Heinrich, 88
Klarer, Josef, 292
Klemm, LeRoy H., 396
Kleppe, Kjell, 462
Klesper, Ernst, 396
Krantz, Gene, 434
Kraus, Charles A., 336
Krebs, Hans (Krebs cycle), 308
Kroto, Harold W., 472
Kurti, Nicholas, 324
Kwolek, Stephanie L., 412

Lagrange, Joseph-Louis, 86
Langmuir, Irving, 262, 508
Latimer, Wendell, M., 268
Laue, Max T. F. von, 48, 246
Lavoisier, Antoine, 82, 84, 86–87, 100
Lawes, John Bennet, 132
Lawrence, Ernest O., 256
LC/MS, 464–465
Lead contamination, 418–419
Lead, tetraethyl, 270–271, 352

Learning, advancement of, 58–59
Le Bel, Joseph-Achille, 136, 164, 182
Leblanc, Nicolas, 156, 196
Le Chatelier, Louis, 60
Le Châtelier's principle (Henry-Louis Le Châtelier), 184–185
Lehmann, Otto, 192
Leidenfrost, Johan Gottlob, 180
Le Rossignol, Robert, 242
Leroux, Pierre-Joseph, 216
Lewis, Gilbert N., 272, 286
Lewis, Winford L., 260
Ley, Steven V., 498
Liebig, Justus von, 118–119, 132, 144, 294
Linde, Carl von, 212
Lind, James, 288
Lipmann, Fritz A., 308
Lipscomb, William, 252
Liquid air, 212–213
Liquid chromatography/mass spectrometry (LC/MS), 464–465
Liquid crystals, 108, 192–193
Liquid nitrogen, 180–181
Lockyer, Joseph Norman, 160
Lohmann, Karl, 308
Long, Crawford W., 54
Lonsdale, Kathleen, 158
Loschmidt, Josef, 154
Lovell, James A, Jr., 434
Lowry, Thomas Martin, 272
LSD, 330–331
Luciferin, 390–391

MacArthur, John Stewart, 190
Macquer, Pierre, 76
Magnetic stirring, 338–339
Maillard, Louis-Camille, 248
Maillard reaction, 248–249
Makogon, Yuri F., 420
MALDI, 474–475
Marckwald, Willy, 208
Marker, Russell, 326, 366
Martin, Archer J. P., 364, 374, 440–441
Mass, conservation of, 86–87
Mass spectrometry, 256–257
Mauve, Perkin's, 142–143
Maxwell-Boltzmann distribution, 168–169
Maxwell, James Clerk, 168
Mayer Maria G., 346
McCord, Joseph, 296
McCormick, Katherine, 366
McElroy, William D., 390
McFarland, David Ford, 160
McMillan, Edwin M., 372
Meldrum, Norman U., 288
Mendeleev, Dmitri I., 162
Mercury, 30–31
Merrifield Robert B., 408
Merrifield synthesis, 408–409
Meselson, Matthew, 392
Metal-catalyzed couplings, 506–507
Metals, electroplating, 94–95
Meteorite, Murchison, 430–431
Methane hydrate, 420–421
Meyer, Gerhard, 508
Meyerhof, Otto Fritz, 308
Meyer, Lothar, 162
Michel Helen V., 456
Midgley, Thomas, Jr., 270, 282, 352
Mietzsch, Fritz, 292

Miller, Samuel A., 370
Miller, Stanley, 376
Miller-Urey experiment, 376–377
Min Chueh Chang, 366
Mirror silvering, 144–145
Mitchell, Peter, 308
Mitsuya, Hiroaki, 466
Modern drug discovery, 476–477
Mohr, Ernst, 358
Moissan, Ferdinand-Frédéric-Henri, 186
Molecular disease, 354–355
Molecule (single) images, 508–509
Mole, the, 206–207
Moore, Richard E., 484
Morehead, James Turner, 198
Morphine, 92–93
Moseley, Henry Gwyn Jeffries, 162
MPTP, 460–461
Müller, Paul H., 332
Mullis, Kary B., 462
Murchison Meteorite, 430–431

Nanotubes, carbon, 482–483
Natta, Giulio, 406, 496
Natural product chemistry, 22
Natural products, 32–33
Neel, James Van Gundia, 354
Negishi, Ei-ichi, 506
Neon, 222–223
Nerve gas, 304–305
Neuman, Edward W., 296
Neumann, Caspar, 72
Newlands, John A. R., 162
Niépce, Nicéphore (Joseph), 126
Nitrogen, liquid, 180–181
Nitroglycerine, 134–135, 166
NMR, 398–399
Nobel, Alfred, 134
Noble gas compounds, 402–403
Nocera, Daniel G., 512
Noddack, Ida T., 306
Noddack, Walter, 306
Nonclassical ion controversy, 356–357
Notation, chemical, 104–105
Novoselov, Konstantin, 492
Nozaki coupling, 452–453
Nozaki, Hitosi, 452
Nuclear magnetic resonance (NMR), 398–399
Nylon, 302–303

Okamoto, Yoshio, 396
Olah, George, 356
Olefin metathesis, 496–497
Olszewski, Karol, 180
Ostwald, Friedrich Wilhelm, 206
Oxidation states, 150–151
Oxygen: history/chemistry of, 84–85; hottest flame and, 388; hydrogen, water and, 82, 102; liquid, 212; ozone and, 130, 446; photosynthesis and, 348, 512; superoxide and, 296–297; weight of, 148
Ozone, 130–131, 446–447

Packard, Martin E., 398
Palmieri, Luigi, 160
Palytoxin, 452, 484–485
Paquette, Leo, 458
Paracelsus, 52, 54, 64
Paris green, 106–107

Parquet, Paul, 176–177
Parsons, Charles A., 384
Pasteur, Louis, 136, 218
Patterson, Clair C., 418
Pauling, Linus C., 254, 268, 296, 300, 320–321, 354, 368, 398, 402, 468
Pauson, Peter L., 370
PCR, 462–463
Pechmann, Hans von, 210–211, 294
Pelouze, Théophile-Jules, 134
Penicillin, 340–341
PEPCON explosion, 478–479
Perey, Marguerite, 318
Periodic table, 26, 150, 162–163, 222, 258
Perkin, William Henry, 142–143, 176, 282
Perrier, Carlo, 306
Perrin, Michael W., 294
Pestka, Sidney, 440
PET imaging, 450–451
Peyrone, Michele, 416
Pfann, William G., 378
Phillips, David C., 414
Phillips, Peregrine, 74
Philosopher's stone, 40–41, 56, 68
Phlogiston, 66–67, 84
Phosphate fertilizer, 132–133
Phosphorus, 68–69
Photochemistry, 122–123
Photosynthesis, 348–349, 512
Photosynthesis, artificial, 512–513
pH and indicators, 240–241
The pill, 366–367
Pincus Gregory G., 366
Pincus, Gregory Goodwin, 366
Pirkle, William, 396
Platinum, 48, 220, 352, 416
Plato, 26
Pliny the Elder, 34
Plunkett, Roy J., 316–317
Polanyi, Michael, 300
Polonium/radium, 232–233
Polyethylene, 294–295
Polymerase chain reaction, 462–463
Polymers and polymerization, 124–125
Polywater, 424–425
Pople, John A., 436
Porcelain, 36–37
Porto, Sérgio, 424
Positron emission tomography (PET), 450–451
Potier, Pierre, 480
Poulletier de la Salle, François, 108
Prasher, Douglas, 400
Priestley, Joseph, 78, 84
Protein crystallography, 414–415
Protein, green fluorescent (GFP), 400–401
Prussian blue, 72–73, 76, 380
Purification, 22–23

Qin Shi Huang, 30
Quasicrystals, 468–469
Quinine, 62–63

Rabe, Paul, 62
Radithor, 264–265
Radium/polonium, 232–233
Ramazzini, Bernardino, 70
Ramsay, William, 222
Rao, Yellapragada S., 344
Rapamycin, 442–443
Reaction mechanisms, 312–313

Recrystallization and polymorphs, 488–489
Reduction, Birch, 336–337
Reduction, electrochemical, 98–99, 510
Reductionism, 26, 28
Refining, zone, 378–379
Reichstein, Tadeus, 360
Reinitzer, Friedrich, 192
Reppe chemistry, 280–281
Resolution and chiral chromatography, 396–397
Respiration, cellular, 308–309
Reverse-phase chromatography, 440–441
Richards, Rex E., 398
Roberts, John D., 398
Robinson, Robert, 312
Robiquet, Pierre-Jean, 96
Rodebush, Worth Huff, 268
Roebuck, John, 74
Rohloff, John C., 494
Roman concrete, 34–35
Rosenberg, Barnett, 416
Rosenkranz, George, 366
Rosinger, Arthur, 338
Rotary evaporator, 22, 362–363
Roughton, Francis, 288
Rousseau, Denis L., 424
Royal water (aqua regia), 48–49
Rubber, 128–129
Runge, Friedlieb Ferdinand, 110
Rutherford, Ernest, 258

Sabatier, Paul, 220
Sachse, Hermann, 358
Sagan, Carl, 454
Salvarsan, 244–245
Sanger, Frederick, 364
Sanger sequencing, 364–365
Sarett, Lewis, 360
Savile, Christopher K., 504
The Sceptical Chymist (Boyle), 64
Schatz, Albert, 332
Scheele, Carl W., 70, 76, 190
Scheuer, Paul J., 484
Schönbein, Christian F., 130, 134
Schott, Friedrich O., 202
Schrader, Gerhard, 304
Schrock, Richard R., 496
Seaborg, Glenn T., 372–373
Segrè, Emilio G., 306
Sehgal, Suren, 442
Separatory/sep funnel, 140–141
Sertürner, Friedrich W. A., 92
Sharpless, Karl B., 490
Shechtman, Dan, 468
Sheehan, John C., 340
Shikimic acid shortage, 494–495
Shimomura, Osamu, 400
Shoolery, James N., 398
Shukhov, Vladimir, 194
Sigma/pi bonding, 284–285
Silicones, 228–229
Silvering, mirror, 144–145
Simon, Eduard, 124
Simon, Francis, 324
Singer, Seymour J., 354
Single-molecule images, 508–509
Smalley, Richard, 472
Smithies, Oliver, 386
Smylie, Robert E., 434
Soap, 18–19

Sobrero, Ascanio, 134
Soddy, Frederick, 258
Sokoloff, Louis, 450
Solvay, Ernest, 156, 196
Solvay process, 156–157
Sørensen, Søren, 240
Soret, Jacques-Louis, 130
Soxhlet extractor, 174–175
Spectrometry, mass, 256–257
Spectroscopy, flame, 146–147
Spectroscopy, infrared, 234–235
Spider silk, 238–239
Stadie, William C., 288
Stahl, Franklin, 392
Stahl, Georg E., 72
Stainless steel, 250–251
Stark, David Dewey, 266
Staudinger, Hermann, 124
Steel, Viking, 42–43
Steinhardt, Paul, 468
Steroid chemistry, 108, 326–327
Stevenson, James, 310
Stock, Alfred, 252
Stokes, George G., 138
Stone, Edward, 216
Stork, Gilbert, 62
Strehler, Bernard L., 390
Streptomycin, 332–333
Structural formula, 154–155
Strutt, John W., 222, 262
Sugars, Fischer and, 182–183
Sulfanilamide, 292–293
Sulfanilamide, elixir, 310–311
Sulfuric acid, 74–75
Sun, Eugene, 488
Supercritical fluids, 112–113
Superoxide, 296–297
Surface chemistry, 262–263
Suzuki, Akira, 506
Swigert, John L., Jr., 434–435
Swords, Ulfberht, 42
Synge, Richard L. M., 364
Synthetic diamond, 384–385
Synthetic (unnatural) products, 458–459
Szent-Györgyi, Albert, 288, 308
Szily, Pál, 240

Tanaka, Koichi, 474
Tartaric acid crystals, 136–137, 164
Taxol, 480–481
Technetium, 306–307
Teflon, 316–317
Tennant, Smithson, 48
Terra-cotta warriors, 30–31
Tetraethyl lead, 270–271, 352
Tetrahedral carbon atoms, 164–165
Thalidomide, 394–395
Thallium poisoning, 380–381
Thénard, Louis-Jacques, 186
Theophanes the Confessor, 38
Thermal cracking, 194–195
Thermite, 200–201
Tholin, 454–455
Thomson, Joseph John, 256, 286
Thomson, William, 212
Thurlow, Nathaniel, 236
Tiselius, Arne, W. K., 386
Tishler, Max, 360
Titanium/titanium dioxide, 88–89, 511
Tollens, Bernhard, 144
Toxicology, 52–53

Transition state theory, 300–301
Transuranic elements, 372–373
Travers, Morris, 222
Trommsdorff, Hermann, 122
Tropsch, Hans, 276
Tschirnhaus, Ehrenfried Walther von, 36
Tsien Roger Y., 400
Tsvet, Mikhail, 230, 426, 440
Tuppy, Hans, 364
Turin, Alan M., 428

Uemura, Daisuke, 484
Ulfberht swords, 42
Unnatural products, 458–459
Urea synthesis, 116–117
Urey, Harold, 286, 376

Vacuum-line technique, boranes and, 252–253

Van den Broek, Antonius, 162
Vauquelin, Nicolas-Louis, 96, 114
Venable, Francis Preston, 198
Viking steel, 42–43
Vitamin B_{12} synthesis, 444–445
Vitamin C, 116, 290–291
Volta, Alessandro, 94
Vorländer, Daniel, 192

Waksman, Selman A., 332–333
Walker, John Ernest, 308
Wall, Monroe Eliot, 480
Walsh, Alan, 146
Wani, Mansukh C., 480
Warfare, chemical, 260–261. See also Nerve gas
Watkins, Harold C., 310
Watson, James D., 382, 392
Werner, Alfred, 204, 416

Westheimer, Frank H., 448
White, Emil H., 390
Wieland, Otto, 108
Wildman, Samuel G., 348
Wilkins, Maurice H. F., 382
Wilkinson, Geoffrey, 370
Willson, Thomas Leopold, 198
Windaus, Adolf Otto Reinhold, 108
Winstein, Saul, 356
Wöhler, Friedrich, 114, 116, 118, 198
Wolf, Alfred P., 450
Wollaston, William H., 146
Woodfill, Jerry, 434
Woodward-Hoffman rules, 422–423
Woodward, Robert B., 62, 340, 370, 422, 444–445, 458
Wróblewski, Zygmunt, 180

X-ray crystallography, 246–247

Yaghi, Omar, 486
Yarchoan, Robert, 466
Yorkshire alum, 60–61
Ytterby, 90–91

Zeidler, Othmar, 332
Zhabotinsky, Anatol, 428
Ziegler, Karl W., 406, 496
Ziegler-Natta catalysis, 406–407
Zone refining, 378–379
Zymase fermentation, 218–219

Image Credits

akg-images: 357; © Bruni Meya 43

Bridgeman: Biblioteca Nacional, Madrid 39; Biblioteque des Arts Decoratifs, Paris, France/Archives Carmet 45; © The Fine Art Society, London, UK 107

© C. Gerhardt GmbH & Co. KG Germany: 175

Corbis: Bettman 335, 351, 471

CSIRO: © Daryl Peroni 277

Deposit Photos: © ZanozaRU 151; ©vshivkova 491

Fairfax Syndication: 381

Courtesy Felix R. Fischer: 509

Fundamental Photos: © Richard Megna 103, 336

Getty Images: ©AFP 471; © Carsten Peter/Speleoresearch & Films 15; ©Tim Graham 133

Harvard University Archives: HUGFP 68.38.1p Box 2 445

© The Israel Museum, Jerusalem: Elie Posner from the exhibition *White Gold Revealing the World's Earliest* 25

Courtesy Imageing Technology Group at Beckman Institute: Dr. Larry Millet and Janet Sinn-Hanlon at the University of Illinois 475

iStock: © 36clicks 513; © alacatr 299; © alantobey 31; © amete 173; © amitus 393; © Aneb 89; © chinaface 21; © DavidMednick 237; © Jim DeLillo 19; © Diane Diederich 97; © Shunya Fan 473; © Garsya 317; © Hailshadow 105; © Hanis 67; © HultonArchive 437; © RobertKacpura 453; © Savas Keskiner 343; © Joakim Leroy 197; © LPETTET 223; © pedrosala 153; © PICSUNV 439; © proximinder 405; © raclro 159;

© RapidEye 331; © Remains 75; © Scharvik 219; © slovegrove 191; © styfz22 129; © Thomas_EyeDesign 181; © toddtaulman 263; © Trout55 269; © westphalia 179; © yurok 301

Lawrence Berkeley National Lab: 373

Library of Congress: 73, 83, 127, 135, 305, 333, 425

© Derek B. Lowe: 155

Courtesy Ulrich Lüning: 279

NASA: 435; Ames 377; JPL 457; Ozone Watch 511

National Cancer Institute: Larry Otsby 417, 467

National Institutes of Health: 361

National Library of Medicine: 183

National Science Foundation: 193; Joseph McConnell, Desert Research Institute 419

NIST: 499, 513

NOAA: PMEL EOI Program 113

Courtesy North York Moors National Park Authority: 61

Courtesy Oak Ridge Associated University: 264

Courtesy Simon Park/Exploring the Invisible: 117

Rijksmuseum: 29

Science Source: © Andrew Brookes/National Physical Laboratory 465; © Martyn F. Chillmaid 145; © Crown Copyright courtesy of Central Science Laboratory 399; © ER Degginger 403; © Kenneth Eward 285; © Eye of Science 397, 507; © Simon Fraser/Medical Physics, RVI, Newcastle Upon Tyne 475; © Tom Hollyman 321; ©

Andrew Lambert Photography 121; © Lawrence Berkeley National Lab 273; © New York Public Library 441; © Alfred Pasieka 109, 307; © Philippe Plailly 429; © Power and Syred 409; © SPL 143, 247; © Chris Taylor/CSIRO 235; © Sheila Terry 187; © Charles D. Winters 171, 185, 207, 231, 489

Shutterstock: © Alegria 2; © Leonid Andronov 365; © Antonio S 165; © areeya_ann 367; © Alexander Bark 221; © axz700 503; © bluecrayola 137, 227; © BlueRingMedia 319; © Thum Chia Chieh 141; © Chrisferra 229; © concept w 163; © Cousin_Avi 481; © danilo ducak 297; © DnD-Production.com 379; © EastVillageImages 463; © Elenamiv 85; © evantravels 401; © Everett Historical 195, 261, 283; © Tyler Fox 485; © Vladimir Gjorgiev 389; © GROGL 443; © Peter Gudella 111; © hacohab 213; © Johann Heigason 427; © Inna_77 323; © Jhaz Photography 131; © Bernd Juergens 249; © Sabine Kappel 241; Sebastian Kaulitzki 309, 355; © Pitsanu Kraichana 201; © longtaildog 443; © Luis Louro 295; © Robyn Mackenzie 495; © magnetix 205; © Marioner 349; © Mikhail Markovskiy 76; © Mediagram 407; © molekuul.be 371; © Mopic 383; © nikkytok 77; © Oleksiy Mark 493; © Production Perig 27; © petarg 369; © Albert Russ 99, 167; © Yenyu Shih 363; © swissmacky 169; © tanewpix 375; © taro911 Photography 353; © Tischenko Irina 203; © Ursa Studio 339; © vetkit 123; © Vittoriano Junior 215; © Vshivkova 345; © wasanajai 209; © Peter Weber 501; © XXLPhoto 125; © yanqiu 17

Solvay Photolibrary: © Lisa Means 157

U.S. Department of Energy, National Energy Technology Laboratory: 421

U.S. Navy: 161

Wellcome Library, London: 33, 41, 47, 53, 57, 65, 87, 101, 119, 170, 217, 225, 233, 245, 265, 477; Science Museum, London 63, 81, 95, 293

Courtesy Wikimedia Foundation: 59, 93, 126, 148, 176, 192, 211; Armand69 35; Bin im Garten 359; Biodiversity Heritage Library 411; Tony Boehle 487; Art Bromage 431; Chemical Heritage Foundation 51, 341; Department of Energy 347; Doc RNDr. Josef Reischig, CSc. 291; Dr EG 251; Dschwen 5; Dutch National Archives 255; Ed Wescott/US Army/Manhattan Engineering District 257; Simon A. Eugster 449; Fae 313; FDA 311, 395; Hannes Grobe/AWI 139; Daniel Grohman 49; Hallwyl 37; Inaglory 239; Irri Images 387; Julien Bobrof, Frederic Bouquet/LPS, Orsay, France 91; J.W. Photography from Annapolis 147; Jynto 459; Roman Kohler 385; Library of Congress 149; Douglas A. Lockhard 423; Patrick J. Lynch 289; Jens Maus 451; Mfomich 267; mikulova 315; Ben Mills 497; Mstroeck 483; NASA 71, 253, 455, 479, 505; AURA/Hubble Heritage Team STScI/ESA/NASA 287; National Archive and Records Administration 189, 199, 243, 271, 303, 329; National Library of Norway 214; Mari-Lan Nguyen 23; NIST 469; Oak Ridge National Laboratory/US Department of Energy 259, 325; Osmotheque 177; Quit007 391; Suraj Rajan 461; Rillke 327; Wolfgang Sauber 115; Taavi.Ivan 415; US Department of Defense 413; Sander van der Wel 281; Wellcome 69

Courtesy Yale University: Beinecke Rare Book and Manuscript Library 166